AF576059

Interdisciplinary Researches for Cultural Heritage Conservation

Interdisciplinary Researches for Cultural Heritage Conservation

Editor

Cesareo Saiz-Jimenez

MDPI • Basel • Beijing • Wuhan • Barcelona • Belgrade • Manchester • Tokyo • Cluj • Tianjin

Editor
Cesareo Saiz-Jimenez
Instituto de Recursos
Naturales y Agrobiologia,
IRNAS-CSIC
Sevilla
Spain

Editorial Office
MDPI
St. Alban-Anlage 66
4052 Basel, Switzerland

This is a reprint of articles from the Special Issue published online in the open access journal *Applied Sciences* (ISSN 2076-3417) (available at: https://www.mdpi.com/journal/applsci/special_issues/Cultural_Heritage_Conservation).

For citation purposes, cite each article independently as indicated on the article page online and as indicated below:

LastName, A.A.; LastName, B.B.; LastName, C.C. Article Title. *Journal Name* **Year**, *Volume Number*, Page Range.

ISBN 978-3-0365-6818-8 (Hbk)
ISBN 978-3-0365-6819-5 (PDF)

Cover image courtesy of Agora S.L. Juan Aguilar

Contents

About the Editor

Cesareo Saiz-Jimenez

Cesareo Saiz-Jimenez is a Research Professor at the Instituto de Recursos Naturales y Agrobiologia, CSIC, Seville, Spain. Ph D in Biology (Complutense University of Madrid, Spain) and Ph.D. in Chemical Engineering and Material Sciences (Technical University of Delft, The Netherlands). He coordinated European Commission projects in the fields of urban atmospheric pollution, biodeterioration of monuments, geomicrobiology of subterranean environments, environmental microbiology, etc. He was the founder and the coordinator of the CSIC Thematic Network on Cultural Heritage and the Spanish Thematic Network for Science and Technology for the Conservation of Cultural Heritage. More than 500 papers and books have been published (e.g. "Molecular Biology and Cultural Heritage", "Air Pollution and Cultural Heritage", "Écologie Microbienne de la Grotte de Lascaux", "The Conservation of Subterranean Cultural Heritage", etc.).

Editorial

Special Issue on Interdisciplinary Researches for Cultural Heritage Conservation

Cesareo Saiz-Jimenez

Instituto de Recursos Naturales y Agrobiologia, IRNAS-CSIC, 41012 Sevilla, Spain; saiz@irnase.csic.es

Citation: Saiz-Jimenez, C. Special Issue on Interdisciplinary Researches for Cultural Heritage Conservation. *Appl. Sci.* **2023**, *13*, 1824. https://doi.org/10.3390/app13031824

Received: 22 January 2023
Accepted: 28 January 2023
Published: 31 January 2023

1. Introduction

UNESCO defines cultural heritage as "the legacy of physical artefacts and intangible attributes of a group or society that are inherited from past generations, maintained in the present and bestowed for the benefit of future generations".

The roots of each country and region are anchored in its own cultural heritage. Cultural heritage is an invaluable legacy and is integral to our future, but due to its fragile and finite nature we sought to identify the best, most sustainable means of preserving cultural heritage. This implies the conservation of movable (paintings, sculptures, artifacts) and immovable (monuments, archaeological sites, industrial archaeology) properties. Today, cultural heritage is exposed to air pollution, flooding, earthquakes, wrong management activities, etc., which threaten its integrity. To mitigate damages, research should focus on analyzing and alleviating deterioration and provide technological solutions for enhancing the conservation of cultural heritage. These goals can be achieved through interdisciplinary and transdisciplinary studies.

This Special Issue on Interdisciplinary Researches for Cultural Heritage Conservation in Applied Sciences aims to bring together some of the latest researches in this field. A total of 10 papers focusing on different aspects of cultural heritage are categorized into the three topics detailed below and summarized.

2. Rock Art, Mural Paintings and Stone Monuments

Some of the earliest forms of rock art are cave paintings and petroglyphs, carved or scratched into the rock surface. The most famous examples of rock art include Altamira and Lascaux caves, well-known examples of flawed management that produced substantial biodeterioration phenomena, widely discussed in the scientific literature. The pioneer studies on these two caves promoted the extension to other caves affected by similar problems.

Altamira, Lascaux and many other caves presented different types of colonization by phototrophic biofilms, a severe alteration produced by artificial lighting. The most noticeable effect of the lighting was the growth of a dense phototrophic community of cyanobacteria, algae and bryophytes on the speleothems, walls and ground of Tesoro Cave, Rincon de la Victoria, Spain. Jurado et al. [1] tested the application of different mechanical and chemical cleaning procedures and showed the effectiveness of hydrogen peroxide and sodium hypochlorite for the successful cleaning of green biofilms from the entire cave.

Cañaveras et al. [2] studied the deterioration processes affecting the sandstones supporting the paintings and carvings of the southern Spanish shelters of *Tajo de las Figuras* (Cádiz) and *Peñas Cabrera* (Málaga). The data showed that the most common deterioration mechanisms corresponded to natural physical and chemical phenomena, including mechanical (frost) and chemical (carbo-hydrolysis) weathering, in addition to biological colonization, for which corrective measures are difficult to apply without producing unforeseen consequences, mainly when the state of conservation of the paintings is acceptable.

Rabbachin et al. [3] investigated the factors involved in the degradation of petroglyphs in the Negev desert of Israel, with a focus on biodeterioration processes. The multi-analytical approach used in this study allowed the characterization of the microbiomes of limestones covered with black varnish, in close proximity to the petroglyphs, as well as the weathering of the limestones, thus unveiling mineral–microbial interactions.

Earliest known examples of wall paintings are found in Crete and in Egyptian tombs that were continued in Etruscan necropolis, Roman, medieval and Renaissance monuments. The *Salón de Reinos* is a remnant of the 17th century *Palacio del Buen Retiro* (Madrid, Spain) built between 1632 and 1640. The *Salón* housed famous paintings, now deposited in the *Museo Nacional del Prado*, and mural paintings located in the ceiling and window vaults with a poor conservation status, evidenced by fissures, water filtration and fungal colonizations. Jurado et al. [4] stated that most of the isolated bacteria and fungi from the paintings originate from plants, saprophytes or pathogens, data consistent with the abundant vegetation in parks and gardens near the *Salón de Reinos* (Retiro Park and Botanical Garden). Likewise, the presence of a few bacteria and fungi with biodeteriorative activities, previously detected in other murals and frescoes, was also noteworthy.

The monuments have not only suffered the ravages of times, but are also damaged by vandalism in modern times. An example is the Cross of the Inquisition located in the Seville City Hall, Spain. The Cross was vandalized in September 2019 and the restoration concluded in September 2021. Along with the restoration, a mineralogical and microbiological study was carried out in a few small fragments. The exposure of the Cross of the Inquisition to an urban environment for more than 100 years promoted the colonization of the limestone by a lichenic community, associated with bacteria, fungi and bryophytes, which were detrimental for the stone. Appropriate biocides and consolidants were used in the restoration [5].

3. Stained-Glass Windows

Two papers deal with materials used in stained-glass windows. Palomar et al. [6] determined the composition of flashed glasses that have been used since medieval times in stained-glass windows. They used optical microscopy, field emission scanning electron microscopy, and linear energy dispersive spectroscopy to determine the thickness of the colored layers, and laser-induced breakdown spectroscopy for the analysis of major and minor components, which provided valuable information on chronology, provenance, and manufacturing processes of these types of glasses.

Machado et al. [7] studied the production of grisaille, the first glass-based paint to be used in stained-glass windows spread throughout Europe from the 12th century onwards and still used today. The authors determined that raw materials directly affect the properties and appearance of grisailles and the proportions between the different grisaille components (base glasses, coloring agents, firing processes, etc.) were key factors in their stability.

4. Miscellaneous Papers

Morales-Martín et al. [8] explored the use of optical sol–gel environmental pH sensors for air evaluation in exhibition halls of the *Museo Nacional de Ciencias Naturales* (Madrid, Spain). The effects of Madrid air pollution affected the exhibition halls near the entrance with slightly low pH values, while halls located far from the entrance showed neutral pH values. Environmental pH sensors allowed the museum to make decisions on preventive conservation strategies.

Bernabéu-Larena et al. [9] analysed the case of Canal de Isabel II, a water supply public work and its historic infrastructure built in the second half of the 19th century and the early 20th century. The study showed that the water supply that maintained its original use has also taken on new functions through the conversion of some of its parts and added new functions to the existing ones, such as the enhancement of its heritage value.

Wang et al. [10] proposed Lunyu ontology to formally represent concepts concerning morality in The Analects of Confucius. In this research, the authors explored ways to develop a class hierarchy from discrete and abstract ideas and also adopted a term-and-characteristic-guided methodology derived from taking into account the ISO principles on Terminology, whereby "a term is a verbal designation of a concept" and "a concept is a unique combination of (essential) characteristics".

Overall, the main themes of the papers published in this Special Issue cover important achievements on diverse Cultural Heritage aspects that were also complemented by other Applied Sciences Special Issues devoted to Cultural Heritage.

Funding: This research received no external funding.

Conflicts of Interest: The author declares no conflict of interest.

References

1. Jurado, V.; Hernandez-Marine, M.; Rogerio-Candelera, M.A.; Ruano, F.; Aguilar, C.; Aguilar, J.; Saiz-Jimenez, C. Cleaning of Phototrophic Biofilms in a Show Cave: The Case of Tesoro Cave, Spain. *Appl. Sci.* **2022**, *12*, 7357. [CrossRef]
2. Cañaveras, J.C.; Sanz-Rubio, E.; Sánchez-Moral, S. Weathering Processes on Sandstone Painting and Carving Surfaces at Prehistoric Rock Sites in Southern Spain. *Appl. Sci.* **2022**, *12*, 5330. [CrossRef]
3. Rabbachin, L.; Piñar, G.; Nir, I.; Kushmaro, A.; Pavan, M.J.; Eitenberger, E.; Waldherr, M.; Graf, A.; Sterflinger, K. A Multi-Analytical Approach to Infer Mineral–Microbial Interactions Applied to Petroglyph Sites in the Negev Desert of Israel. *Appl. Sci.* **2022**, *12*, 6936. [CrossRef]
4. Jurado, V.; Gonzalez-Pimentel, J.L.; Hermosin, B.; Saiz-Jimenez, C. Biodeterioration of Salón de Reinos, Museo Nacional del Prado, Madrid, Spain. *Appl. Sci.* **2021**, *11*, 8858. [CrossRef]
5. Jurado, V.; Cañaveras, J.C.; Gomez-Bolea, A.; Gonzalez-Pimentel, J.L.; Sanchez-Moral, S.; Costa, C.; Saiz-Jimenez, C. Holistic Approach to the Restoration of a Vandalized Monument: The Cross of the Inquisition, Seville City Hall, Spain. *Appl. Sci.* **2022**, *12*, 6222. [CrossRef]
6. Palomar, T.; Martínez- Weinbaum, M.; Aparicio, M.; Maestro-Guijarro, L.; Castillejo, M.; Oujja, M. Spectroscopic and Microscopic Characterization of Flashed Glasses from Stained Glass Windows. *Appl. Sci.* **2022**, *12*, 5760. [CrossRef]
7. Machado, C.; Vilarigues, M.; Pinto, J.V.; Palomar, T. The Influence of Raw Materials on the Stability of Grisaille Paint Layers. *Appl. Sci.* **2022**, *12*, 10515. [CrossRef]
8. Morales-Martín, D.; Agua, F.; Barreiro, J.; Garvía, A.L.; García-Heras, M.; Villegas, M.A. Environmental pH Evaluation in Exhibition Halls of Museo Nacional de Ciencias Naturales (CSIC, Madrid). *Appl. Sci.* **2021**, *11*, 10091. [CrossRef]
9. Bernabéu-Larena, J.; Cabau-Anchuelo, B.; Plasencia-Lozano, P.; Hernández-Lamas, P. Use and Management in the Heritage Conservation of the Historic Water Supply of Canal de Isabel II, Madrid. *Appl. Sci.* **2022**, *12*, 6731. [CrossRef]
10. Wang, F.; Wei, T.; Wang, J. Ontology-Driven Cultural Heritage Conservation: A Case of The Analects of Confucius. *Appl. Sci.* **2022**, *12*, 287. [CrossRef]

MDPI

Article

Cleaning of Phototrophic Biofilms in a Show Cave: The Case of Tesoro Cave, Spain

Valme Jurado [1,*], Mariona Hernandez-Marine [2], Miguel Angel Rogerio-Candelera [1], Francisco Ruano [3], Clara Aguilar [3], Juan Aguilar [3] and Cesareo Saiz-Jimenez [1]

1 Instituto de Recursos Naturales y Agrobiologia, IRNAS-CSIC, 41012 Sevilla, Spain; marogerio@irnase.csic.es (M.A.R.-C.); saiz@irnase.csic.es (C.S.-J.)
2 Facultad de Farmacia, Universidad de Barcelona, 08028 Barcelona, Spain; marionahernandez@ub.edu
3 Agora S.L., 41928 Sevilla, Spain; frgdepau@hotmail.com (F.R.); aguilarlinaresclara@yahoo.es (C.A.); agora2@telefonica.net (J.A.)
* Correspondence: v.jurado@csic.es

Abstract: Show caves have different grades of colonization by phototrophic biofilms. They may receive a varied number of visits, from a few thousand to hundreds of thousands of visitors annually. Among them, Tesoro Cave, Rincon de la Victoria, Spain, showed severe anthropic alterations, including artificial lighting. The most noticeable effect of the lighting was the growth of a dense phototrophic community of cyanobacteria, algae and bryophytes on the speleothems, walls and ground. The biofilms were dominated by the cyanobacterium *Phormidium* sp., the chlorophyte *Myrmecia israelensis*, and the rhodophyte *Cyanidium* sp. In many cases, the biofilms also showed an abundance of the bryophyte *Eucladium verticillatum*. Other cyanobacteria observed in different biofilms along the cave were: *Chroococcidiopsis* sp., *Synechocystis* sp. and *Nostoc* cf. *edaphicum*, the green microalgae *Pseudococcomyxa simplex*, *Chlorella* sp. and the diatom *Diadesmis contenta*. Preliminary cleaning tests on selected areas showed the effectiveness of hydrogen peroxide and sodium hypochlorite. A physicochemical treatment involving the mechanical removal of the thickest layers of biofilms was followed by chemical treatments. In total, 94% of the surface was cleaned with hydrogen peroxide, with a subsequent treatment with sodium hypochlorite in only 1% of cases. The remaining 5% was cleaned with sodium hypochlorite in areas where the biofilms were entrapped into a calcite layer and in sandy surfaces with little physical compaction. The green biofilms from the entire cave were successfully cleaned.

Keywords: cyanobacteria; chlorophytes; bryophytes; surface cleaning; hydrogen peroxide; sodium hypochlorite

Citation: Jurado, V.; Hernandez-Marine, M.; Rogerio-Candelera, M.A.; Ruano, F.; Aguilar, C.; Aguilar, J.; Saiz-Jimenez, C. Cleaning of Phototrophic Biofilms in a Show Cave: The Case of Tesoro Cave, Spain. *Appl. Sci.* **2022**, *12*, 7357. https://doi.org/10.3390/app12157357

Academic Editors: Ramona Iseppi and Gyungsoon Park

Received: 3 June 2022
Accepted: 21 July 2022
Published: 22 July 2022

1. Introduction

Caves are heterogeneous karst systems composed of different compartments. As one progresses from the outside to inside the cave, it can be observed that the microbial populations colonizing the walls and ceilings are different, and separate into different zones. In consequence, there is a transition between adjacent ecological systems that involves a change in the structure and composition of the community and is limited by their spatial distribution. Phototrophic communities (cyanobacteria, algae, lichens, bryophytes and lower plants) are usually abundant in the area closest to the outside, generally illuminated by natural light, and disappear towards the deepest area, depending on the decrease in the light gradient.

Cyanobacteria and algae are especially successful microorganisms in subterranean environments due to the high humidity and carbon dioxide concentration and the very low light intensity in which they thrive. Some authors call these biofilms lampenflora [1–3]. The lighting in show caves and other underground environments generates a series of problems, derived from the growth of phototrophic communities, which generally represents a serious ecological distortion and an aesthetic problem due to the presence of biofilms with green to black colorations colonizing the speleothems and walls.

The problem of the appearance of green biofilms in caves with Paleolithic paintings is not new. A few cases have been documented in the scientific literature and the phototrophic communities threaten the Paleolithic paintings.

Lascaux Cave in France had to be closed at the beginning of the 1960s due to the invasion of the alga *Bracteacoccus minor* as a result of the intense flow of visitors and hours of artificial light. The biofilms were removed with antibiotics and formaldehyde [4,5]. Altamira Cave was closed in 2002 due to the occurrence of cyanobacteria and algae on the paintings [6,7]. A third case was recently reported in El Castillo Cave, with the occurrence of biofilms in the Polychromes Panel [8]. In this cave, chlorophytes, specifically *Jenufa aeroterrestrica*, a new species described in 2015 [9], as well as *Coccomyxa* sp. were found in the biofilms, but no cyanobacteria. In Altamira and El Castillo caves, with initial stages of phototrophic colonization, lighting was turned off.

Other caves showed extensive green biofilms on speleothems, but far from the paintings, such as in Tito Bustillo Cave, Asturias, whose stalactites and stalagmites were abundantly colonized by phototrophic microorganisms and, in particular, the sediments showed an exuberant colonization by the calcifying cyanobacterium *Scytonema julianum*, common in caves, tombs and catacombs [10–12].

Del Rosal et al. [13,14] found *Jenufa* sp. in Nerja Cave, with abundant phototrophic biofilms on the speleothems and walls. In areas with water percolation (occasional or habitual), chlorophytes (*Jenufa* sp. and *Desmococcus endolithicus*) were commonly identified, while cyanobacteria and the red alga *Cyanidium* sp. abounded in drier areas.

The Andalusian caves, Gruta de las Maravillas, Murcielagos, Tesoro and Nerja, to mention a few caves, exhibit different grades of phototrophic biofilm development. All of them can be accessed and receive a varied number of visits, from a few thousand to hundreds of thousands of visitors annually. Among them, Tesoro and Nerja caves stand out due to the abundance of phototrophic biofilms. While Nerja Cave biofilms have been thoroughly studied [15], the composition of the biofilms from Tesoro Cave is relatively unknown.

Tesoro Cave (Figure 1) is situated in Rincon de la Victoria, about 11 km from Malaga, Spain. The cave is located in a calcareous massif (Cantal Alto), formed on a thick bed of about 70 m of limestone. The cave is about 80 m above sea level. The Cantal hill is widely karstified with an abundance of small cavities and two main caves, Tesoro and Rincon de la Victoria, the first located in the central area of Cantal Alto massif, and the second more to the south.

The climate is semiarid and mesothermal, with small surplus of water in winter. The water that reaches the karst is scarce; a significant quantity is lost by evapotranspiration, and the remainder is lost in the urbanized area surrounding the cave entrance. The only water supply to Cantal karst is reduced to direct rain infiltration with a few dripping points in the cave; the response time to rain has 2–3 h of delay. This explains why the generation of speleothems is so scarce in the cave [16].

The cave measurements of CO_2, temperature and relative humidity showed particular features. In Sala del Lago and Sala del Volcan, the CO_2 concentrations were from 200 to 550 ppm, the temperature 14.8–16.8 °C, and the relative humidity 80% to 90%, all of which are normal values for a karstic cavity [5–8,13–15]. However, the non-visited area (Galeria de Breuil) shows anomalous concentrations of CO_2 [16], such as 20,000 ppm, the temperature rises to 22.7 °C, and the relative humidity reaches 100% saturation. These values are characteristic of environments with poor ventilation; in other words, the karstic system seems to behave, in this Galeria, as a confined and almost impermeable enclosure that presents a small renewal of air.

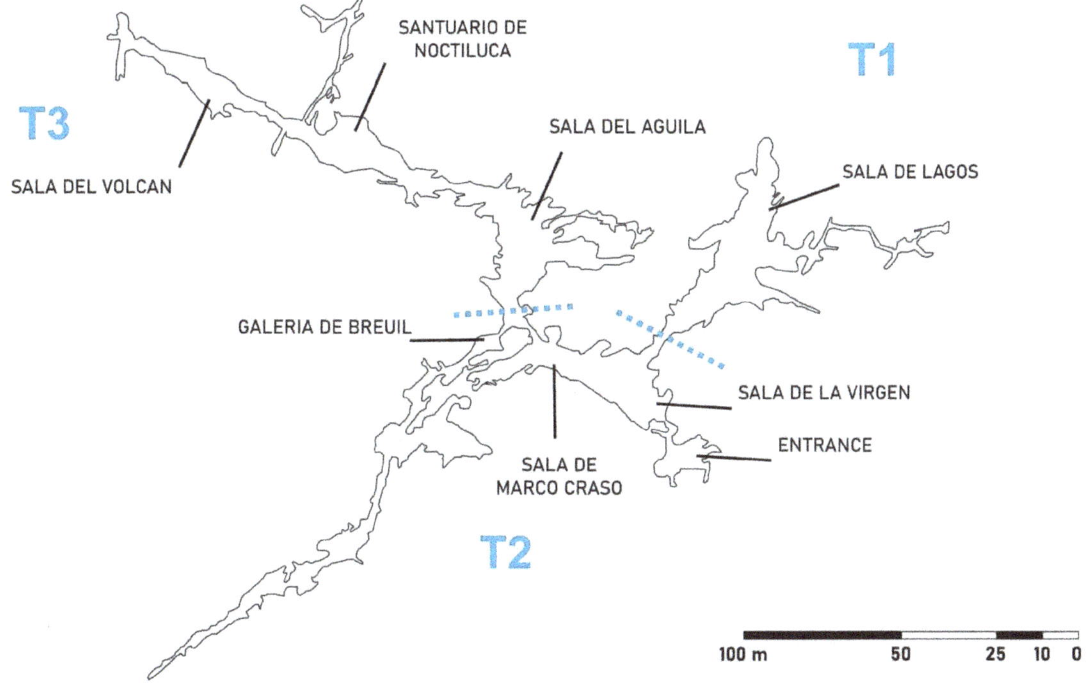

Figure 1. Map of Tesoro Cave. T1–T3 represent the three sectors into which the cave was divided for study purposes.

Tesoro Cave has experienced severe anthropic alterations. On the surface, the area was urbanized with roads, parking plots and buildings. Inside the cave, strong alterations for facilitating visits are observed, including the removal of sediments, building materials extraneous to the karst, artificial lighting, the construction of an artificial lake, an elevator, and a wide staircase [16]. Surface corrosion was favored by the microenvironmental conditions in some galleries. The most noticeable effect of lighting was the growth of a dense phototrophic community of cyanobacteria, green algae and bryophytes on the speleothems, walls and ground, near the lighting points and wet areas. The remains of wood and an old electric installation are also evident.

For study purposes, the cave was divided into three sectors: T1, mainly formed by a wide hall, Sala de Lagos, with an artificial lake; T2, centered by the main entrance to the cave with an elevator and staircase as well as halls and galleries at both sides to give access to zones T1 and T3; and T3, ending in a deep hall, Sala del Volcan (Figure 1). Here, we report a study on the phototrophic community in Tesoro Cave, the cleaning of biofilms and the used protocols.

2. Materials and Methods

Phototrophic biofilm samples were scraped off with a sterile scalpel, placed into 1.5 mL sterile Eppendorf tubes and stored at 4 °C until arrival at the laboratory.

Taxonomic identifications were based on the morphological characters of the whole biofilm and on colonies grown on 15% agar plates with medium BG11 (Sigma-Aldrich, Steinheim, Germany) or Bold's Basal Medium (BBM) (Sigma-Aldrich, Steinheim, Germany) and incubated at room temperature and a low light intensity (semi-closed window oriented to the north) [17,18]. The identification of the different microalgae and cyanobacteria was based on the available taxonomic keys and confirmed in AlgaeBase [19–24]. Molecular biology techniques were applied for the identification of *Cyanidium* sp. DNA was extracted using the FastDNASPIN for Soil Kit (MP Biomedicals, Illkirch, France). The amplification of 16S gene sequences was performed using the cyanobacteria-specific primer pair, Cya106F (5′-CGGACGGGTGAGTAACGCGTGA-3′) and Cya 781R (5′-GACTACTGGGGTATCTAATCCCWTT-3′) [25]. The PCR amplification protocol consisted of the following thermal conditions: 95 °C for 2 min; 35 cycles of 95 °C for 15 s, 60 °C for 15 s, 72 °C for 2 min; and a final step of 72 °C for 10 min. DNA libraries of PCR-amplified products were constructed using the TOPO-TA cloning kit (Invitrogen, Carlsbad, CA, USA) according to the manufacturer's recommendations. Plasmids were extracted with the JetQuick Plasmid Miniprep Spin kit (Genomed, Löhne, Germany), following the manufacturer's protocol, and sequenced by Secugen Sequencing Services (CSIC, Madrid, Spain). The identification of phylogenetic neighbors was determined using the BLASTN algorithm [26]. The 16S rRNA gene sequences of clones and its closest related sequences were multiplied and aligned using CLUSTAL W [27]. The phylogenetic tree was constructed using the maximum-likelihood method [28] and Kimura's two-parameter model with a discrete Gamma distribution in MEGA version 11 [29]. A bootstrap analysis of 1000 re-samplings was used to evaluate the robustness of the tree. All sequences were submitted to GenBank with consecutive accession numbers ON619559–ON619566.

The direct observation of phototrophic biofilms and isolate colonies from enrichment cultures was carried out using an Axioplan microscope (Carl Zeiss, Oberkochen, Germany) and the images captured with an AxioCam MRc5 digital camera and processed with Axiolan LE software and a scanning electron microscope (Quanta 200 FEI Φ EDAX) [13,14]. The samples were subjected to pre-fixation with acrolein vapor and post-fixation with osmium tetroxide vapor. The samples were double coated with carbon and sputtered gold.

Several treatments were tested for biofilm removal and cleaning. The treatments applied for cleaning the whole cave included mechanical removal by brush and cleaning with liquid nitrogen, with hydrogen peroxide and/or sodium hypochlorite.

Physicochemical cleaning was a two-phase treatment: the first physical and the second chemical. The first consisted of the mechanical removal of the thickest layers of biofilms using a spatula and a brush. This treatment was conditioned by the type of biofilm and the rock surface. In fact, this was applied to dense biofilms showing abundance of bryophytes and/or algae that could easily be removed from the rock surfaces. The application was never done dry, but rather by means of previous hydration of the surfaces. Once finished, a chemical treatment was applied, which was conditioned by the typology of the surface (solid rock surfaces), the type of biofilms (residual algae and/or bryophyte protonema not removed by brushing) and the environmental conditions (dry or wet surfaces). The products used were as follows.

Cleaning with hydrogen peroxide: hydrogen peroxide at 10% in water was very effective in dry areas. It was applied by impregnation with a brush or by spraying with nebulizers. In the case of more humid areas and with more irregular textures, it was applied in a percentage of 50% in aqueous solution. This was done by first performing various tests. We observed that in areas with high humidity (due to condensation or seepages), the product was diluted, lowering the concentration and losing effectiveness. This chemical was the most widely used.

Cleaning with sodium hypochlorite: cleaning with 10% sodium hypochlorite in aqueous solution was used in very porous or textured areas where cracks abounded and the conditions, in all ways, were very extreme, and in areas of very high biofilm development, especially in areas located around light sources. It was applied by impregnation with a brush.

Subsequent chemical cleaning: this treatment was mainly applied in areas where the biofilms were entrapped into a calcite layer and on sandy surfaces with little physical compaction. Two different scenarios were considered, namely, fragile and calcified zones.

Fragile zones: small stalactites, or very fragile surfaces. The treatment was applied with a nebulizer in areas where hydrogen peroxide was not effective enough.

Calcified zones: applied in areas where the biofilm was covered by calcite crystals, showing an intense green hue. To access the biofilms, pads impregnated with sodium hypochlorite were placed on the mineral layer, which reduced the evaporation of the product and facilitated its action.

3. Results and Discussion

3.1. Composition of Biofilms in Different Cave Sectors

3.1.1. Sector T1

The cave was divided into three sectors with different ecological and environmental characteristics. Sector T1, largely represented by Sala del Lago, was influenced by the artificial lake occupying almost the entire sector, which provided high relative humidity, near saturation. In addition, the connection with the exterior was relevant through the water. This and the lighting produced abundant phototrophic biofilms composed of cyanobacteria, microalgae and bryophytes near the lighting lamps. The biofilms were dominated by the cyanobacterium *Phormidium* sp. (Figure 2B), the chlorophyte *Myrmecia israelensis* (Figure 2A) and the rhodophyte *Cyanidium* sp. It is remarkable that *Cyanidium* sp. was only found in this sector and not in the biofilms from the rest of the cave.

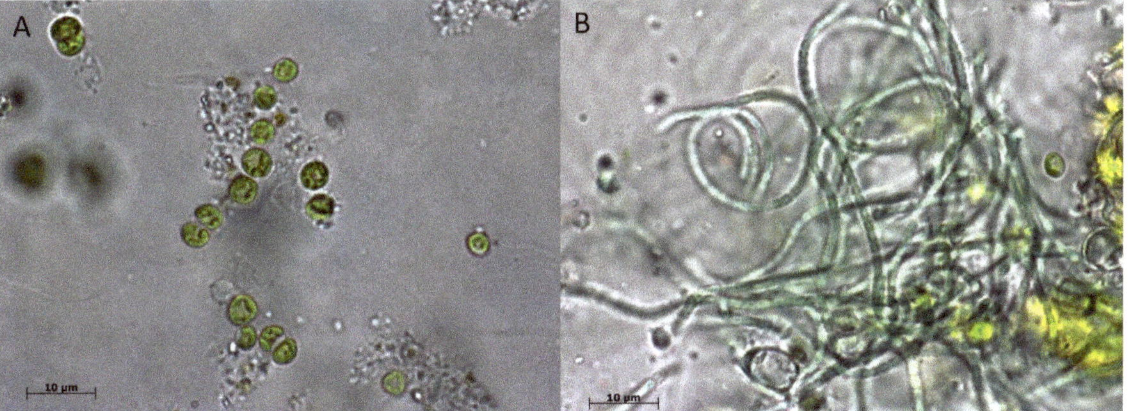

Figure 2. (**A**) Cells of *Myrmecia israelensis*. (**B**) Filaments of *Phormidium* sp. Scale bar: 10 µm.

The presence of *Cyanidium* sp. in Tesoro Cave, which was also observed in Nerja Cave some 50 km away [14], deserves a special mention. The reports on the genus *Cyanidium*, identified by microscopy and morphology and retrieved from non-extremophilic habitats, are very scarce, but not those from extreme and acidophilic environments, where *Cyanidium caldarium* is well represented [30,31]. *Cyanidium* sp. would not correspond to *C. caldarium* due to its very different habitat and ecology. In fact, *Cyanidium* strains from caves cluster into a separate monophyletic lineage within the class *Cyanidiophyceae* [32].

The first morphospecies of the genus *Cyanidium* that did not inhabit acidic ecosystems or volcanic areas were described from samples from caves (*Cyanidium chilense*) and from fissures in coastal rocks, both in Chile [33]. Subsequently, by means of molecular analysis, three species of non-extremophilic aerophytic *Cyanidium* have been described, two of them *Cyanidium* sp. Monte Rotaro and *Cyanidium* sp. Sybil Cave, inhabitants of Italian caves [34,35], and the third, *Cyanidium* sp. Atacama, in a cave on the Chilean coast [36].

The sequences of clones related to the genus *Cyanidium* identified in Tesoro Cave present identities of 95.09%, 94.75% and 92.76% with the strains from Atacama, Monte Rotaro and Sybil Cave, respectively (Figure 3). In the phylogenetic tree, the Tesoro Cave sequences form a robust clade far from previous *Cyanidium* isolates, suggesting that they could represent a different species of the genus.

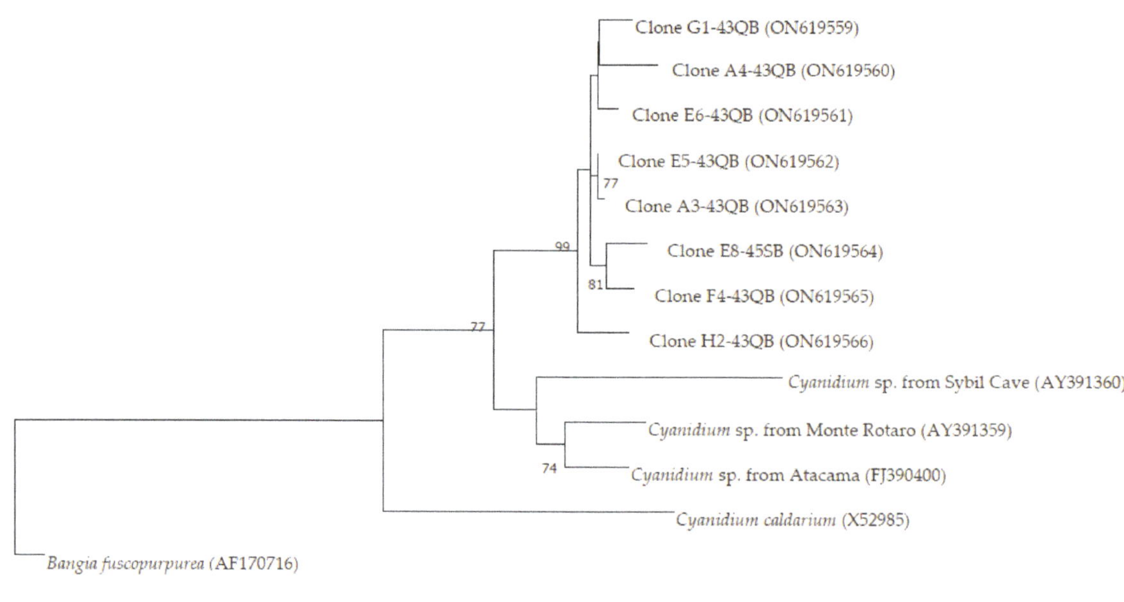

Figure 3. Maximum-likelihood phylogenetic tree based on 16S rRNA gene sequences showing the relationship of clones from phototrophic biofilms collected in Tesoro Cave to isolated species of the genus *Cyanidium*. Bootstrap values (>50%) are expressed as percentages of 1000 replicates. The 16S rRNA gene sequence of *Bangia fuscopurpurea* was used as the outgroup. Bar: 0.02 substitutions per nucleotide position.

In many cases, the biofilms also showed abundance of the bryophyte *Eucladium verticillatum*. The cause was the intense illumination by means of incandescent lamps, as shown in Figure 4. *Eucladium verticillatum* usually live outdoors in very pronounced shaded areas and produce rhizoid buds. However, this bryophyte is common in limestone caves, where it grows at varied ranges of light irradiation [37,38]. Whitehouse [39] reported the presence of *E. verticillatum* in Altamira Cave in 1955. We also find it in Encajero Cave, in Quesada, Jaen, with a thick spongy growth giving rise to considerable tufa formations, associated with *Stichococcus bacillaris* [40].

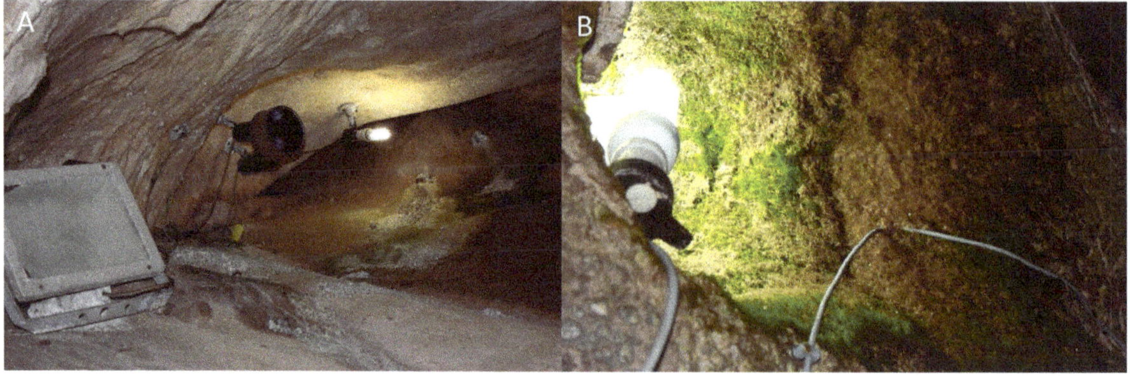

Figure 4. Growth of the bryophyte *Eucladium verticillatum*. (**A**) Sala del Volcan. (**B**) Sala del Aguila.

In the studied cases, cells of *M. israelensis*, *Phormidium* sp. and *E. verticillatum* formed a dense biofilm. Scanning electron microscopy images of the biofilms showed a dense network of *M. israelensis* cells (Figure 5A) and filaments of *Phormidium* sp. (Figure 5B). The cells of *M. israelensis* and *Phormidium* sp. are usually attached to each other by appendages. Some cells are lodged in depressions that suggest an active mechanism of calcite dissolution.

Figure 5. Scanning electron images of biofilms. (**A**) Cells of *Myrmecia israelensis*. (**B**) Filaments of *Phormidium* sp. Scale bar: 5 μm.

3.1.2. Sector T2

This sector is characterized by the location of an elevator and the staircase in the center, connecting with the entrance, and with halls and galleries at both sides connecting with sectors T1 and T3. Therefore, this sector was anthropically modified and showed high environmental changes with respect to the rest of the cave.

In Sala de Marco Craso, there was a phototrophic colonization occupying large extensions. The light microscope as well as the scanning electron microscope study revealed that the phototrophic community was almost exclusively formed by *M. israelensis*. The substratum was heavily corroded, and on it had been deposited masses of the algae.

In other biofilms from Sala de Marco Craso, several species of cyanobacteria were observed: *Synechocystis* sp. and *Nostoc* cf. *edaphicum* (Figure 6A,B) and the microalgae *Pseudococcomyxa simplex*, *Chlorella* sp. and sporocysts of *M. israelensis* (Figure 6C). Sporocysts are cells within which spores are formed. The identification of a species of *Nostoc* is interesting, since it indicated that there is nitrogen fixation from the air in the cave by cyanobacteria, which would enrich the niche and provide this element for the growth of non-nitrogen-fixing heterotrophic microorganisms. The violet color of Figure 6A is due to the pigment phycocyanin that produces *Nostoc* and that adheres to membranes.

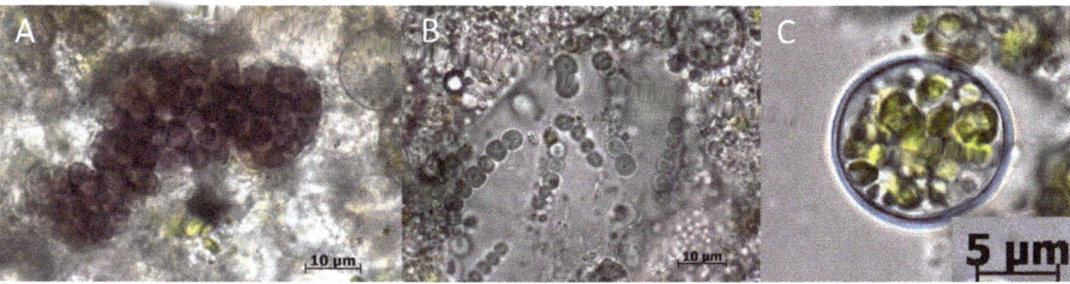

Figure 6. (**A**,**B**) *Nostoc* cf. *edaphicum*. (**C**) Sporocysts of *Myrmecia israelensis*. Scale bar: 10 μm in (**A**,**B**), and 5 μm in (**C**).

3.1.3. Sector T3

In sector T3, Sala del Aguila, the biofilms developed significantly. Phototrophic colonization adopts different morphologies in the distribution on the mineral substratum, from circular colonies to a homogeneous biofilm covering the rock surface.

The study of the cultured sample in the laboratory from Sala del Aguila biofilms showed cells of *Synechocystis* sp., *Diadesmis contenta*, *Phormidium* sp., *Chroococcidiopsis* sp., *Pseudococcomyxa simplex* and *M. israelensis*, as well as the bryophyte *E. verticillatum*.

In the samples from the circular colonies, a whitish mass formed by organisms that retained color, as well as many empty sheaths and dead organisms were revealed under the light microscope. This may be due to the alternation of dry and wet seasons.

We did not study fungi, because given the amount of bacteria and cyanobacteria in the biofilms, which usually produce bioactive (antifungal) substances, and in light of previous studies in other caves, it does not seem that fungi can attain great relevance in these phototrophic communities, as reported for Nerja Cave [15].

In the gallery accessing to Santuario de Noctiluca, the biofilms were also abundant. Under the light microscope, *M. israelensis* and *Phormidium* sp. were evidenced. The scanning electron microscope study showed a dense web of filaments of both cyanobacteria and associated filamentous bacteria.

The phototrophic organisms found in Tesoro Cave have been reported in many other European caves [14,37,41–49].

3.2. *Preliminary Cleaning Testing*

After reviewing the different cleaning procedures reported in the literature [50–53], we selected three methods: (i) mechanical cleaning with liquid nitrogen, (ii) cleaning with sodium hypochlorite, and (iii) cleaning with hydrogen peroxide. They were tested in biofilms from Sala del Aguila.

The cleaning with liquid nitrogen on biofilms was performed with a brush. Liquid nitrogen provided a cleaning technique that combined mechanical removal with freezing. This protocol, in the case of a biofilm composed mainly of the bryophyte *E. verticillatum*, or cyanobacteria and green algae, was much less effective than when using chemicals, because it was not able to completely remove the green biofilms.

The cleaning with sodium hypochlorite on the biofilm of cyanobacteria and algae was effective, and was maintained for a long time (Figure 7). This treatment has been applied in many caves all over the world for removing lampenflora [52]. Two months later, the rock presented a whiter hue than at the end of the cleaning, indicating that the chemical had a residual effect.

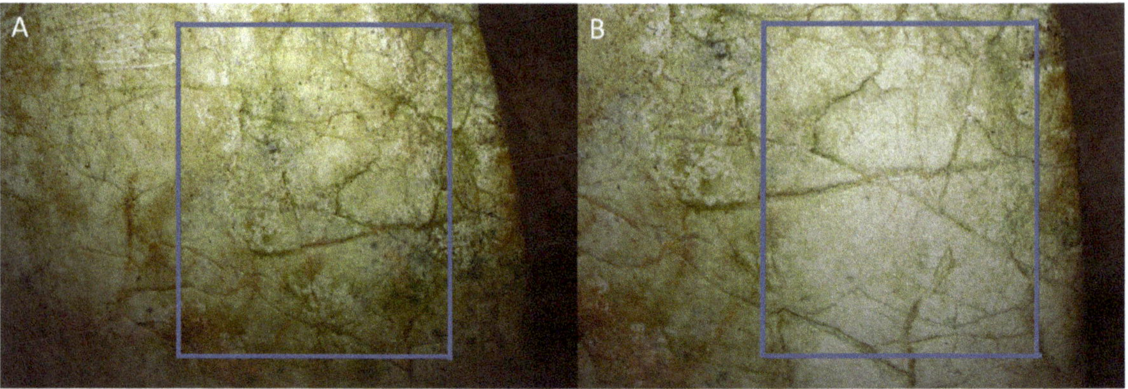

Figure 7. Cleaning of phototrophic biofilms with sodium hypochlorite. (**A**) Before treatment. (**B**) After treatment.

Cleaning with hydrogen peroxide on *E. verticillatum* showed that the hydrogen peroxide applied directly with a brush was effective. In the case of abundant colonization of bryophytes, it is preferable to use the mechanical removal of the organisms, that can be done easily, and to only apply hydrogen peroxide to the residual protonema and rhizoids that remain on the rock. Likewise, in the areas on which bryophytes are growing, lighting should be removed (Figure 8).

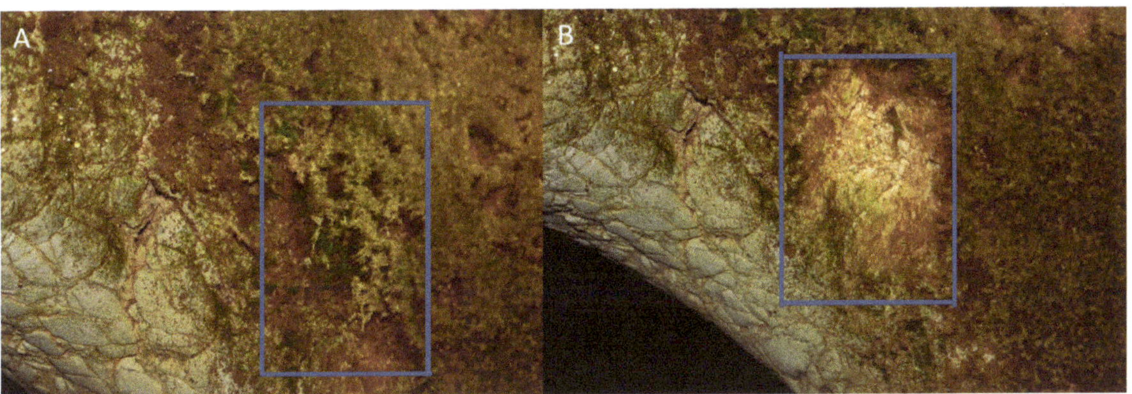

Figure 8. Cleaning with hydrogen peroxide on *Eucladium verticillatum*. (**A**) Before cleaning. (**B**) After cleaning.

The cleaning with hydrogen peroxide on biofilms of cyanobacteria, algae and bryophytes was very effective as it eliminated the phototrophic community without leaving residues (Figure 9). This was considered an environmentally acceptable procedure for removing cave biofilms [51]. Cleaning efficiency can be maintained after treatment if the lighting is removed after cleaning. In fact, under these environmental conditions, the cleaned areas should not develop new biofilms and will guarantee the persistence of the cleaning effect.

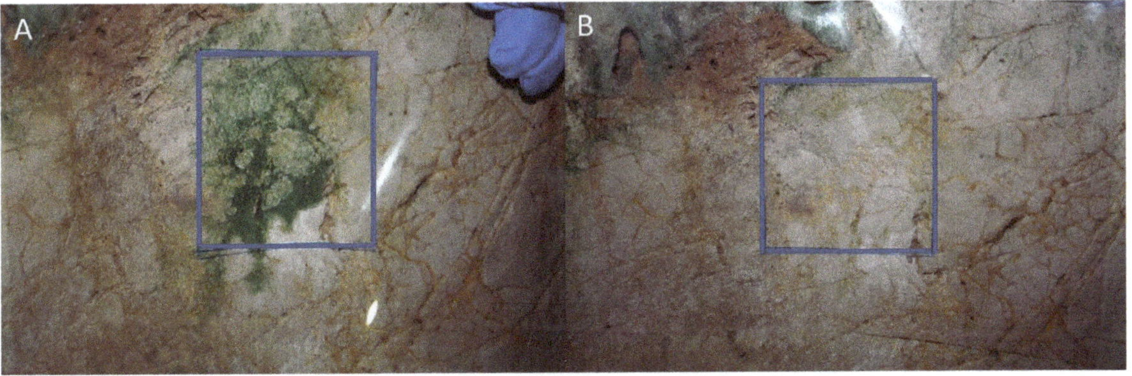

Figure 9. Cleaning of biofilms with hydrogen peroxide. (**A**) Before cleaning. (**B**) After cleaning.

From the preliminary tests carried out in the cave, it can be inferred that the most suitable methods for the cleaning and removal of the extensive green biofilms coating the cave surface were the use of hydrogen peroxide and sodium hypochlorite. The first should be used preferentially for its innocuousness. The second can leave residue, and odor can remain for a short time. Because of its complicated transport and execution, as well as possible danger to the applicator and for not having shown greater effectiveness than the other two treatments, the use of liquid nitrogen is not recommended.

3.3. Recommended Cleaning Protocols

The recommended procedure is first to carry out mechanical cleaning in those cases where the biofilm is thick, such as, for example, in cases of bryophytes and dense algae biofilms. Subsequently, hydrogen peroxide must be used to remove the adhered residues, and not be mechanically removed. This allows the use of less of the chemical and optimization of the results.

Cases where the green color is not completely eliminated can be treated exceptionally well with sodium hypochlorite, first testing a small rock area to check its effectiveness. If the green color does not disappear with the hydrogen peroxide, or subsequently with sodium hypochlorite, it is likely because the cells with the chlorophyll pigment are embedded in a calcite matrix.

Most of the rock art paintings are in a gallery not accessible to visitors (Galeria de Breuil, Figure 1), and therefore without lighting. No green biofilms were observed in this part of the cave. However, Sala del Aguila has a marked problem of colonization by phototrophic microorganisms, associated with the presence of a lighting source, which is very close to some of the rock art painting. This converts the cave into no longer a natural monument, but a site of cultural interest, and therefore the precautions that the current legislation provides for this site were adopted. The occurrence of a lighting point relatively near a painted horse protome, a typology from the Upper Paleolithic, threatened the painted area (Figure 10). The occurrence of bryophyte protonema and/or phototrophic microorganisms represents a real danger for the conservation of the painting. Considering these conditions, mechanical cleaning was proposed for the bryophytes and active phototrophic colonization, which was supervised at all times by the archaeological direction of the cave throughout the duration of the intervention.

Figure 10. Lighting point in Sala del Aguila with phototrophic biofilms next to the Paleolithic painting. On the right, tracing of a Paleolithic horse protome using digital image analysis.

3.4. Effective Cleaning of Biofilms in Tesoro Cave by Agora S.L.

After the preliminary testing of the cleaning procedures, the treatments applied in cleaning the whole cave included mechanical removal and cleaning with hydrogen peroxide and/or sodium hypochlorite. The distribution of cleaning treatments was as follows: physicochemical cleaning, 95% (hydrogen peroxide 94%, sodium hypochlorite 1%); only chemical cleaning, 5%: sodium hypochlorite. The products were applied according to the needs of the area to be cleaned. The cleaning lasted two months for three full-time restorers (Figure 11).

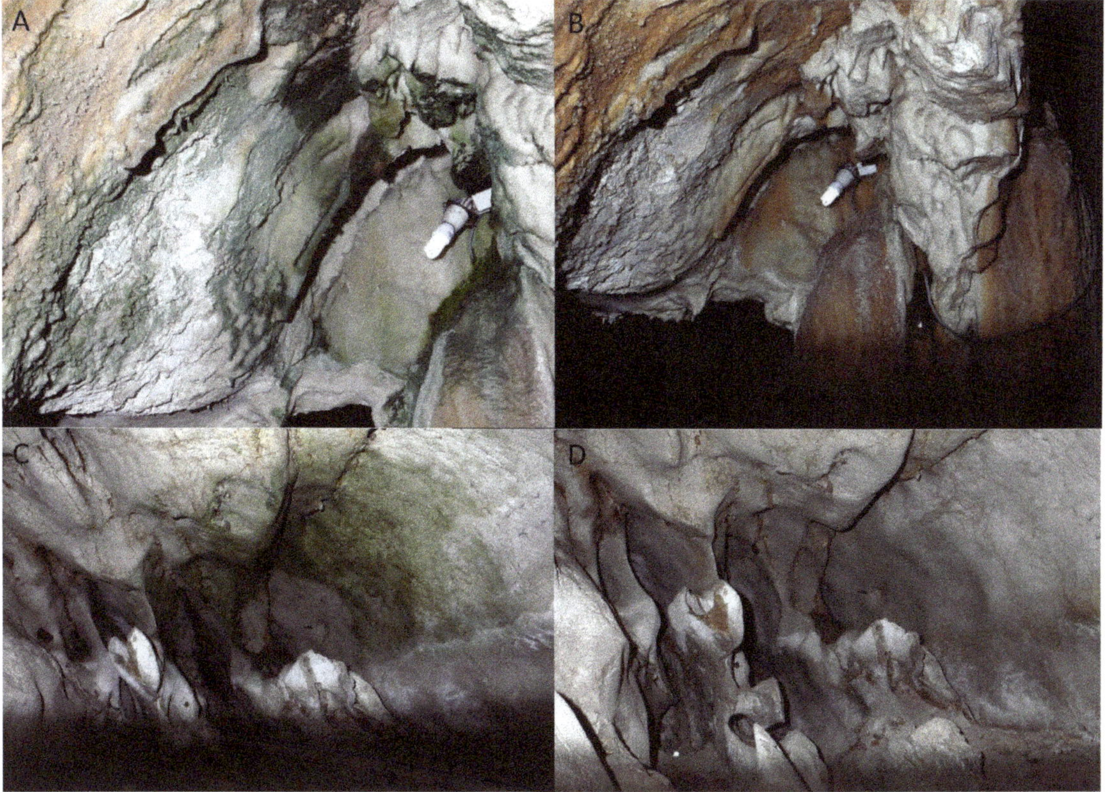

Figure 11. (**A**) Biofilms in Sala del Lago (T1) before cleaning. (**B**) After cleaning. (**C**) Biofilms in Sala del Aguila (T3) before cleaning. (**D**) After cleaning. Cleaning consisted of mechanical removal of biofilms and treatment of the residues with hydrogen peroxide.

4. Conclusions

The phototrophic biofilms covering the speleothems, walls and ground in Tesoro Cave, Rincon de la Victoria, Spain, were successfully cleaned using a protocol involving the mechanical removal of dense biomass, treatment with hydrogen peroxide and, when needed, it was followed by sodium hypochlorite. To guarantee long-term efficiency, the old lighting system should be replaced with a new design with monochrome or LED lamps to be located in the driest area of the cave, or alternatively, implementing visits with electric lanterns, which is possible due to the easy walkable cave trail.

Author Contributions: Conceptualization, C.S.-J.; investigation, V.J., M.H.-M., M.A.R.-C. and C.S.-J.; cleaning and restoration, F.R., C.A. and J.A.; writing—original draft preparation, C.S.-J.; writing—review and editing, C.S.-J. All authors have read and agreed to the published version of the manuscript.

Funding: The research and cave cleaning were funded by Rincon de la Victoria City Hall.

Institutional Review Board Statement: Not applicable.

Informed Consent Statement: Not applicable.

Data Availability Statement: Data supporting reported results on taxonomical and morphological identification of organisms are available from the authors.

Acknowledgments: The authors wish to acknowledge the professional support of the CSIC Interdisciplinary Thematic Platform Open Heritage: Research and Society (PTI-PAIS) as well as the facilities provided by Rincon de la Victoria City Hall for this study. The authors thank Jesus Muñoz Fuente, Real Jardin Botanico, CSIC, Madrid, for the identification of *Eucladium verticillatum*.

Conflicts of Interest: The authors declare no conflict of interest. The funders had no role in the design of the study; in the collection, analyses, or interpretation of data; in the writing of the manuscript, or in the decision to publish the results.

References

1. Mulec, J. Lampenflora. In *Encyclopedia of Caves*; White, W.B., Culver, D.C., Eds.; Academic Press: Waltham, MA, USA, 2012; pp. 451–456.
2. Trinh, D.A.; Trinh, Q.H.; Tran, N.; Guinea, J.G.; Mattey, D. Eco-friendly remediation of lampenflora on speleothems in tropical karst caves. *J. Cave Karst Stud.* **2018**, *80*, 1–12. [CrossRef]
3. Nikolić, N.; Zarubica, N.; Gavrilović, B.; Predojević, D.; Trbojević, I.; Simić, G.S.; Popović, S. Lampenflora and the entrance biofilm in two show caves: Comparison of microbial community, environmental, and biofilm parameters. *J. Cave Karst Stud.* **2020**, *82*, 69–81. [CrossRef]
4. Lefèvre, M. La maladie verte de Lascaux. *Stud. Conserv.* **1974**, *19*, 126–156.
5. Martin-Sanchez, P.; Miller, A.Z.; Saiz-Jimenez, C. Lascaux Cave: An example of fragile ecological balance in subterranean environments. In *Microbial Life of Cave Systems*; Engel, A.S., Ed.; DeGruiter: Berlin/Heidelberg, Germany, 2015; pp. 280–301.
6. Saiz-Jimenez, C.; Cuezva, S.; Jurado, V.; Fernandez-Cortes, A.; Porca, E.; Benavente, D.; Cañaveras, J.C.; Sanchez-Moral, S. Paleolithic art in peril: Policy and science collide at Altamira Cave. *Science* **2011**, *334*, 42–43. [CrossRef]
7. Cuezva, S.; Jurado, V.; Fernandez-Cortes, A.; Garcia-Anton, E.; Rogerio-Candelera, M.A.; Ariño, X.; Benavente, D.; Cañaveras, J.C.; Saiz-Jimenez, C.; Sanchez-Moral, S. Scientific data suggest Altamira Cave should remain closed. In *Microbial Life of Cave Systems*; Engel, A.S., Ed.; DeGruiter: Berlin/Heidelberg, Germany, 2015; pp. 303–320.
8. Jurado, V.; Gonzalez-Pimentel, J.L.; Fernandez-Cortes, A.; Martin-Pozas, T.; Ontañon, R.; Palacio, E.; Hermosin, B.; Sanchez-Moral, S.; Saiz-Jimenez, C. Early detection of phototrophic biofilms in the Polychrome Panel, El Castillo Cave, Spain. *Appl. Biosci.* **2022**, *1*, 40–63. [CrossRef]
9. Procházková, K.; Nemcová, Y.; Neustupa, J. A new species *Jenufa aeroterrestrica* (*Chlorophyceae incertae sedis*, *Viridiplantae*), described from Europe. *Preslia* **2015**, *87*, 403–416.
10. Ariño, X.; Hernandez-Marine, M.; Saiz-Jimenez, C. Colonization of Roman tombs by calcifying cyanobacteria. *Phycologia* **1997**, *36*, 366–373. [CrossRef]
11. Saiz-Jimenez, C. Biogeochemistry of weathering processes in monuments. *Geomicrobiol. J.* **1999**, *16*, 27–37. [CrossRef]
12. Albertano, P. Cyanobacterial biofilms in monuments and caves. In *Ecology of Cyanobacteria II: Their Diversity in Space and Time*; Whitton, B.A., Ed.; Springer: Dordrecht, The Netherlands, 2012; pp. 317–343.
13. Del Rosal Padial, Y.; Jurado Lobo, V.; Hernández-Mariné, M.; Roldán Molina, M.; Sáiz-Jiménez, C. Biofilms en cuevas turísticas: La Cueva de Nerja y la Cueva del Tesoro. In *El Karst y el Hombre: Las Cuevas Como Patrimonio Mundial*; Andreo, B., Durán, J.J., Eds.; Asociación de Cuevas Turísticas Españolas: Madrid, Spain, 2016; pp. 103–114.
14. Del Rosal, Y.; Jurado, V.; Roldán, M.; Hernández-Mariné, M.; Saiz-Jiménez, C. Cyanidium sp. colonizadora de cuevas turísticas. In *Estudio y Conservación del Patrimonio Cultural. Actas*; Moreno Oliva, M., Rogerio Candelera, M.A., López Navarrete, J.T., Hernández Jolín, V., Eds.; Universidad de Málaga: Málaga, Spain, 2015; pp. 170–173.
15. Jurado, V.; del Rosal, Y.; Gonzalez-Pimentel, J.L.; Hermosin, B.; Saiz-Jimenez, C. Biological control of phototrophic biofilms in a show cave: The case of Nerja Cave. *Appl. Sci.* **2020**, *10*, 3448. [CrossRef]
16. Jurado, V.; Novakova, A.; Hernandez-Marine, M.; Saiz-Jimenez, C. Cueva del Tesoro, Rincon de la Victoria, Malaga: A treasure of biodiversity. In *The Conservation of Subterranean Cultural Heritage*; Saiz-Jimenez, C., Ed.; CRC Press/Balkema: Leiden, The Netherlands, 2014; pp. 207–214.
17. Rippka, R.; Deruelles, J.; Waterbury, J.W.; Herdman, M.; Stanier, R.G. Generic assignments, strain histories and properties of pure cultures of cyanobacteria. *J. Gen. Microbiol.* **1979**, *111*, 1–61. [CrossRef]
18. Bischoff, H.W.; Bold, H.C. *Phycological Studies IV. Some Soil Algae from Enchanted Rock and Related Algal Species*; University of Texas Publication: Austin, TX, USA, 1963; Volume 6318, pp. 1–95.
19. Ettl, H.; Gärtner, G. *Syllabus der Boden, Luft und Flechtenalgen*; Fischer, G., Ed.; Springer: Berlin/Heidelberg, Germany, 1995.
20. Komárek, J.; Anagnostidis, K. Cyanoprokaryota I: Chroococcales. In *Süsswasserflora von Mitteleuropa 19/1*; Ettl, H., Gärtner, G., Heynig, H., Mollenhauer, D., Eds.; Gustav Fischer: Jena, Germany, 1998.
21. Komárek, J.; Anagnostidis, K. Cyanoprokaryota 2: Oscillatoriales. In *Süsswasserflora von Mitteleuropa 19/2*; Büdel, B., Krienitz, L., Gärtner, G., Schagerl, M., Eds.; Spektrum Akademischer Verlag: Berlin/Heidelberg, Germany, 2005.

22. Komárek, J. Cyanoprokaryota 3: Heterocytous Genera. In *Süsswasserflora von Mitteleuropa 19/3*; Büdel, B., Gärtner, G., Krienitz, L., Schagerl, M., Eds.; Spektrum Akademischer Verlag: Berlin/Heidelberg, Germany, 2013.
23. Albertano, P.; Ciniglia, C.; Pinto, G.; Pollio, A. The taxonomic position of *Cyanidium*, *Cyanidioschyzon* and *Galdieria*: An update. *Hydrobiologia* **2000**, *433*, 137–143. [CrossRef]
24. Guiry, M.D.; Guiry, G.M. AlgaeBase. World-Wide Electronic Publication. Galway: National University of Ireland. Available online: https://www.algaebase.org (accessed on 1 March 2022).
25. Nübel, U.; Garcia-Pichel, F.; Muyzer, G. PCR primers to amplify 16S rRNA genes from cyanobacteria. *Appl. Environ. Microbiol.* **1997**, *63*, 3327–3332. [CrossRef] [PubMed]
26. Altschul, S.F.; Gish, W.; Miller, W.; Myers, E.W.; Lipman, D.J. Basic local alignment search tool. *J. Mol. Biol.* **1990**, *215*, 403–410. [CrossRef]
27. Thompson, J.D.; Higgins, D.G.; Gibson, T.J. CLUSTAL W: Improving the sensitivity of progressive multiple sequence alignment through sequence weighting, position-specific gap penalties and weight matrix choice. *Nucleic Acids Res.* **1994**, *22*, 4673–4680. [CrossRef] [PubMed]
28. Felsenstein, J. Evolutionary trees from DNA sequences: A maximum likelihood approach. *J. Mol. Evol.* **1981**, *17*, 368–376. [CrossRef] [PubMed]
29. Tamura, K.; Stecher, G.; Kumar, S. MEGA11: Molecular Evolutionary Genetics Analysis Version 11. *Mol. Biol. Evol.* **2021**, *38*, 3022–3027. [CrossRef]
30. Gross, W.; Heilmann, I.; Lenze, D.; Schnarrenberger, C. Biogeography of the Cyanidiaceae (Rhodophyta) based on 18S ribosomal RNA sequence data. *Eur. J. Phycol.* **2001**, *36*, 275–280. [CrossRef]
31. Walker, J.J.; Spear, J.R.; Pace, N.R. Geobiology of a microbial endolithic community in the Yellowstone geothermal environment. *Nature* **2005**, *434*, 1011–1014. [CrossRef]
32. Iovinella, N.; Eren, A.; Pinto, G.; Pollio, A.; Davis, S.J.; Cennamo, P.; Ciniglia, C. Cryptic dispersal of Cyanidiophytina (Rhodophyta) in non-acidic environments from Turkey. *Extremophiles* **2018**, *22*, 713–723. [CrossRef]
33. Ciniglia, C.; Pinto, G.; Pollio, A. Cyanidium from caves: A reinstatement of *Cyanidium chilense* Schwabe (Cyanidiophytina, Rhodophyta). *Phytotaxa* **2017**, *295*, 86–88. [CrossRef]
34. Ciniglia, C.; Yoon, H.S.; Pollio, A.; Pinto, G.; Bhattacharya, D. Hidden biodiversity of the extremophilic Cyanidiales red algae. *Mol. Ecol.* **2004**, *13*, 1827–1838. [CrossRef] [PubMed]
35. Ciniglia, C.; Cennamo, P.; De Natale, A.; De Stefano, M.; Sirakov, M.; Iovinella, M.; Yoon, H.S.; Pollio, A. *Cyanidium chilense* (Cyanidiophyceae, Rhodophyta) from tuff rocks of the archeological site of Cuma, Italy. *Phycol. Res.* **2019**, *67*, 311–319. [CrossRef]
36. Azúa-Bustos, A.; González-Silva, C.; Mancilla, R.A.; Salas, L.; Palma, R.E.; Wynne, J.J.; McKay, C.P.; Vicuña, R. Ancient photosynthetic eukaryote biofilms in an Atacama Desert coastal cave. *Microb. Ecol.* **2009**, *58*, 485–496. [CrossRef] [PubMed]
37. Mulec, J.; Kubesová, S. Diversity of bryophytes in show caves in Slovenia and relation to light intensities. *Acta Carsologica* **2010**, *39*, 587–596. [CrossRef]
38. Glime, J.M. Caves—Overall bryophyte flora. In *Bryophyte Ecology. Habitat and Role*; Glime, J.M., Ed.; Michigan Technological University: Houghton, MI, USA, 2021; Chapter 18-2; Volume 4, pp. 1–30.
39. Whitehouse, H.L.K. The production of protonemal gemmae by mosses growing in deep shade. *J. Bryol.* **1980**, *11*, 133–138. [CrossRef]
40. Saiz-Jimenez, C.; Hermosin, B. Thermally assisted hydrolysis and methylation of the black deposit coating the ceiling and walls of Cueva del Encajero, Quesada, Spain. *J. Anal. Appl. Pyrol.* **1999**, *49*, 349–357. [CrossRef]
41. Kol, E. Algal growth experiments in the Baradla cave at Aggletek. *Int. J. Speleol.* **1967**, *2*, 457–474. [CrossRef]
42. Mulec, J.; Kosi, G.G.; Vrhovsek, D. Characterization of cave aerophytic algal communities and effects of irradiance levels on production of pigments. *J. Cave Karst Stud.* **2008**, *70*, 3–12.
43. Czerwik-Marcinkowska, J.; Mrozińska, T. Algae and cyanobacteria in caves of the Polish Jura. *Pol. Bot. J.* **2011**, *56*, 203–243.
44. Cennamo, P.; Marzano, C.; Ciniglia, C.; Pinto, G.; Cappelletti, P.; Caputo, P.; Pollio, A. A survey of the algal flora of anthropogenic caves of Campi Flegrei (Naples, Italy) archeological district. *J. Cave Karst Stud.* **2012**, *74*, 243–250. [CrossRef]
45. Czerwik-Marcinkowska, J. Observations on aerophytic cyanobacteria and algae from ten caves in the Ojców National Park. *Acta Agrobot.* **2013**, *66*, 39–52. [CrossRef]
46. Czerwik-Marcinkowska, J.; Wojciechowska, A.; Massalski, A. Biodiversity of limestone caves: Aggregations of aerophytic algae and cyanobacteria in relation to site factors. *Pol. J. Ecol.* **2015**, *63*, 481–499. [CrossRef]
47. Cennamo, P.; Montuori, N.; Trojsi, G.; Fatigati, G.; Moretti, A. Biofilms in churches built in grottoes. *Sci. Total Environ.* **2016**, *543*, [illegible]. [CrossRef]
48. Popkova, A.; Mazina, S. Microbiota of hypogean habitats in Otap Head Cave. *J. Environ. Res. Eng. Manag.* **[illegible]**, [illegible]. [CrossRef]
49. Popović, S.; Nikolić, N.; Jovanović, J.; Predojević, D.; Trbojević, I.; Manić, L.; Simić, G.S. Cyanobacterial and algal abundance and biomass in cave biofilms and relation to environmental and biofilm parameters. *Int. J. Speleol.* **2019**, *48*, 49–61. [CrossRef]
50. Mulec, J.; Kosi, G. Lampenflora algae and methods of growth control. *J. Cave Karst Stud.* **2009**, *71*, 109–115.
51. Faimon, J.; Stelcl, J.; Kubesová, S.; Zimák, J. Environmentally acceptable effect of hydrogen peroxide on cave "lamp-flora", calcite speleothems and limestones. *Environ. Pollut.* **2003**, *122*, 417–422. [CrossRef]

52. Esteban Pérez, E. Study and remediation of environmental problems caused due to the growth of algae in speleothems of calcareous caves adapted for tourism—A case of success in Spain. *J. Environ. Geol.* **2018**, *2*, 20–28.
53. Baquedano Estévez, C.; Moreno Merino, L.; de la Losa Román, A.; Durán Valsero, J.J. The lampenflora in show caves and its treatment: An emerging ecological problem. *Int. J. Speleol.* **2019**, *48*, 249–277. [CrossRef]

applied sciences

Article

Weathering Processes on Sandstone Painting and Carving Surfaces at Prehistoric Rock Sites in Southern Spain

Juan Carlos Cañaveras [1,*], Enrique Sanz-Rubio [2] and Sergio Sánchez-Moral [3]

[1] Department of Environmental and Earth Sciences, University of Alicante, Campus San Vicente del Raspeig, E-03690 Alicante, Spain
[2] GEOMNIA Natural Resources SLNE, C/Cea Bermúdez, 14, E-28003 Madrid, Spain; esanz@geomnia.es
[3] Department of Geology, National Museum of Natural Sciences, MNCN, CSIC, C/José Gutiérrez Abascal, 2, E-28006 Madrid, Spain; ssmilk@mncn.csic.es
* Correspondence: jc.canaveras@ua.es

Abstract: The sandstones which constitute the host rock for the prehistoric artwork in the Rock Groups of Tajo de las Figuras and Peñas de Cabrera (southern Spain) show a serious degree of alteration, due both to natural processes and those related to anthropogenic and animal activity. A detailed study was carried out on the petrological and compositional characteristics of the sandstones (fresh and altered rock) in both rock groups, and on the geological and climatological characteristics of the area in which they are located. The sandstones have very similar petrological and compositional characteristics in both areas. This likeness causes the nature of the natural weathering processes to be similar in the rock areas studied. These processes can be divided in terms of the predominant mechanisms of alteration into three inter-related categories: mechanical weathering, chemical weathering, and bio-induced alteration processes. However, the different climatic conditions of the areas in which the two rock areas are located directly influences the intensity of these processes. The precipitation and the range of temperature variation with heavy winter frosts in the area of El Tajo de las Figuras are significantly higher than in the area of Peñas de Cabrera; this translates into a higher rate of weathering at El Tajo de las Figuras. Regarding the anthropogenic action, two types of influence on the deterioration can be distinguished: a direct one, which consists of scouring and wetting of the walls in order to increase the chromatic contrast; and an indirect one, which is the extraction of blocks of sandstone in the upper part of rock shelters, which in turn encourages the development of the chemical weathering processes.

Keywords: sandstone weathering; chemical weathering; rock shelter; rock art

Citation: Cañaveras, J.C.; Sanz-Rubio, E.; Sánchez-Moral, S. Weathering Processes on Sandstone Painting and Carving Surfaces at Prehistoric Rock Sites in Southern Spain. *Appl. Sci.* **2022**, *12*, 5330. https://doi.org/10.3390/app12115330

Academic Editor: Asterios Bakolas

Received: 9 May 2022
Accepted: 23 May 2022
Published: 25 May 2022

1. Introduction

Rock materials have been employed as a host rock and/or frame for artistic representations, paintings, sculpture, and architecture since prehistoric times. In the last decades, great progress has been made in investigating the processes of alteration in rock used for architectural monuments ([1–11], among others). However, works on to the state of alteration of the host rock for artistic representations located in natural rock outcroppings, both karstic cavities and rock shelters, are much less common. Recent works address this problem in art representations executed in natural rock shelters. Campbell (1991) [12] classified the intensity of the weathering which has affected the sandstone host rock for the historic carvings in the Writing-on-Stone Park in Canada. Pentecost (1991) [13] studied the rate of weathering of the Cretaceous Ardingly Sandstone in the Weald (England) where there are abundant pictorial representations. Benito et al. (1993) [14] analysed the state of conservation of the paintings and carvings in different rock groups in NE Spain together with the processes of weathering in their sandstone host rock. Sjöberg (1994) [15] carried out a study on the alteration of surfaces with carvings from the Bronze Age in southwestern Sweden. Hall et al. (2007) [16] and Meiklejohn et al. (2009) [17] suggested that rock moisture and thermal regimes exert the most damaging influence on indigenous rock art painted on porous sandstones in southern Africa. Díez-Herrero et al. (2009) [18] studied the influence of direct insolation in the weathering of cave paintings on Triassic sandstones at Central Spain. More recently, Peña-Monné et al. (2022) [19] highlighted the role of hydroclastic and haloclastic processes on weathering processes of the sandstones in the painted rock shelters of Cerro Colorado (Argentina).

Paintings and carvings on rock surfaces become closely linked with the natural evolution of the host rock from the moment that they are executed. Because of this, and in order to carry out an extensive study of the state of conservation of these artistic and cultural features, it is first necessary to establish the processes and mechanisms by which the rock has been altered. The purpose of the present work is to establish the alteration processes which affect the sandstones that constitute host rocks for the paintings and carvings of the so-called Rock Groups of Tajo de las Figuras (Cádiz) and Peñas de Cabrera (Málaga), southern Spain.

2. Study Area

The Rock Group of Tajo de las Figuras is located within the municipal boundaries of Benalup (Cádiz province, southwestern Spain) (Figure 1A), in the extreme southeast of the Sierra Momia, at an altitude which ranges between 110 and 170 m above sea level. These sierras, which have little relief, are formed of siliceous sandstone belonging to the Aljibe Formation (lower Miocene) that grades downwards into reddish clays which range in age from Oligocene to Aquitanian. The Aljibe Fm., a 1500 to 2000 m thick succession, consists of meter thick planar-tabular or cross-stratified beds of medium to coarse-grained, pale orange to yellowish brown sandstones. The quartzose sandstones of Aljibe Fm. are strongly tectonized, the basic directions of the diaclase network being E–W. Its structure is in the form of piled up thrust sheets which provides important relief in the form of sierras with an average altitude of 400 to 500 m. It also has an abundance of structures due to tectonic instability, such as landslides and sand dykes [20]. The Rock Group of Tajo de las Figuras includes a total of seven natural rock shelters, mainly facing the south, termed Cueva del Tajo de las Figuras, Cueva del Arco, Cueva Cimera, Cueva Negra, Cueva Alta, Cueva del Tesoro, and Cueva de los Pilones. These host a set of engravings which can be attributed to the upper Palaeolithic and post-Palaeolithic (Neolithic–Chalcolithic–Bronze Age) [21]. These paintings are of zoomorphic and anthropomorphic motives and signs [22] and form their own artistic style within the so-called "schematic" style, which can be differentiated from the rest of the artistic rock works in the Iberian Peninsula [23,24].

The Rock Group of Peñas de Cabrera is located within the municipal boundary of Casabermeja, in the interior of the Málaga province (southern Spain) (Figure 1A). Most rock shelters are sited on the northern flank of a thick succession of sandstones which ends in a small hill (570–620 m above sea level) which is located on the southern side of the Lower Miocene Colmenar Depression. The sandstones are arranged in meter-thick beds which generally are folded with dips that slightly exceed 30°. However, at some points, such as where rock shelters with paintings are found, sandstones appear as blocks with chaotic distribution, being arranged vertically and even seem to be inverted with some frequency. The degree of tectonic fracturing of sandstones is very high, showing a dense network of faults and diaclases with preferential N260° E and N orientation. Given their affinity with the Germanic facies and the lack of fauna in the study area, the Permo-Triassic age has been assigned to these sandstone beds [25]; thus placing them in the Andalusian domain precisely in contact with the flysch of the Colmenar Unit. The Rock Group of Peñas de Cabrera is formed of a total of 23 well-known rock shelters of smaller dimensions facing N, NW, and W. It contains a series of post-Palaeolithic paintings (Chalcolithic) which can also be assigned to the "schematic" style, with an abundance of anthropomorphic and idiomorphic representations, idols, tectiforms, and arboriforms [26].

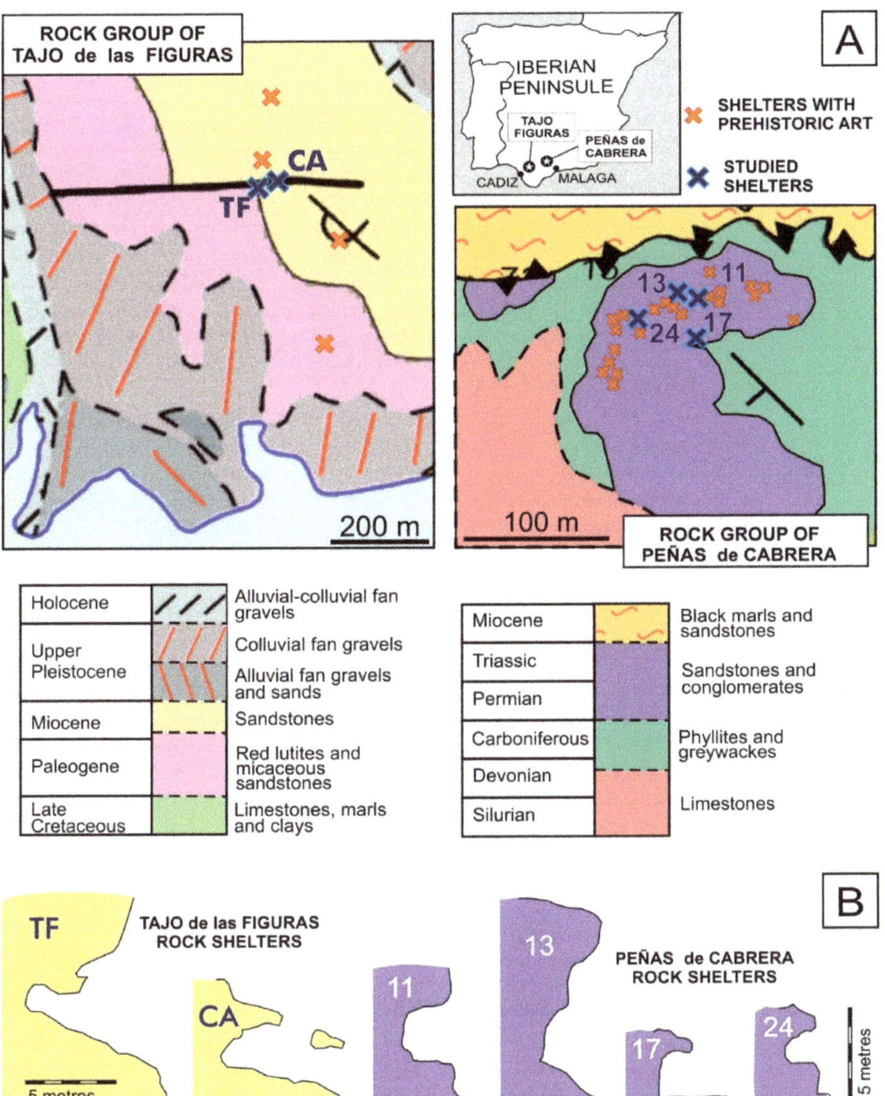

Figure 1. (**A**) Location map. Small crosses indicate locations of rock shelters with prehistoric paintings and engravings. Large crosses indicate locations of the studied rock shelters: Cueva del Tajo de las Figuras (TF), Cueva del Arco (CA), Shelter n° 11, Shelter n° 13, Shelter n° 17, and Shelter n° 24. (**B**) Cross-sections of the studied rock-shelters.

Both Rock Groups (Tajo de las Figuras and Peñas de Cabrera) show a high degree of deterioration, both of natural and anthropic origin, which has promoted various actions aimed at their protection by the relevant authorities, including their temporary closure to tourism [27,28].

As the climatic conditions have a determining role in the development of rock weathering processes [29], the main climatic parameters of the study areas were collected for the period 1995–2014 from data supplied by the Antequera meteorological station (Z = 408, 370146 N, 044454 W) for Peñas de Cabrera and the Grazalema station (Z = 913, 364538 N,

052227 W) for Tajo de las Figuras. Regarding temperatures, both areas have a mesothermal to hot climate, but with a greater seasonal contrast in the area in which Tajo de las Figuras is located with relatively frequent frosts in December, January, and February, and a greater annual thermal fluctuation can be observed. The mean annual temperature values for both Tajo de las Figuras and Peñas de Cabrera areas range from 15.2 to 17.3 °C, and from 14.7 to 16.5 °C, respectively. In the case of precipitations, the differences between the two areas are of much more importance; Grazalema (Tajo de las Figuras) shows indices of very high rainfall with mean annual ranging from 1800 to 2200 mm, which are considerably greater than that in Málaga (Peñas de Cabrera), with mean annual precipitation values ranging from 3310 to 660 mm. The average annual wind speed in Grazalema is 13.86 km/h with gusts of over 130 km/h. In Antequera, the average annual value of the wind speed is slightly lower, 11.99 km/h with some gusts exceeding 80 km/h. At both areas the climate is Csa (hot summer Mediterranean climate) according to the Köppen-Geiger classification [30].

3. Materials and Methods

The geomorphological disposition of the sandstone beds which act as host rock for paintings and carvings was determined through a photo-geological study of the area, a revision of the existing geological cartographies, and field work in which structural and textural data were collected. The study concentrated on six representative rock shelters from the two archaeological groups (Figure 1). Two shelters belong to the Rock Group of Tajo de las Figuras (Cueva del Tajo de las Figuras and Cueva del Arco) and four to the Rock Group of Peñas de Cabrera (rock shelters 11, 13, 17 and 24), as these are the ones that have the largest number of artistic representations. To assess the deterioration phenomena, detailed and comprehensive in situ visual analyses were performed before the sampling campaigns. The ICOMOS ICS glossary (2008) [31] was used for weathering form terminology. A total of 80 samples were taken from these six rock shelters in order to carry out a petrological, mineralogical, and geochemical characterization of alteration processes which affect the sites. These samples correspond both to fresh rock and to altered rock (flaking, filling with joints, oxide patinas, detritus of alveoli, and concretions). Analytical procedures were performed at Museo Nacional de Ciencias Naturales (MNCN-CSIC, Madrid) laboratories. Optical microscopy and SEM techniques were used for the petrographic study. Qualitative and semi-quantitative data were obtained on the chemical composition of the different mineral components present in the sample using an EDS analyser. The mineralogical analyses were performed using X-ray diffractometry, with quartz as the internal standard. The geochemical composition of the host rock and of the alteration products were determined using standard procedures with an X-ray fluorescence spectrometer.

4. Geological Characteristics of Rock Shelters

4.1. Morphology

The general morphological characteristics of the six rock shelters studied are provided below. The main alteration products, whose mechanisms of formation are explained later, are schematically indicated:

- The Cueva del Tajo de las Figuras is a rock shelter opening towards the south, 5.8 m in height, 4.2 m wide, and 8 m deep (Figure 1B). It is located on a vertical incision corresponding to a fault plane with an E–W direction (Figure 2A). It has alveolar structures formed by one centimetre to decimetre-width alveoli with a spherical to ellipsoidal section particularly located in the back wall and the ceiling of the shelter. Holes are used by animals (wasps and birds) for their nests. Patinas of reddish to blackish tones are abundant on the walls and ceiling. Whitish patinas on the external part of the shelter are also present. These stains are closely related to the colonisation of plants and the defecation of animals. The floor of the shelter has a smooth, polished, shining surface. Fallen rock flakes are abundant, particularly in the more external part of the shelter;

- The Cueva del Arco is a rock shelter with a domed roof and an overhanging rock ledge that has partially collapsed. Consequently, a large arch with an elliptical section has been formed at the entrance of the cavity (Figure 2B). Its approximate dimensions are 2.3 m in height, 10 m wide, and 7.5 m deep (Figure 1B). Alveolar surfaces throughout the stratification are abundant on the walls and ceiling. The alveoli are one centimetre in width and have an ellipsoidal shape. Both the walls and the ceiling and floor show patinas of reddish and orange tones and an abundant colony of fungus, lichens, and other plants. Remains of fallen rock flakes are numerous;
- Shelter 11 (Peñas de Cabrera) is morphologically similar to the Cueva del Tajo de las Figuras, but not as deep, with approximate dimensions of 5 m in height, 4 m in width, and 2.5 m deep (Figure 1B). This shelter is orientated to the north and has developed on a practically vertical E–W fracture plane. Alveolar structures similar to those previously described can be seen on the ceiling and upper part of the walls of the shelter (Figure 2C). The floor is covered with a patina of an intense red colour, with bright and dark areas. The red tones of the floor appear to be related to the traces of Neolithic pictorial activity. A well-defined surface blackening can also be observed on walls and ceilings;
- Shelter 13 (Peñas de Cabrera) faces the north according to fracture planes in a N100°E direction and a dip of 50–60° S (Figure 2D), with the development of one meter-width tafone morphologies on its back wall (Figure 2E). This is a shelter with little depth (less than 3 m) and a maximum height of 8 m (Figure 1B). Honeycomb structures formed by alveoli of the order of one millimetre can be seen on the back wall of the shelter together with brownish orange patinas, fallen rock flakes, and white marks from the defecation of animals. Traces of biological communities, such us fungi or lichen, on the inner walls of the shelter are practically absent;
- Shelter 17 (Peñas de Cabrera) is 2.5 m in height, 6 m wide, and 1.8 m maximum in depth (Figure 1B). It faces north and is located very near ground level, thus making a very clear ground-level ledge. This shelter is almost masked by abundant vegetation. Reddish patinas and signs of fallen flakes are very abundant. The lower part of the shelter, up to a height of 50 cm, does not show alteration patinas and is completely devoid of flakes (Figure 2F);
- Shelter 24 (Peñas de Cabrera) faces west and is small (1.8 m high by 4 m wide and a maximum depth of 1.5 m) (Figure 1B). This shelter has a great development of alveolar morphologies in the interior, particularly towards the ceiling. Colonisation of fungi, lichens and other plants is abundant. The walls of the shelter have a very extensive blackish patina.

The degree of tectonic fracturing of the sandstone host rock in both rock groups is very high as they have been affected both by faults and a dense network of joints (Figure 2A,D). There are older recrystallized, hardened, and patinated fractures in the diaclase systems which are more resistant to weathering than the sandstone itself. The fracture network stands out inside the shelters; the sandstone is compartmentalized into patina-encased rock masses. The most recent fractures, on the other hand, are in a period of enlargement. It must be taken into account that this is a tectonically active area [20,25] and therefore the more recent joints must have originated through neo-tectonic activity.

Figure 2. Morphological features of the studied rock-shelters (see text for explanation). (**A**,**B**) Rock Group of Tajo de las Figuras. (**C**–**F**) Rock Group of Peñas de Cabrera. (**A**) View of the entrance of the Cueva del Tajo de las Figuras, which is located in a fault plane; note the high number of fissures and fractures affecting the host rock. (**B**) General view of the Cueva del Arco. (**C**) View of the interior of the shelter 11; honeycomb weathering is developed on walls and ceiling. (**D**) Lateral view of shelter 13, which is developed on a fault plane. (**E**) Frontal view of shelter 13; decimetre-width tafone are developed in the uppermost part of the shelter wall. (**F**) Frontal view of shelter 17; selective weathering due to proximity to ground level is observed.

4.2. Petrology and Geochemistry

The sandstone host rock of the Tajo de las Figuras site is mainly composed of medium to coarse-grained, poorly sorted sub-arkoses. The framework grains show a relatively high degree of roundness (Figure 3A). It is mainly composed of: quartz (80–90%); feldspar (0–15%), with a predominance of plagioclase; mica (<5%), mainly muscovite; tourmaline (<2%); and opaque minerals (<2%). The matrix is very scarce (<5%) and mainly consists of kaolinite masses (to a lesser extent illite). Syntaxial siliceous cements, ferruginous rims and kaolinite pore-linings and pore-fillings were observed. The chemical composition of the Tajo de la Figuras fresh rock samples (Table 1) is characterised by the high content of SiO_2 (>95%) and low content of Al_2O_3 (<1.8%) and other elements. In most of the samples, the MgO and MnO content was below the sensitivity limit of the analytical technique employed (XRF). This fact appears to be related to the low proportion of phyllosilicates observed in the petrographic study.

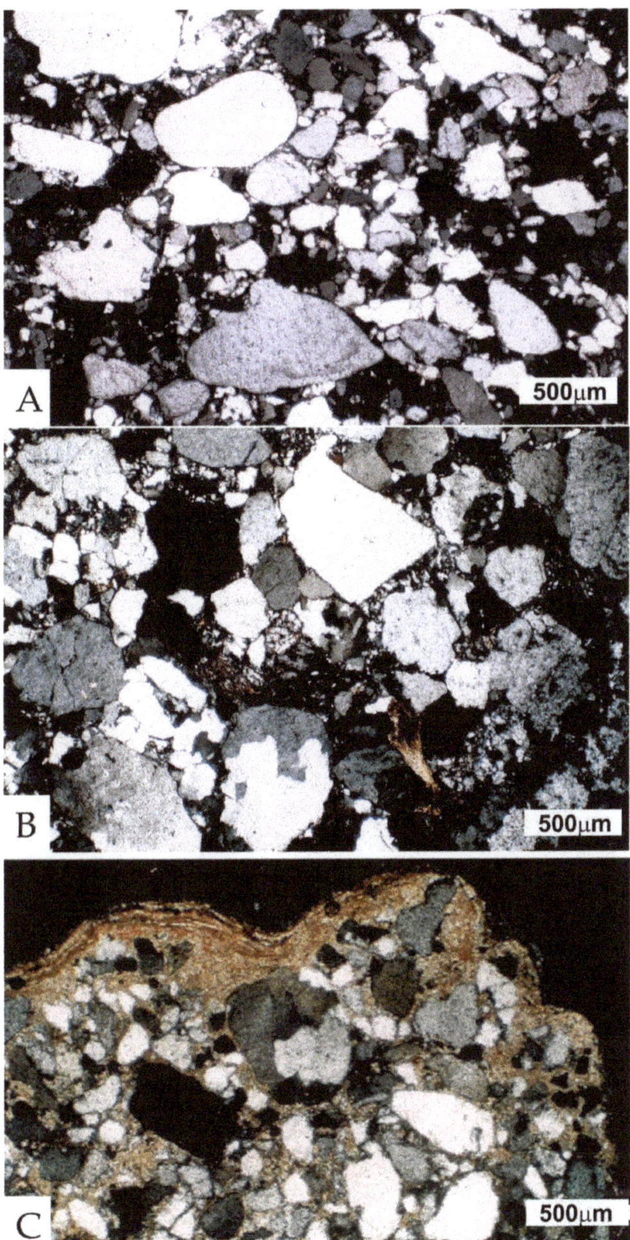

Figure 3. Thin section photomicrographs. (**A**) Poorly sorted sub-arkose showing clast-supported fabric and grains with high degree of roundness (Cueva del Arco). (**B**) Sub-arkose with feldspar and biotite grains showing fine-grained alteration and pore-filling kaolinite cement (Shelter 13, Peñas de Cabrera). (**C**) Stromatolite-like crust surrounding sandstone surface (Cueva del Tajo de las Figuras). Calcite cements filling intergranular pores and partially replacing both siliceous framework grains and cements. All photographs were taken under cross polarized light.

Table 1. Average chemical composition (in weight %) of fresh sandstone host rocks.

	Tajo de las Figuras	Peñas de Cabrera
SiO_2	96.3	91.53
Al_2O_3	1.4	5.5
$FeO + Fe_2O_3$	0.5	0.45
TiO_2	0.25	0.28
MgO	0.10	0.13
CaO	0.07	0.04
MnO	<0.01	<0.01
Na_2O	0.1	0.04
K_2O	0.3	0.98
P_2O_5	0.06	0.1
L.O.I.	0.99	0.90

Peñas de Cabrera sandstones are very similar to those of Tajo de las Figuras. They are medium to coarse-grained, poorly sorted quartz-rich sandstones with grains showing a moderate to high degree of roundness and sphericity. The sandstone is composed mainly of: quartz (80–90%); mica (5–20%); feldspar (mainly plagioclase) (0–15%); and metamorphic rock fragments (0–10%). The matrix is scarce (<5%) and consists of masses of illites and kaolinites. The cements, although scarce, are mainly siliceous and ferruginous. Most of these sandstones can be classified as sub-arkoses, although locally, sub-lithoarenites are relatively abundant (Figure 3B). The chemical composition is also similar to Tajo de las Figuras sandstones (Table 1). The main difference is clearly the greater concentration of Al_2O_3 and K_2O due to the presence of phyllosilicates (mainly mica and illites).

The superficial alteration rinds on the shelters' walls and of the fallen flakes are formed by a compact mass, which is not very porous and has reddish tones. The chemical analysis for these alteration rinds in the two rock groups indicate a marked enrichment in $FeO + Fe_2O_3$ (4.9% $\pm$ 0.5%), MgO (0.42% $\pm$ 0.17%), and in the loss on ignition (L.O.I.) (1.8% $\pm$ 0.3%). This fact is consistent with the petrographic observations made and the mineralogical analyses performed. Framework components show weathering features such as: (1) strongly altered feldspars and biotites; (2) masses of authigenic kaolinites; (3) quartz grains with signs of dissolution and syntaxial overgrowth; (4) and muscovites fragmented into flakes. Fe and Mn oxides/hydroxides in the form of amorphous masses, which commonly appear to be fragmented and re-cemented, are relatively abundant. Traces of biological activity, such as lichen thalli, fungal hyphae, and the remains of insect skeletons were also recognised.

Micrometre to millimetre-thick laminar carbonate crusts (Figures 3C and 4A) formed by 5 to 50 μm-thick layers mainly composed of calcite crystals (micrite and microsparite) also occur. Lenticular gypsum crystals (Figure 4B), Fe and/or Mn oxides/hydroxides grains, clays, and organic elements (biofilms, algae, fungi, and spores) were also recognised. In the areas where these crusts are developed, it was observed that the calcium carbonate crystals penetrated the sandstone replacing and/or disintegrating both the framework components and the matrix/cements (Figure 3C).

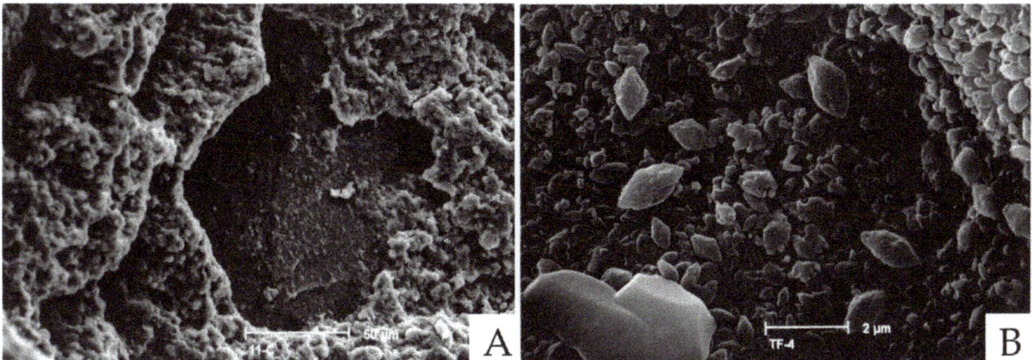

Figure 4. SEM micrographs. (**A**) Detail of stromatolitic calcite layers enveloping grain surfaces (Shelter 11, Peñas de Cabrera). (**B**) Small lenticular gypsum crystals within the stromatolite crust (Cueva del Tajo de las Figuras).

5. Weathering Processes

A simple visual examination of the shelters which form the two rock groups denotes the major effectiveness of weathering processes in the Tajo de las Figuras Group, as is revealed by a major development of alteration rinds, oxide-rich patinas, falling of flakes, etc. A detailed field and laboratory examination of the alteration products and forms allowed us to differentiate between natural weathering processes and those induced by human activity.

5.1. Natural Weathering Processes

The natural weathering processes, observed with different intensity in each of the studied sites, can be divided in terms of the predominant alteration mechanisms in three inter-related groups: mechanical weathering, chemical weathering processes, and bio-induced alteration.

5.1.1. Chemical Weathering Processes

Petrological and geochemical analyses of rock samples from the weathered surface of the walls of the shelters indicate the existence of chemical alteration processes which include dissolution and/or transformation of minerals and the redistribution of ionic substances. The main effect observed was the formation of hardened superficial layers. These layers are enriched in Fe and/or Mn oxides, clays, and silica. This process has been widely referred in the literature as case hardening (e.g., [32–34]). The driving mechanisms for case hardening on sandstones are the hydrolysis and carbo-hydrolysis reactions which affect the aluminosilicate minerals. The carbo-hydrolysis phenomenon constitutes the mechanism whose action has the greatest effectiveness, in a coherent manner, with warm and humid climatic conditions. These conditions favour the adsorption of water and a chemical attack on the rock components through the action of meteoric water which is slightly acidic due to the dissolved CO_2. The CO_2 source may be atmospheric or organic (vegetable covering and soils rich in organic material in the upper area of the shelters). Next is detailed the geochemical expression that summarises feldspar alteration and clay authigenesis:

$6KAlSi_3O_8 + 4CO_2 + 4H_2O \rightarrow K_2Al_4(Al_2Si_6O_{20})(OH)_4 + 4K^+ + 12SiO_2 + 4HCO_3^-$
K-feldspar Illite

$2KAlSi_3O_8 + 2H^+ + 9H_2O \rightarrow Al_2Si_2O_5(OH)_4 + 2K^+ + 4SiO_2 + 8H_2O$
K-feldspar Kaolinite

$2CaAlSi_3O_8 + 11H_2O + 2CO_2 \rightarrow Al_2Si_2O_5(OH)_4 + 2Ca^{2+} + 2HCO_3^- + 4H_4SiO_4$
Anorthite Kaolinite

An example of this phenomenon can be observed in Figure 5A,B, with feldspar crystals altered by carbo-hydrolysis and kaolinite crystals neoformed as the result of the alteration. It is precisely the abundant presence of kaolinite (Figure 5C) in the Tajo de las Figuras samples which corresponds to the high humidity level in this location since this authigenic mineral has been commonly employed as an indicator of wet climates [35]. The availability of water is very important in the development of the chemical weathering processes [36], since the solubility of the feldspars is very low (e.g., 3×10^{-7} mol for the K-feldspar and 6×10^{-7} mol/L for the Na-feldspar) [37] its saturation is reached rapidly and the dissolution of feldspars needs a continuous renewal of under-saturated water.

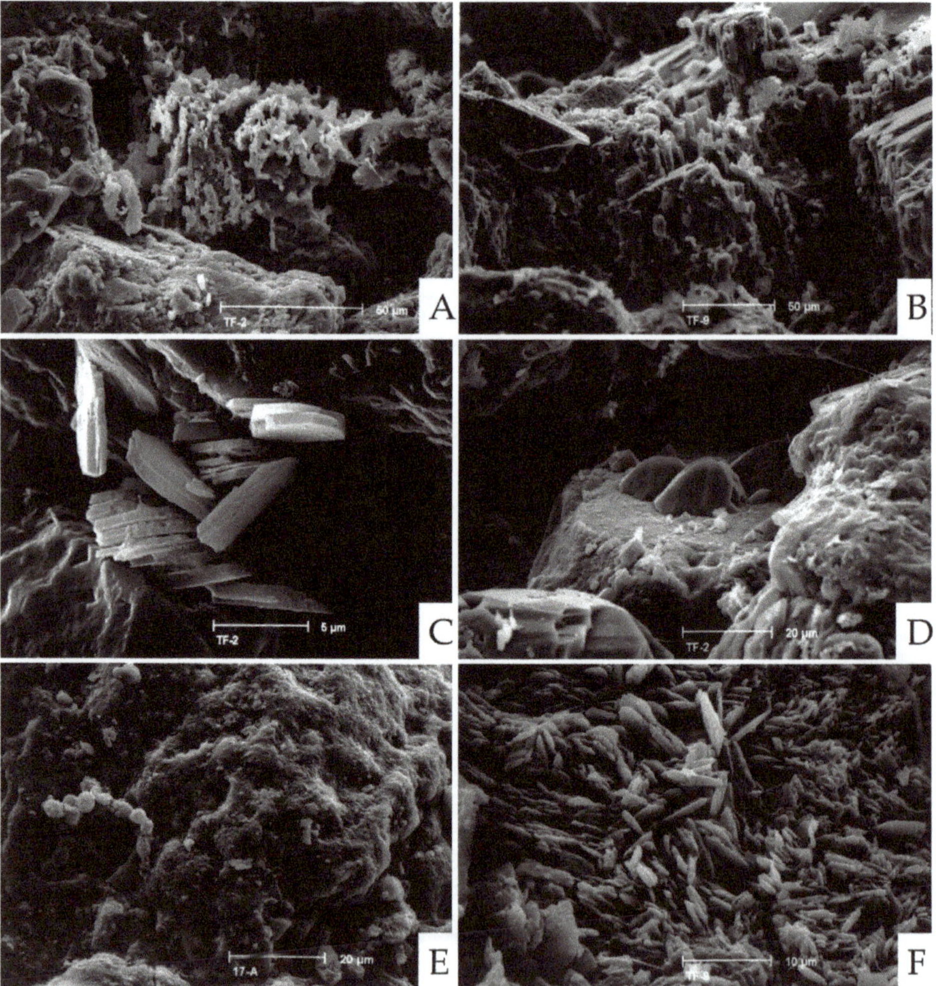

Figure 5. SEM micrographs. (**A**) Altered Na-feldspar crystal; on right, authigenic kaolinite cement. (**B**) Altered K-feldspar; note the development of intracrystalline fractures and etching features on crystal surfaces. (**C**) Detail of authigenic kaolinite aggregates showing booklet morphology. (**D**) Siliceous overgrowth on quartz crystal. (**E**) Organic biofilm coating quartz grain surfaces. (**F**) Brucite aggregates.

The silica released (initially as H_4SiO_4) migrates to the exterior of the rock according to the preferential directions of water flow and starts to form the superficial hardened rinds and fillings in the system of joints and fractures (Figure 5D). A similar chemical attack, even more rapid and efficient, is produced simultaneously on the micaceous ferromagnesian minerals, such as biotite ($2K(Mg,Fe,Mn)_3AlSi_3O_{10}(OH)_2$)). The products are the neoformation of kaolinite, the release of Fe-oxides (limonite and haematite by oxidation) (Figure 5E) and, in smaller proportions, of Mg-oxides (brucite) (Figure 5F) and Mn-oxides (pyrolusite, manganite). These oxides migrate to the outer layers of the rock in a similar manner to the silica and, together, form alteration rinds with reddish tones, which are common in all the studied shelters. EDS chemical analyses clearly show the Fe and Mn enrichment of these alteration rinds. The simple dissolution of quartz and feldspar along their crystalline edges may be added to the processes, favouring the redistribution of cations in the superficial parts of the rock.

The formation of these hardened surfaces inhibits the progress of superficial alteration processes since these have low permeability and prevent the ingress of water to the rock interior. A crumbly area of mineral impoverishment developed below these hardened rinds. Flakes detachment takes place in favour of these areas (Figure 6A). When a detaching of part of the alteration rinds is produced in the form of flakes, it can be seen that the migration of the silica continues to progress and takes the preferred route between the contact and discontinuity areas, i.e., between these rinds and the fresh rock. Once a piece of hardened rind is broken or separated, the heterogeneity favours the development of microclimatological differences, which numerous authors have evoked as the cause of cavernous weathering features (e.g., [38–40]). Case hardening, simultaneous with the removal of sand grains by dissolution or salt weathering are the responsible for the formation of alveoli and honeycomb patterns on rock surfaces [41].

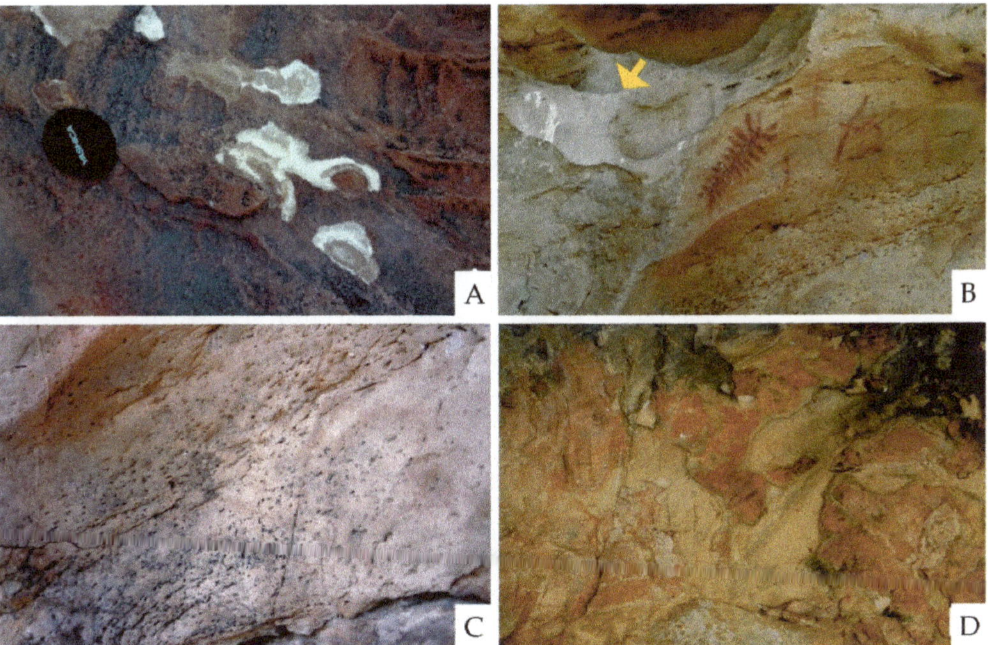

Figure 6. Weathering products. (**A**) Flaking processes and bleaching; Cueva del Arco. (**B**) Incipient alveolar weathering affecting schematic Palaeolithic rock paintings; on top left, bird droppings (arrows); Shelter 13. (**C**) Detail of alveolar weathering; Shelter 13. (**D**) Flaking; Shelter 17.

These chemical alteration processes can favour the development of mechanical weathering processes, whose main effect is granular disintegration through the superficial dissolution of the grains [34,42] or by the effect of increasing the volume and the pressure between crystals/grains through the alteration of biotites and feldspars [32].

5.1.2. Mechanical Weathering Processes

In the two cases studied, two different but inter-related mechanisms were observed:

(a) Thermal weathering due to the effect of changes in temperature suffered by rocks, both seasonally and in night-day cycles. This temperature variation may lead to the development of a thermal gradient between the surface and the interior of the rock, given its low thermal conductivity, the effect being their final breaking [17,43–46]. This process is clearly seen in the case of the Peñas de Cabrera shelters, where the metamorphic rock grains, with dark tones, can become relatively abundant (10%) in some parts of the shelters. Given their colouring and their textural characteristics, these components have higher specific heat and expansion coefficients than quartz and feldspar (majority components in the host rock). This leads to differential expansion processes and, therefore, variations in greater magnitude through dilatation-retraction phenomena, which in the long term cause their separation. As a consequence, on many occasions, lines of holes were observed in the rock. These alignments, which originally corresponded to laminations enriched in slaty grains, produce incipient alveolar surfaces (Figure 6B,C). In the studied cases, the heating agent is the sunshine, for which reason the term 'insolation weathering' can be used [47]. Effects of thermal expansion due to fire [48] have not been recognised. The effects of lighting fires have only been translated into the formation of blackish patinas with abundant sooty particles;

(b) Frost weathering as a consequence of successive freeze-thaw cycles. The water present in the rock, whether through the infiltration of meteoric water or through the nocturnal condensation phenomena, is introduced in the rock through weakness planes (fractures, joints, and bedding planes) and may freeze in the winter periods. In many instances, these weakness planes constitute the contact between the superficial alteration rinds and the fresh rock. In this case, the effectiveness of the weathering phenomena is increased by its simultaneous action with chemical and biological weathering processes. The final result is the detaching of large superficial rock flakes (Figure 6D) and the granular disintegration of the rock. Structural (fractures, joints) and textural (laminations, abundance of phyllosilicates) features of the host rock have an effect on its frost resistance and its response to other weathering processes such as salt weathering [49,50]. In the two studied rock groups it was possible to check that the phenomena of formation and shedding of flakes are very common and intense. Nevertheless, it must be stressed that in the Tajo de las Figuras site a greater effectiveness was appreciated as a consequence of the greater thermal oscillation.

5.1.3. Bio-Induced Alteration Processes

The structure of the natural shelters studied favours the colonisation of the surface part of the rock by diverse biological communities, including various colonies of mosses, nitrophilous lichens and blue-green algae, particularly associated with fissures and cracks. The walls with paintings are not much affected. Small black stains observed in the Peñas de Cabrera shelters can be attributed to the activity of fungi [51]. In Tajo de las Figuras shelters a greater profusion of nitrophilous lichens, fungi, and crypto-endolithic cyanobacteria were recognised, particularly in the Cueva del Arco, given the major incidence of direct light on the walls of this shelter. The influence of these organisms on the alteration processes of host rock is much contrasted [52–54]. Previous information has been obtained on the action of lichens both in biochemical alteration processes [55–57] and in the biophysical processes [58,59] of fungi [60,61] and crypto-endolithic cyanobacteria [3,62,63]. The action of micro-organisms on Fe-rich rinds [64] has also been recognised. These biological communities facilitate water adsorption and the maintenance of a certain level of humidity and they produce CO_2 and organic compounds which are soluble in water. These waters can accelerate the chemical alteration processes and induce the precipitation of $CaCO_3$ in the form of thin crusts. In the Tajo de las Figuras site, these lining alteration crusts show great development, which is significantly higher than in the Peñas de Cabrera site. This fact can be related to the high indices of rainfall in the location area with its own more closed shelter structure and possibly with the higher deterioration through anthropic actions (paintings getting wet) presented in this rock group. These crusts have different types of action: (1) covering the rock surface, including the areas where the paintings are present; (2) favouring the retention of oxides from the chemical alteration processes; and (3) favouring the precipitation of salts, such as gypsum, whose crystallization pressure in many cases causes mechanical tensions which aid the flaking of these crusts (haloclasty).

5.2. Human and Aninal-Induced Deterioration Processes

The previous paragraphs focus on the main natural weathering processes detected in the host sandstones. Nevertheless, it is important to emphasise the influence of anthropogenic action on the state of conservation of the paintings. Both in the Tajo de las Figuras and Peñas de Cabrera sites, the signs left by visitors when wetting and rubbing the paintings in order to bring up the chromatic contrasts in the artistic features are evident. This action considerably increases the effect of the wetting-drying cycles of the host rock. The effect produced by the continuous rubbing of the walls and floors is unquestionable; this is particularly noticeable in the floor of the Cueva del Tajo de las Figuras [26–28]. Other deterioration effects resulting from human activity are: (1) fires lit in the shelters, which have caused the rock surfaces to become black; and (2) removal of blocks of sandstone (Figure 7A), which has left despoiled areas where the water accumulates and where organic soils are produced. These actions can considerably favour the natural processes of rock alteration (e.g., shelter 13, Peñas de Cabrera). Animal activity has also caused deterioration both in the host rock and in the paintings. The presence of wasps' nests (Figure 7B) and the influence of birds, bats, goats, and deer is evident with defecations in the interiors of the shelters (Figure 6B) and rubbing against the walls.

Figure 7. Human and animal-induced deterioration. (**A**) Sandstone excavation near the top of shelters 11 and 13, the floors of these artificial depressions commonly dam water, and the development of organic soils is favoured. (**B**) Wasp nests on the ceiling of the Cueva del Tajo de las Figuras.

6. Discussion and Conclusions

The results of field work and the laboratory analyses showed that the most common mechanisms of alteration in the study area are: carbo-hydrolysis (chemical weathering), frost weathering (mechanical weathering) and bio-induced encrustation (biotic weathering). Other mechanisms such as haloclasty, abrasion, or slaking are considered to be of lesser importance. Haloclasty, combined with other mechanisms, has been interpreted as being the most important in coastal environments ([33,34,41,65], among others) and even desert areas [66]. However, in our case study, it is not easy to evoke a source of airborne salts in the form of an aerosol [67,68] because of the long distance to marine waters and industrial centres [69]. The very scarce gypsum crystals in the alteration rind and in the bio-induced crust are due to interstitial fluids which migrate outwards controlled by environmental conditions [67,70]. This is consistent with the existence of levels of gypsum in the underlying facies below sandstones beds [20,25]. Hydro-aeolian abrasion processes participate mainly in the removal of weathering debris in tafoni and alveoli [14,19,40]. Both wind-blown and water-driven deposits, including organic matter remains and salts, partially cover the shelter surfaces and/or infiltrate through the porous host rock. The slaking phenomena must have more influence on some shelters (e.g., shelter 17), where repeated cycles of wetting and drying facilitate extensive flaking by means of ordered-water molecular pressure mechanisms [47,71].

The similar structural and lithological characteristics of the studied sandstones mean that, qualitatively, the alteration processes that affect them are similar. In this sense, this has only been observed as the presence of a greater proportion of slaty grains in the Peñas de Cabrera sandstones and their separation by thermal weathering processes causes the formation of lines of small alveoli which was not observed in Tajo de las Figuras. However, it was proved that the different climatic conditions in the location of both rock groups directly influence the intensity of physicochemical weathering of sandstones. The Tajo de las Figuras area shows a rainfall index three times higher than in the Peñas de Cabrera area. In addition, the Tajo de las Figuras area has a greater seasonal contrast, with a wider thermal oscillation range and frequent frosts during the winter months. These differences in the climatic features cause the recognised alteration processes and the interaction between them to be more effective in the Tajo de las Figuras rock group, since:

- The greater availability of water leads to greater intensity in the chemical alteration processes. The most obvious effects of which are: (1) a higher degree of alteration in the feldspars, with kaolinite neoformation; (2) a greater development in case hardening processes with the formation of superficial alteration rinds, which are enriched in silica, Al and Fe, and Mg and Mn oxides;
- This hardened rind, a priori, inhibits the development of mechanical erosion processes but favours the retention of the water infiltrated in the new rock-rind contact area. When cracking or flaking occurs, the existence of a discontinuous rind produces differences in the distribution of permeability, humidity, etc., favouring the alteration processes. This is particularly noticeable in the discontinuity planes which constitute the contact areas between the superficial alteration rind and the underlying rock. The latter is generally bleached, relatively poor in cations, such as Si, Al, Fe, and Mn, and is crumbly. These discontinuity planes act similar to preferential water accumulation areas, in such a way that the action of the winter freeze-thaw cycles becomes more effective and the flakes formation and separation phenomena are more common;
- At the same time, the splitting away of the flakes leaves areas of fresh rock once more exposed to chemical alteration;
- The releasing of Ca^{2+} and HCO_3^- by the feldspars carbo-hydrolysis favours a greater development of the bio-induced carbonate crusts. This in turn has an incrusting effect and favours the physical and chemical attack processes towards the interior of the rock.

With respect to the rates of weathering, the preservation of rock paintings denotes low rates of weathering, at least in the immediate host rock. Although no data are available in the form of old photographs, for example to quantify the alteration, the existence of hardened rinds lining a large part of the walls of the shelters of alveolar structures on the walls and ceilings of same and biological colonisation are indicative of slow rates of alteration in the shelters [13,36,72]. The rates of alteration would vary as a function of the climatic and geographical contexts and the lithological and geomorphic characteristics of the host rock, generally ranging from microns to a few millimetres a year [13,72–74].

In the present case, considering both art sites, climate is the most important controlling factor in the rates of alteration as indicated previously. However, considering each shelter separately, the major influence of other local factors can be observed, both in the types and rates of alteration. In this way, it may be stated that in shelters 24 (facing W) and the Cueva del Arco, given that they have a greater effect of natural light, the rates of alteration (in particular, biotic) are relatively greater. On the other hand, in the case of the Cueva del Tajo de las Figuras, the biotic alteration through micro-organisms is restricted, given its nature of a more enclosed shelter. For shelter 17, its closeness to floor level causes the alteration in the lower part to accelerate with regard to the rest of the shelter due to the greater degree of humidity [32].

It is necessary to note that these relatively low rates of alteration correspond to the more recent evolution of the shelters, at least that which is clearly later than the time that the paintings were executed (Epi-Palaeolithic-Neolithic). The genesis of rock shelters probably dates from the humid and glacial cold stages of the Quaternary, when mechanical alteration processes (freeze-thaw cycles) were predominant and the rates of alteration were considerably greater.

The alteration features found in the different shelters mainly correspond to natural physical and chemical phenomena. For this reason, corrective measures are difficult to apply without producing unforeseen consequences. Presently, the state of conservation of the paintings is not bad. Natural weathering processes can be considerably accelerated by both anthropic and animal activity. Anthropic action can be direct (wetting and rubbing against the paintings, lighting of fires) or indirect (removal of blocks of sandstone from the ceilings of the shelters). Large animals can produce alteration by different actions (rubbing, defecation, building of nests, etc.) which are particularly harmful for this type of artwork.

Author Contributions: Conceptualisation, J.C.C. and S.S.-M.; Investigation, J.C.C., S.S.-M. and E.S.-R.; Writing–original draft, J.C.C., S.S.-M. and E.S.-R. All authors have read and agreed to the published version of the manuscript.

Funding: This study forms part of the research agreement "Study and diagnosis of the archaeological groups and deposits of the Autonomous Community of Andalusia in their geomorphological, biological and climatic contexts" financed by the Culture Council, Junta of Andalusia (Spain).

Institutional Review Board Statement: Not applicable.

Informed Consent Statement: Not applicable.

Data Availability Statement: The study did not report any data.

Acknowledgments: Mineralogical and geochemical analyses were conducted with the assistance of R. González, M. Vallejo and M.I. Ruiz (MNCN-CSIC). J. Bedoya (MNCN-CSIC) assisted with the SEM. The authors wish to acknowledge the professional support of the CSIC Interdisciplinary Thematic Platform Open Heritage: Research and Society (PTI-PAIS).

Conflicts of Interest: The authors declare no conflict of interest.

References

1. Chowdhury, A.N.; Punuru, A.R.; Gauri, K.L. Weathering of limestone beds at the Great Sphinx. *Environ. Geol. Water Sci.* **1990**, *15*, 217–223. [CrossRef]
2. Saíz-Jiménez, C.; García-Rowe, J.; García del Cura, M.A.; Ortega-Calvo, J.J.; Roekens, E.; Van Grieken, R. Endolithic cyanobacteria in Maastricht limestone. *Sci. Total Environ.* **1990**, *94*, 209–220. [CrossRef]
3. Ortega-Calvo, J.J.; Ariño, X.; Hernández-Marine, M.; Saíz-Jiménez, C. Factors affecting the weathering and colonization of monuments by phototrophic microorganisms. *Sci. Total Environ.* **1995**, *167*, 329–341. [CrossRef]
4. Zezza, F. Marine spray and polluted atmosphere as factors of damage to monuments in the Mediterranean coastal environment. In Proceedings of the 3th International Symposium on the Conservation of Monuments in the Mediterranean, Venice, Italy, 22–25 June 1994; pp. 269–273.
5. Heinrichs, K.; Fitzner, B. Deterioration of rock monuments in Petra/Jordan. In Proceedings of the 9th International Congress of the Deterioration and Conservation of Stone, Amsterdam, The Netherlands, 19–24 June 2000; Volume 2, pp. 53–61.
6. Paradise, T.R. Sandstone weathering and aspect in Petra, Jordan. *Z. Geomorphol.* **2002**, *46*, 1–17. [CrossRef]

7. Sánchez-Moral, S.; Luque, L.; Cuezva, S.; Soler, V.; Benavente, D.; Laiz, L.; Gonzalez, J.M.; Saiz-Jimenez, C. Deterioration of building materials in Roman catacombs: The influence of visitors. *Sci. Total Environ.* **2005**, *349*, 260–276. [CrossRef] [PubMed]
8. Sánchez-Moral, S.; Cañaveras, J.C.; Benavente, D.; Fernández-Cortés, A.; Cuezva, S.; Elez, J.; Jurado, V.; Rogerio-Candelaria, M.A.; Saiz-Jiménez, C. A study on the state of conservation of the Roman Necropolis of Carmona (Sevilla, Spain). *J. Cult. Herit.* **2018**, *34*, 185–197. [CrossRef]
9. André, M.F.; Etienne, S.; Mercier, D.; Vautier, F.; Voldoire, O. Assessment of sandstone deterioration at Ta Keo temple (Angkor): First results and future prospects. *Environ. Geol.* **2008**, *56*, 677–688. [CrossRef]
10. André, M.F.; Voldoire, O.; Roussel, E.; Vautier, F.; Phalip, B.; Peou, H. Contrasting weathering and climate regimes in forested and cleared sandstone temples of the Angkor region. *Earth Surf. Process. Landf.* **2012**, *27*, 519–532. [CrossRef]
11. Zhang, J.; Huang, J.; Liu, J.; Jiang, S.; Li, L.; Shao, M. Surface weathering characteristics and degree of Niche of Sakyamuni Entering Nirvana at Dazu Rock Carvings, China. *Bull. Eng. Geol. Environ.* **2019**, *78*, 3891–3899. [CrossRef]
12. Campbell, I.A. Classification of rock weathering at Writing-on-Stone Provincial Park, Alberta, Canada. *Earth Surf. Process. Landf.* **1991**, *16*, 701–711. [CrossRef]
13. Pentecost, A. The weathering rates of some sandstone cliffs, central Weald, England. *Earth Surf. Process. Landf.* **1991**, *16*, 83–91. [CrossRef]
14. Benito, G.; Machado, M.J.; Sancho, C. Sandstone weathering processes damaging prehistoric paintings at the Albarracín Cultural Park, NE Spain. *Environ. Geol.* **1993**, *22*, 71–79. [CrossRef]
15. Sjöberg, R. Diagnosis of weathering on rock carving surfaces in northern Bohuslän, Southwest Sweden. In *Rock Weathering and Landform Evolution*; Robinson, D.A., Williams, R.B.G., Eds.; Wiley: Chichester, UK, 1994; pp. 223–241.
16. Hall, K.; Meiklejohn, I.; Arocena, J. The thermal responses of rock art pigments: Implications for rock art weathering in southern Africa. *Geomorphology* **2007**, *91*, 132–145. [CrossRef]
17. Meiklejohn, K.I.; Hall, K.; Davis, J.K. Weathering of rock art at two sites in the KwaZulu-Natal Drakensberg, southern Africa. *J. Archaeol. Sci.* **2009**, *36*, 973–979. [CrossRef]
18. Díez-Herrero, A.; Gutiérrez-Pérez, I.; Lario, J.; Cañaveras, J.C.; Benavente, D.; Sánchez-Moral, S.; Alonso-Azcárate, J. Analysis of potential direct insolation as a degradation factor of cave paintings in Villar del Humo, Cuenca, Central Spain. *Geoarchaeology* **2009**, *24*, 450–465. [CrossRef]
19. Peña-Monné, J.L.; Sampietro-Vattuone, M.M.; Ariel Báez, W.; García-Giménez, R.; Matía Stábile, F.; Martínez Satgnaro, S.Y.; Tissera, L.E. Sandstone weathering processes in the painted rock shelters of Cerro Colorado (Córdoba, Argentina). *Geoarchaeology* **2022**, *37*, 332–349. [CrossRef]
20. Ménanteau, L.; Vanney, J.R.; Zazo, C. Belo et son environnement (Détroit de Gibraltar). Étude physique d'un site antique. Belo II. *Archéologie* **1983**, *4*, 39–221.
21. Ripoll, S.; Mas, M.; Torra, G. Grabados paleolíticos en la Cueva del Tajo de las Figuras (Benalup, Cádiz). *Espac. Tiempo y Forma* **1991**, *4*, 111–126. [CrossRef]
22. Cabré, J.; Hernández-Pacheco, E. Avance al estudio de las pinturas prehistóricas del extremo sur de España (laguna de Janda), Trabajos de la Comisión de Investigaciones Paleontológicas y Prehistóricas. *Mus. Nac. Cienc. Nat.* **1914**, *3*, 35.
23. Baldellou, V. "II Reunión de Prehistoria Aragonesa": La terminología en el arte rupestre Post-Paleolítico. In *Bolskan Revista de Arqueología Oscense*; Instituto de Estudios Altoaragonenes: Huesca, Spain, 1989; Volume 6, pp. 5–14.
24. Breuil, H.; Burkitt, M.C. *Rock Paintings of Southern Andalusia. A Description of a Neolithic and Copper Age Art Group*; Clarendon Press: Oxford, UK, 1929.
25. Barba, A.; Martín, A.; Piles, E. *Memoria del Mapa Geológico de España 1:50,000 de la Hoja n° 1039 de Colmenar*; Instituto Geológico y Minero de España: Madrid, Spain, 1979; p. 88.
26. Barroso, C.; Medina, F. El conjunto rupestre de arte post-paleolítico de Peñas Cabreras (Casabermeja, Málaga). *Anu. Arqueol. Andal.* **1989**, *II*, 333–345.
27. Rodríguez Guzman, S.; Santana, I.; Martínez, J. La gestión del arte rupestre en Andalucia. Actuaciones en materia de protección y conservación. *Panel Rev. Arte Rupestre* **2001**, *1*, 32–43.
28. Mas, M. *La Cueva del Tajo de las Figuras (Cádiz)*; UNED: Madrid, Spain, 2005; 256p.
29. Peltier, C. The geographic cycle in periglacial regions as it is related to climatic geomorphology. *Ann. Assoc. Am. Geogr.* **1950**, *40*, 214–236. [CrossRef]
30. Geiger, R. Classificação climática de Köppen-Geiger, Great. In *Commons Attribution-ShareAlike 3.0 Unported*; 1936. Available online: https://creativecommons.org (accessed on 12 January 2020).
31. ICOMOS-ICS. Illustrated Glossary on Stone Deterioration Pattern. 2008. Available online: http://international.icomos.org/publications/monuments_and_sites/15/pdf/Monuments_and_Sites_15_ISCS_Glossary_Stone.pdf (accessed on 12 January 2020).
32. Dragovich, D. The origin of cavernous surfaces (tafoni) in granitic rocks of southern south Australia. *Z. Geomorphol.* **1969**, *13*, 163–181.
33. Mustoe, G.E. The origin of honeycomb weathering. *Geol. Soc. Am. Bull.* **1982**, *93*, 108–115. [CrossRef]
34. Mottershead, D.N.; Pye, K. Tafoni on coastal slopes, South Devon, U.K. *Earth Surf. Process. Landf.* **1994**, *19*, 543–563. [CrossRef]
35. Drever, J.L. *The Geochemistry of Natural Waters*; Prentice-Hall: Englewood Cliffs, NJ, USA, 1982; 388p.
36. Friedmann, E.I. Endolithic microorganisms in the Antartic cold desert. *Science* **1982**, *215*, 1045–1053. [CrossRef]
37. Berner, R.A. Rate controls of mineral dissolution under earth surface conditions. *Am. J. Sci.* **1978**, *278*, 1235–1252. [CrossRef]

38. Martini, I.P. Tafoni weathering, with examples from Tuscany, Italy. *Z. Geomorphol.* **1978**, *22*, 4–67.
39. Conca, J.L.; Astor, N.M. Capillary moisture flow and the origin of cavernous weathering in dolerite of Bull Pass, Antartica. *J. Geol.* **1987**, *15*, 151–154. [CrossRef]
40. Sancho, C.; Benito, G. Factors controlling tafoni weathering in the Ebro Basin (NE Spain). *Z. Geomorphol.* **1990**, *34*, 165–177. [CrossRef]
41. McBride, E.F.; Picard, D. Origin of honeycombs and related weathering forms in Oligocene Macgno Sandstone, Tuscan coast near Liborno, Italy. *Earth Surf. Process. Landf.* **2004**, *29*, 713–735. [CrossRef]
42. Wray, R.A.L.; Sauro, F. An updated global review of solutional weathering processes and forms in quartz sandstones and quartzites. *Earth-Sci. Rev.* **2017**, *171*, 520–557. [CrossRef]
43. Ollier, C. *Weathering*, 2nd ed.; Longman: London, UK, 1984; 270p.
44. Warke, P.A.; Smith, B.J. Short-term rock temperature fluctuations under simulated hot desert conditions: Some preliminary data. In *Rock Weathering and Landform Evolution*; Robinson, D.A., Williams, R.B.G., Eds.; Wiley: Chichester, UK, 1994; pp. 57–70.
45. Hoerlé, S.; Salmon, A. Microclimatic data and rock art conservation at Game Pass Shelter in the Kamberg Nature Reserve, KwaZulu-Natal. *S. Afr. J. Sci.* **2004**, *100*, 340–341.
46. Hoerlé, S. Rock temperatures as an indicator of weathering processes affecting rock art. *Earth Surf. Process. Landf.* **2006**, *31*, 383–389. [CrossRef]
47. Ollier, C.; Pain, C. *Regolith, Soils and Landforms*; John Wiley & Sons: Hoboken, NJ, USA, 1996; 316p.
48. Allison, R.J.; Goudie, A.S. The effects of fire on rock weathering: An experimental study. In *Rock Weathering and Landform Evolution*; Robinson, D.A., Williams, R.B.G., Eds.; Wiley: Chichester, UK, 1994; pp. 1–56.
49. Jiménez-González, I.; Rodríguez-Navarro, C.; Scherer, G.W. Role of clay minerals in the physicomechanical deterioration of sandstone. *J. Geophys. Res.* **2008**, *113*, F02021. [CrossRef]
50. Cnudde, V.; De Boever, W.; Dewanckele, J.; De Kock, T.; Boone, M.; Boone, M.N.; Silversmit, G.; Vincze, L.; Van Ranst, E.; Derluyn, H.; et al. Multi-disciplinary characterization and monitoring of sandstone (Kandla Grey) under different external conditions. *Q. J. Eng. Geol. Hydrogeol.* **2013**, *46*, 95–106. [CrossRef]
51. Saíz-Jiménez, C.; Ortega-Calvo, J.J.; de Leeuw, J.W. The chemical structure of fungal melanins and their possible contribution to black stains in stone monuments. *Sci. Total Environ.* **1995**, *167*, 305–314. [CrossRef]
52. Saiz-Jimenez, C. Biogeochemistry of weathering processes in monuments. *Geomicrobiol. J.* **1999**, *16*, 27–37. [CrossRef]
53. Warscheid, T.; Braams, J. Biodeterioration of stone: A review. *Int. Biodeterior. Biodegrad.* **2000**, *46*, 343–368. [CrossRef]
54. Miller, A.Z.; Sanmartin, P.; Pereira-Pardo, L.; Dionisio, A.; Saiz-Jimenez, C.; Macedo, M.F.; Prieto, B. Bioreceptivity of building stones: A review. *Sci. Total Environ.* **2012**, *426*, 1–12. [CrossRef]
55. Galván, J.; Rodriguez, C.; Ascaso, C. The pedogenic action of lichens in metamorphic rocks. *Pedobiologia* **1981**, *21*, 60–73.
56. Cooks, J.; Otto, E. The weathering effects of the lichen *Lecidea aff. sarcogynoides* (Koerb.) on Magaliesburg quartzite. *Earth Surf. Process. Landf.* **1990**, *15*, 491–500. [CrossRef]
57. Ariño, X.; Ortega-Calvo, J.J.; Gómez-Bolea, A.; Sáiz-Jiménez, C. Lichen colonization of the Roman pavement at Baelo Claudia (Cádiz, Spain): Biodeterioration vs. bioprotection. *Sci. Total Environ.* **1995**, *167*, 353–363. [CrossRef]
58. Moses, C.A.; Smith, B.J. A note on the role of the lichen *Collema auriforma* in solution basin development on a carboniferous limestone. *Earth Surf. Process. Landf.* **1993**, *18*, 363–368. [CrossRef]
59. McCarroll, D.; Viles, H.A. Rock-weathering by the lichen *Lecidea auriculata* in an artic alpine environment. *Earth Surf. Process. Landf.* **1995**, *20*, 199–206. [CrossRef]
60. Friedman, G.M.; Sanders, J.E.; Kopaska-Merkel, D.C. *Principles of Sedimentary Deposits*; McMillan Publishing Company: New York, NY, USA, 1992; 717p.
61. Petersen, K.; Kuroczkin, J.; Strzelczyk, A.B.; Krumbein, W.E. Distribution and effects of fungi on and in sandstones. In *Biodeterioration*; Houghton, D.R., Smith, R.N., Eggins, H.O.W., Eds.; Elsevier: London, UK, 1988; Volume 7, pp. 123–128.
62. Bell, R.A.; Athey, P.V.; Sommerfeld, M.R. Crytoendolithic algae communities of the Colorado Plateau. *J. Phycol.* **1986**, *22*, 429–435. [CrossRef]
63. Wu, F.; Zhang, Y.; He, D.; Gu, J.-D.; Guo, Q.; Liu, X.; Duan, Y.; Zhao, J.; Wang, W.; Feng, H. Community structures of bacteria and archaea associated with the biodeterioration of sandstone sculptures at the Beishiku Temple. *Int. Biodeterior. Biodegrad.* **2021**, *164*, 105290. [CrossRef]
64. Adams, J.B.; Palmer, F.; Staley, J.T. Rock weathering in deserts: Mobilization and concentration of ferric iron by microorganisms. *Geomicrobiol. J.* **1992**, *10*, 99–114. [CrossRef]
65. McGreevy, J.P. A preliminary scanning electron microscope study of honeycomb weathering of sandstone in a coastal environment. *Earth Surf. Process. Landf.* **1985**, *10*, 509–518. [CrossRef]
66. Mustoe, G.E. Cavernous weathering in the Capitol Reef Desert, Utah. *Earth Surf. Process. Landf.* **1983**, *8*, 517–526. [CrossRef]
67. Brierly, W.B. Atmospheric sea salts design criteria areas. *J. Environ. Sci.* **1965**, *8*, 15–23.
68. Skarveit, A. Wet scavenning of sea salts and acid compounds in a rainy coastal area. *Atmos. Environ.* **1982**, *16*, 2715–2724. [CrossRef]
69. Malloch, A.J.C. Salt-spray deposition on the maritime cliffs of the Lizard peninsula. *J. Ecol.* **1972**, *60*, 103–112. [CrossRef]
70. Winkler, E.M. *Stone: Properties, Durability in Man's Environment*, 2nd ed.; Springer: New York, NY, USA, 1975; 230p.

71. Weiss, T.; Siegesmund, S.; Kirchner, D.; Sippel, J. Insolation weathering and hygric dilatation: Two competitive factors in stone degradation. *Environ. Geol.* **2004**, *46*, 402–413. [CrossRef]
72. Turkingnton, A.V.; Paradise, T.R. Sandstone weathering: A century of research and innovation. *Geomorphology* **2005**, *67*, 229–253. [CrossRef]
73. Gill, E.D. Rapid honeycomb weathering (tafoni formation) in greywacke, S.E. Australia. *Earth Surf. Process. Landf.* **1981**, *6*, 81–83. [CrossRef]
74. Matsukura, Y.; Matsuoka, N. Rates of tafoni weathering on uplifted shore platforms in Nijima-Zaki, Boso Peninsula, Japan. *Earth Surf. Process. Landf.* **1991**, *16*, 51–56. [CrossRef]

applied sciences

MDPI

Article

A Multi-Analytical Approach to Infer Mineral–Microbial Interactions Applied to Petroglyph Sites in the Negev Desert of Israel

Laura Rabbachin [1,*], Guadalupe Piñar [1], Irit Nir [2], Ariel Kushmaro [2,3], Mariela J. Pavan [3], Elisabeth Eitenberger [4], Monika Waldherr [5], Alexandra Graf [5] and Katja Sterflinger [1]

1 Institute of Natural Sciences and Technology in the Arts (INTK), Academy of Fine Arts Vienna, Schillerplatz 3, 1010 Vienna, Austria; g.pinarlarrubia@akbild.ac.at (G.P.); k.sterflinger@akbild.ac.at (K.S.)
2 Avram and Stella Goldstein-Goren Department of Biotechnology Engineering, Ben-Gurion University of the Negev, Be'er Sheva 8410501, Israel; irin@post.bgu.ac.il (I.N.); arielkus@bgu.ac.il (A.K.)
3 Ilse Katz Institute for Nanoscale Science and Technology, Ben-Gurion University of the Negev, Be'er Sheva 8410501, Israel; marielap@bgu.ac.il
4 Institute of Chemical Technology and Analytics, TU Wien, Getreidemarkt 9/164, 1060 Vienna, Austria; elisabeth.eitenberger@tuwien.ac.at
5 Department of Applied Life Sciences, University of Applied Sciences, FH Campus Wien, Favoritenstraße 226, 1100 Vienna, Austria; monika.waldherr@fh-campuswien.ac.at (M.W.); alexandra.graf@fh-campuswien.ac.at (A.G.)
* Correspondence: l.rabbachin@akbild.ac.at

Abstract: Petroglyph sites exist all over the world. They are one of the earliest forms of mankind's expression and a precursor to art. Despite their outstanding value, comprehensive research on conservation and preservation of rock art is minimal, especially as related to biodeterioration. For this reason, the main objective of this study was to explore the factors involved in the degradation of petroglyph sites in the Negev desert of Israel, with a focus on biodegradation processes. Through the use of culture-independent microbiological methods (metagenomics), we characterized the microbiomes of the samples, finding they were dominated by bacterial communities, in particular taxa of Actinobacteria and Cyanobacteria, with resistance to radiation and desiccation. By means of XRF and Raman spectroscopies, we defined the composition of the stone (calcite and quartz) and the dark crust (clay minerals with Mn and Fe oxides), unveiling the presence of carotenoids, indicative of biological colonization. Optical microscopy and SEM–EDX analyses on thin sections highlighted patterns of weathering, possibly connected to the presence of biodeteriorative microorganisms that leach the calcareous matrix from the bedrock and mobilize metal cations from the black varnish for metabolic processes, slowly weathering it.

Keywords: petroglyphs; Negev desert; biodeterioration; nanopore sequencing technology; metagenomics; analytical techniques

Citation: Rabbachin, L.; Piñar, G.; Nir, I.; Kushmaro, A.; Pavan, M.J.; Eitenberger, E.; Waldherr, M.; Graf, A.; Sterflinger, K. A Multi-Analytical Approach to Infer Mineral–Microbial Interactions Applied to Petroglyph Sites in the Negev Desert of Israel. *Appl. Sci.* **2022**, *12*, 6936. https://doi.org/10.3390/app12146936

Academic Editor: Asterios Bakolas

Received: 25 May 2022
Accepted: 5 July 2022
Published: 8 July 2022

1. Introduction

Rock art, in the form of petroglyphs and pictograms, is found worldwide and has an undoubtedly immense value as it is considered one of the first forms of expression of ancient societies and the prehistoric precursor to art [1–4]. As part of the natural landscape, petroglyphs are constantly exposed to anthropogenic and natural weathering processes [5–7], but despite this, knowledge regarding preservation and conservation of this valuable cultural heritage is limited.

Although there is considerable research and published work focusing on the physical state of rock art sites worldwide [7–9], research focusing on the role of biological agents in the deterioration of rock art is still minimal. Nevertheless, physical and chemical weathering processes initiated by stone-dwelling microorganism (biodeterioration) can play a significant role in the degradation patterns of the stone [10]. Biodeterioration includes surface alterations by crusts and pigments that can change the color of the surface [11], physico–chemical disintegration of the stone material, and dissolution of the lithic substrate by inorganic and organic acids [12]. For example, hyphal growth and biofilm development can cause physical disruptions and cracks in stone materials, while the release of acids can dissolve carbonates, such as limestone, as well as quartz, causing mineral leaching [13]. All these processes contribute to weakening the stone matrix.

In the Negev desert of Israel, the majority of petroglyph sites are spread throughout the Negev central highlands and offer a window into the life and culture of desert people who inhabited the region since prehistoric times [14]. The petroglyphs are considered to be from 3000 BCE [15] and include figurative images (zoomorphic and anthropomorphic), geometric shapes, symbols and inscriptions. The engravings are carved through a dark crust, the so called desert or rock varnish, which coats the local Eocene and late Cretaceous limestone rocks [16]. Desert varnish is defined as a thin (10–200 μm), dark coating highly enriched in manganese that forms on exposed rock surfaces in arid and hyper-arid environments. It is composed of clay minerals and amorphous silica (~70%) in a matrix of poorly crystallized manganese (Mn) and iron (Fe) oxides and hydroxides [17–19]. Despite the fact that geochemical aspects of rock varnish have been thoroughly investigated [19–24] and it is believed that its primary source of material is airborne dust, there is an ongoing debate about the exact process of formation and specifically whether or not microorganisms are involved in this process [5,25,26]. In a recent study, Lingappa et al. [27] proposed a new hypothesis for varnish formation. They state that Cyanobacteria accumulate manganese as a nonenzymatic antioxidant system and that, consequently to the cells death, the manganese-rich residue left behind is oxidized to generate the manganese oxides present in the varnish.

Cyanobacteria are often reported as one of the dominant phyla present in rock varnish, along with Actinobacteria, Proteobacteria, Bacteroidetes and Chloroflexi [16,28]. In fact, despite the harsh environmental conditions, microbial studies revealed a broad spectrum of stone-dwelling microorganism associated with desert varnish [29,30]. Just as it is not fully understood the role played by the microorganisms in the formation of the rock varnish, it is also uncertain if and how the microbial community influences the weathering of the stone. In fact, the literature available about this topic is very limited. A few studies report serious damage in numerous rock art sites due to lichen colonization [31,32]. Nir et al. [33], with a study combining scanning electron microscopy (SEM) and metagenomic sequencing methods, suggest that the stone surface of petroglyphs in the Negev desert is colonized by a complex microbial and lichen community with the biochemical potential to induce biodeterioration. Despite these studies, comprehensive research on deterioration of rock art panels, particularly by microorganisms, is still needed.

For these reasons, the aim of the present study was to investigate the potential microbial involvement in the weathering of petroglyph sites through a multi-analytical approach based on culture-independent microbiological methods (metagenomics) to investigate the microbiome associated with the petroglyph panels, and on microscope observation and physico–chemical methods to analyze the lithic substrate.

2. Materials and Methods

2.1. Study Location and Sampling

Two petroglyph sites in the Negev desert were considered for this study (Figure 1a,b): one is located in the central highlands, close to the ruins of the ancient city of Avdat (N 30″47′10°; E 34″46′20°), and the second one is in the west highlands close to the community settlement of Ezuz (N 30″48′5°; E 34″28′30°).

Figure 1. Map of the Negev desert indicating the location of the two petroglyph sites (**a**); view of the rock art site in Ezuz with examples of rock engravings (**b**); image of a sampling point (**c**); and the sample observed by stereomicroscope (6×) (**d**).

Both areas are considered an arid climatic zone. The annual temperature in this region ranges between −3.5 °C in winter and 40 °C in summer (Israel Meteorological Service, Sde Boker station), while precipitation ranges from 90 to 100 mm, mostly during the winter season [34,35]. There are about 200 nights of dew/year, providing abundant liquid water for microbial lithobiontic colonization in this desert [36].

Sampling was carried out during winter (February and March) 2021. In these months, meteorological measurements were implemented with an in situ monitoring system (situated at Carmey Avdat farm, about 100 m from the petroglyph sites) to define local environmental factors. The monitoring system contained a Campbell CR6 data logger (Campbell Scientific Inc., Logan, UT, USA) with sensors for air temperature (°C), air relative humidity (RH), and rock surface temperature (thermocouples). In addition, it can also measure the amount of rain received in the area. The data are recorded every 15 min and every whole hour. Thus, it was found that the air temperature from January–April 2021 ranged from a minimum of 1.9 °C to a maximum of 38.5 °C, with daily temperature fluctuation up to 20 °C, and the area received 22.5 mm of rain (Supplementary Figure S1a,b). Additionally, previous measurements revealed that the average rock surface temperature was relatively higher than the average air temperature, reaching up to 56.3 °C in summer.

The stone samples were collected from similar rock types (limestone covered with desert varnish) in close proximity to the petroglyphs using a hammer and chisel previously sterilized with 70% ethanol (Figure 1c,d). Seven sampling points were chosen, and from each one different slabs of rock surface, about 3 cm × 3 cm in size, were taken and placed in sterile plastic bags (Table 1). For each sampling point, part of the slabs was transferred to the Ben Gurion University of the Negev, Be'er Sheeva, Israel where Raman and XRF analysis were done, and part to the microbiology laboratory of the Academy of Fine Arts of Vienna, Austria, where the rest of the analyses were performed.

Table 1. List of samples collected at the two petroglyph sites with brief description.

Sample	Petroglyph Site	Brief Description of the Crust
AV1	Avdat	Very dark crust covering a reddish layer
AV2	Avdat	Very dark crust covering a reddish layer
AV3	Avdat	Thinner black crust, intermixed with reddish layer
EZ1	Ezuz	Very dark crust covering a reddish layer
EZ2	Ezuz	Very dark crust covering a reddish layer
EZ3	Ezuz	Dark crust covering a reddish layer
EZ4	Ezuz	Orange thick crust

Six of the samples (AV1, AV2, AV3, EZ1, EZ2, EZ4) were prepared as polished petrographic thin sections. The samples were vacuum-impregnated with epoxy resin from Struers, Denmark. For better visualization of the porosity of clear minerals, the resin was dyed with a blue dye (EpoBlue from Buehler GmbH, Düsseldorf, Germany). After impregnation, the samples were cut into thin sections, mounted on glass slides and polished to approximately 30 μm in thickness. The samples were left uncovered for analysis by transmitted and reflected light optical microscopy and SEM–EDX.

2.2. DNA Extraction, Whole Genome Amplification (WGA), Library Preparation and Sequencing

To assess the composition of the stone microbiome, genomic DNA was extracted from 4 samples using the FastDNA SPIN Kit for soil (MP Biomedicals, Illkrich, France) following the instruction of the manufacturer. For each sample, two DNA extractions with 0.5 g of crushed stone were performed, and the obtained DNA was pooled to proceed with the library preparation. DNA yields were quantified using a Qubit 2.0 fluorometer (Thermo Fisher Scientific, Waltham, MA, USA) with the Qubit dsDNA HS Assay Kit. The "Premium whole genome amplification protocol" available in the online Oxford Nanopore community was followed to perform the WGA and library preparation. All reactions for the WGA were executed in a BioRad C 1000 Thermal Cycler. The REPLI-g Midi Kit (Qiagen, Hilden, Germany) was employed, which uses innovative multiple displacement amplification (MDA) technology. Library preparation was performed following all the steps described by Piñar et al. [37], using the Ligation Sequencing Kit 1D SQK-LSK109 and the Flow cell Priming Kit EXP-FLP001 (Oxford Nanopore Technologies, Oxford, UK).

Once the library was prepared, the MinKNOW™ software 21.02.2 (Oxford Nanopore Technologies, Oxford, UK) was used to perform quality control of the flow cells (SpotOn Flow cell Mk I R9 Version, FLO-MIN 106D) prior to priming and loading of the DNA library. Sequencing was performed in the MinIon Mk1C device for 48 h.

2.3. Sequence Analysis

The latest version of high accuracy basecalling of the fast5 files was conducted with Guppy basecalling software (Oxford Nanopore Technologies, Limited) v5.0.11 + 2b6dbff using the dna_r9.4.1_450bps_hac model. The acquired reads were filtered to obtain Q scores > 9 and to remove adapter sequences (120 bases trimmed at 5′ and 3′ ends). The shortest 5% of all reads were also removed from the analysis. After filtering, the median quality scores were between 13.1 and 13.5, representing a mean error rate of 4.9%. To compensate for the remaining error rate, classification was performed with a more conservative value for the minimum length of partial hits (50).

Taxonomic classification was carried out using Centrifuge v1.0.4 (Johns Hopkins University CCB, Baltimore, MD, USA) with a custom database containing reference genomes for Bacteria, Archaea and Fungi retrieved from NCBI. For Bacteria and Archaea, genomes with complete genome status, and for Fungi, genomes with the complete and draft genome indication were chosen. Addition of draft genomes for Fungi was selected due to their somewhat larger and more complex genomes resulting in a low number of complete genome entries. In addition to the default parameters, the minimum length of partial hits was set to 50, and k, the number of distinct primary assignments per read, was set to 1.

Relative abundances were calculated for all genera with a read count abundance of more than 0.5% in at least one sample and used for the clustered heatmap, which was generated with the R package pheatmap. All data are available at the NCBI BioProject PRJNA847191.

2.4. Optical Microscopy (OM)

For preliminary observation of the samples, a stereomicroscope (Wild M650, Heerbrugg, Switzerland) with magnification up to 40× was employed. The system was equipped with a digital camera connected to a computer, and AmScopeX software allowed acquisition of images at different magnifications.

Detailed morphological analysis of the sample surfaces and observations of thin sections were performed using a VHX-6000 digital microscope (Keyence, Osaka, Japan). Observations were completed with plane polarized light (PPL), cross polarized light (XPL) or with normal transmitted light. The microscope has an LED light source (5700 K). With this microscope, it is possible to obtain fully focused images of uneven surfaces by a depth composition function, which also allows the creation of a 3D image with 3D height information. The pictures were recorded using a VH-Z20 objective with a magnification range from 20× to 200× and a VH-Z100 objective with a magnification range from 100× to 1000×. Measurements of features of interest were performed directly with the microscope software.

2.5. X-ray Fluorescence (XRF)

The stone slabs were examined with a portable ELIO spectrometer (XGLab, Milan, Italy) equipped with a rhodium (Rh) X-ray tube with a maximum power of 4 W at 50 kV and a silicon drift chamber detector (SDD) with a thin beryllium window. The X-ray beam had a diameter of about 1 mm. The analyses were carried out in air; therefore, elements with atomic number lower than silicon (Si) cannot be detected. The measurements were performed on the crust of the samples and on the bedrock, positioning the stone slab in such a way that only the contribution of the rock was measured. The time of acquisition was set at 60 s, the excitation voltage was 40 kV, and the tube current was 60 μA. The results were elaborated using Elio Software v1.6.0.29.

With a Dremel tool, the black varnish, the orange layer and the bedrock of one sample were separated, and the loose powders were analyzed with a wavelength dispersive (WDXRF) spectrometer Axios (1 kW) with SuperQ version 5 software (PANalytical B.V. Almelo, The Netherlands). The special software Omnian, based on the fundamental parameter method, was used for quantitative analysis. All elements (from fluorine onwards) that could be identified by the method were summed and normalized to 100%.

2.6. Scanning Electron Microscopy Coupled with Energy Dispersive X-ray Spectroscopy (SEM–EDX)

Small slabs of the stone samples and the thin sections were analyzed by SEM FEI Quanta 200 scanning electron microscope (SEM) combined with an EDX Ametek Octane Pro system. The stone slabs were placed on a sulfur-free carbon adhesive glued onto an aluminum stub. SEM images of the examined surfaces were obtained in low vacuum mode, without the need of a conductive coating, with a secondary electron LFD detector for morphological analysis of the stone slabs or a dual backscattered electrons (BSE) detector for analysis of the thin sections, at 20 kV voltage, an average working distance of 10 mm and a chamber pressure between 70 and 90 Pa. EDX analyses were performed to gain chemical information about areas of interest both as punctual measurement and elemental maps. For punctual measurements, the acquisition time was set at 50 s.

2.7. Micro-Raman Spectroscopy

Micro-Raman spectroscopy was used on the untreated stone slabs to study both the mineralogical and biological components of the crust and associated rock substrate. The measurements were done with a confocal Horiba LabRam HR Evolution (Kyoto, Japan), equipped with a Syncerity CCD detector (deep-cooled to –60° C, 1024 × 256 pixels). The excitation source was a 532 nm laser with power on the sample between 0.5 and 5 mW. Laser power was kept particularly low (>1 mW) on the black crust since some iron minerals are easily transformed when heated up by the laser.

The laser was focused with a 50× LWD objective (Olympus LMPlanFL-N, NA = 0.5) to a spot of about 1.3 mm. The measurements were taken using a 600 g mm^{-1} grating and a 100 mm confocal hole. Spectral acquisition times were between 3 and 40 s, with variable accumulations from 1 to 10. The spectra were baseline corrected and denoised directly with the instrument's software (LabSpec 6, v6.5.1.24, Horiba, Kyoto, Japan). Further processing of the Raman spectra, such as normalization, was performed with OPUS 7.5 spectroscopy software by Bruker. Identification of the compounds was done either by comparison to spectra obtained with Raman spectral databases (KnowItAll Informatics Systems 2021, Rruff^{TM} Project [38] and Pigments Checker Raman Database [39]) or by comparison with Raman spectra in literature.

3. Results

3.1. Metagenomic Analysis

3.1.1. DNA Yield and Sequencing Analysis

Four sequencing runs were carried out, each in an independent nanopore flow cell loaded with a DNA library prepared from each stone sample. The total reads generated per sequencing run ranged from 1,240,915 to 3,329,384, with total yields between 5.1 to 15.7 Gb. After a first filtering, the total reads per run ranged from 1,080,553 to 2,885,034, and about 2% of the analyzed reads in each sample were phylogenetically assigned. In addition, the mean length of the DNA fragments sequenced was 4678 Kb with an average quality score of 13.2. Details about the sequencing data are summarized in Supplementary Table S1.

3.1.2. Microbiomes of the Stone Samples

The results of the analyzed reads of the DNA sequencing data that could be phylogenetically assigned revealed a similar distribution of superkingdoms among the investigated samples (Figure 2), showing a microbiome dominated by bacteria, with a lower proportion of eukaryotes and a very small proportion of archaea. More specifically, samples AV1 and EZ4 showed the highest relative abundance of bacterial communities (92% and 98%, respectively), while the microbiomes of samples AV3 and EZ1 showed higher proportions of eukaryotes (29 and 35%, respectively). A very small fraction of identified reads was assigned to archaea, which represented less than 1% of the total microbiome of all the samples.

3.1.3. Bacterial Communities

Analysis of the bacterial communities, which dominated over the eukaryotic communities, revealed six main phyla, namely *Cyanobacteria, Actinobacteria, Proteobacteria, Bacteroidetes, Firmicutes* and *Deinococcus-Thermus*, representing together on average 98% of the total bacteria, in addition to some other phyla, each representing less than 0.5% of the total bacterial community and marked as "others" in Figure 3.

The most dominant phylum was *Actinobacteria* (54–61.5% of bacteria), with the exception of sample Ezuz 1, which showed a massive dominance of the phylum *Cyanobacteria* (68% of bacteria) and a lower proportion of *Actinobacteria* (14% of bacteria). The second most abundant phylum was *Cyanobacteria* in samples Avdat 1 and Avdat 3 (26–39% of bacteria), while in sample Ezuz 4, *Cyanobacteria* represented only 0.6% of the bacterial community. The phylum *Proteobacteria* was present with a relative high abundance (21% of bacteria) in Ezuz 4 and with lower proportions in the other samples (1–9%), followed by the phyla *Bacteroidetes* (6–16% of bacteria) and *Firmicutes* (0.5–2%). The last two phyla accounted for less than 0.5% in sample Avdat 1. The phylum *Deinococcus-Thermus* was identified in very low proportions in all the samples, reaching more than 0.5% only in samples Avdat 3 and Ezuz 4 (0.5–0.8% of bacteria).

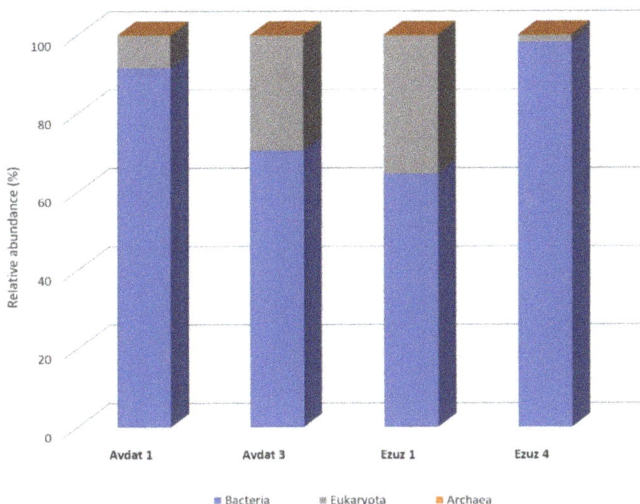

Figure 2. Relative abundance of bacteria, eukaryotes and archaea in the microbiomes of the four samples analyzed.

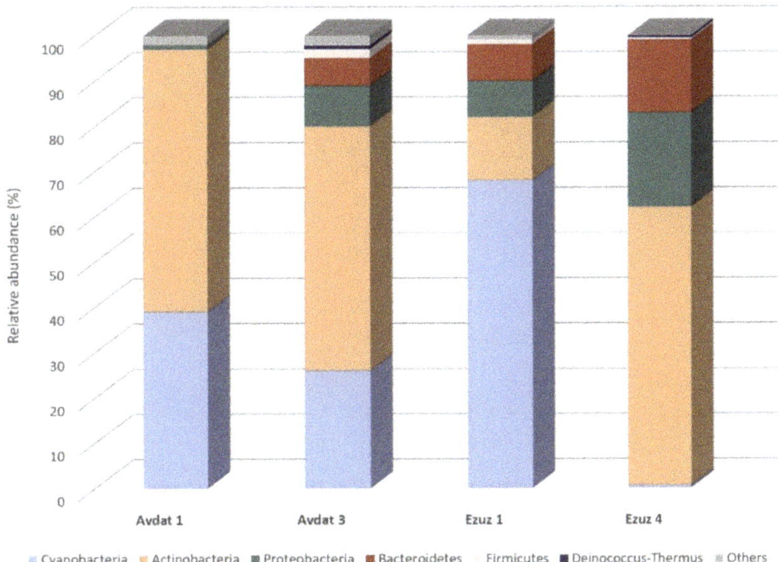

Figure 3. Relative abundance of bacterial communities at the phylum level (cutoff 0.5%).

In Supplementary Figure S2, the relative abundance of the bacterial communities at the class level is presented, while all the genera representing more than 0.5% of the total microbiome in each of the samples are summarized in the form of a heatmap in Figure 4. The heatmap shows the relative abundance of each genus in each sample, highlighting differences and similarities between them.

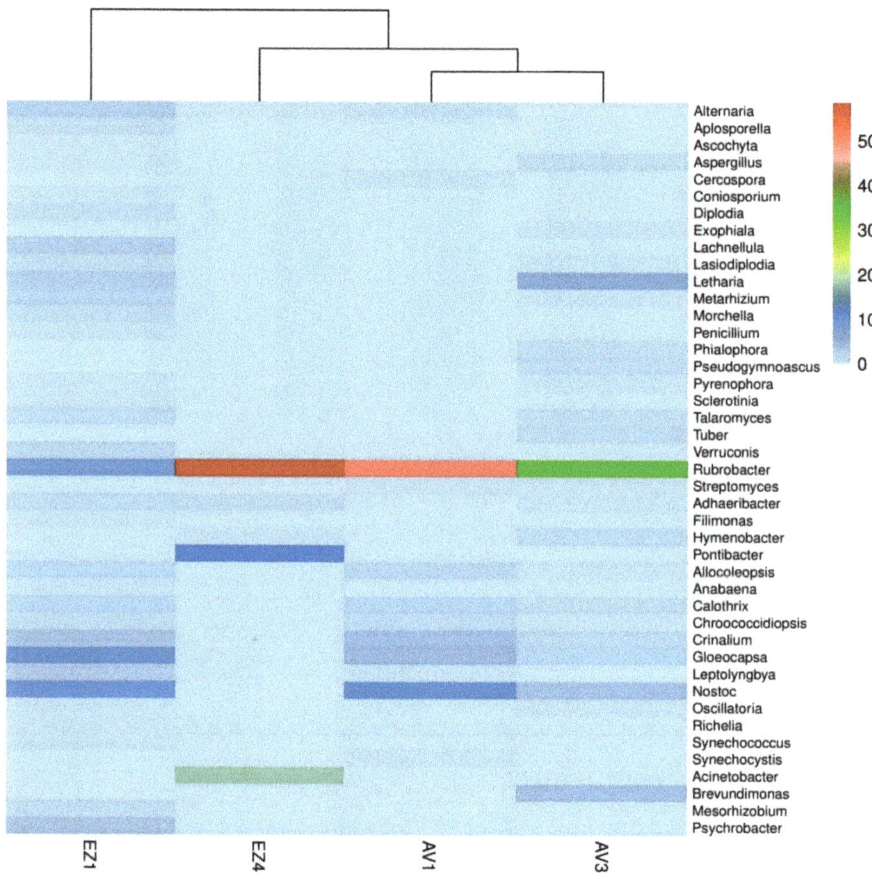

Figure 4. Heatmap displaying the relative abundance (%) of the total microbial community (bacteria and fungi) at the genus level in each sample; the genera represent an abundance greater than 0.5% in at least one sample.

Within the phylum *Actinobacteria*, the dominant class was *Rubrobacteria*, with all the members identified belonging to the *Rubrobacter* genus, present with a very high relative abundance (35–58% of total reads) in all the samples and a lower abundance in Ezuz 1 (8% of total reads) (Figure 4). Members of the class *Actinomycetia* were detected in all the samples with low relative proportions (0.6–4% of bacteria), but *Streptomyces* was the only genus present with more than 0.5% relative abundance in samples Avdat 1 and Avdat 3 (Figure 4).

Cyanobacteria were the second most represented phylum within the bacteria and the one with the highest biodiversity, except in Ezuz 4 where no genus of cyanobacteria was significantly abundant (above 0.5% of total reads). Inside this phylum, in samples AV1, AV2 and EZ1, members of the order *Nostocales* were detected, such as the genus *Nostoc*, which was present with a relatively high abundance in samples AV1 and EZ1 (~10% of total reads) but was also relatively abundant in AV3 (4% of total reads); the genus *Calothrix*, present with a relative low abundance (1–2% of total reads), and the genera *Richelia* and *Anabaena* were found solely in sample EZ1. Other members of the Nostocales, such as the genera *Fischerella*, *Tolypothrix*, *Rivularia*, and *Nodularia* were also identified, but contributed less than 0.5% of the total reads. Among the Oscillatoriales, the genus *Crinalium*, was the most abundant (2–4% of total reads), followed by the genus *Allocoleopsis* (0.5–2% of total reds) and the genus *Oscillatoria*, present only with low abundance in samples AV3 and EZ1 (~0.6% of total reads). Members of the order *Synechococcales* were identified in low proportions only in two samples, being the genus *Synechocystis* only present in AV1, and the genus *Synechococcus* only present in EZ1. Whitin the order *Chroococcales*, the genus *Gloeocapsa* was the only one well-represented in all the samples, with high relative abundance in EZ1 (9% of total reads) and lower proportions in samples AV1 (5% of total reads) and AV3 (2% of total reads). In addition, the genus *Chroococcidiopsis* (order *Chroococcidiopsidales*) was identified in AV1, AV3 and EZ1 in similar proportions (~1% of total reads), while *Leptolyngbya* (order *Pseudanabaenales*) was detected only in sample EZ1 (2% of total reads) and in lower proportion in AV1.

The phylum *Proteobacteria* was present in all the samples, with *Gammaproteobacteria* class being the most abundant in samples AV1, EZ1 and EZ4. Inside this class, members of the order *Moraxellales* were detected, such as the genus *Acinetobacter*, which accounted for over 18% of the total microbiome of EZ4 but was below 0.5% of total reads of the other samples. The genus *Psychrobacter* was instead identified in significant proportion only in sample EZ1 (Figure 4).

In contrast, members of the class *Alphaproteobacteria* were more abundant in sample AV3, with the genus *Brevundimonas* representing 4% of the total reads. Within this class, the genus *Mesorhizobium* was only significantly represented in sample EZ1.

The phylum *Bacteroidetes* was well-represented in all the samples except AV1, in which this phylum was below 0.5% of the bacterial reads. Whitin this phylum, the class *Cytophagia* was the most abundant, with the genus *Adhaeribacter* being present in AV3, EZ1 and EZ4 (0.7–1.5% total reads) and the genus *Hymenobacter* only significantly identified in samples AV3 and EZ4. Interestingly, the *Pontibacter* genus was detected with a relative high abundance (12% of total reads) in sample EZ4 but did not account for more than 0.5% of the total reads in the other samples (Figure 4). The *Chitinophagia* class was present with more than 0.5% of bacterial reads only in samples AV3 and EZ1. Inside this class, only the genus *Filimonas* accounted for more than 0.5% of the total reads and was exclusively in sample EZ1.

The phylum *Firmicutes* was represented in samples AV3, EZ1 and EZ4 with members of the *Bacilli* class, but each of the identified genera within this class contributed less than 0.5% of the total reads.

Finally, the phylum *Deinococcus-Thermus* was only significantly represented in AV3 and EZ4 with the class *Deinococci*, but no genera accounted for more than 0.5% of the total reads.

3.1.4. Eukaryotic Communities

Eukaryotic communities were mainly present in samples AV3 and EZ1 and only represented a low proportion of the microbiome of samples AV1 and EZ4. The eukaryotic sequences were affiliated with the fungi kingdom, in which the phylum *Ascomycota* represented more than 98% of the total fungi reads, with the phylum *Basidiomycota* around 1%. The phylum *Mucoromycota* was present only in two samples with very low relative abundance, less than to 0.1%.

The relative abundance of fungal communities at the class level is reported in Supplementary Figure S3. Within the phylum *Ascomycota*, the *Dothideomycetes* class was dominant in samples AV1, EZ1 and EZ4 with a very high relative abundance (42–58% of fungi) and accounted for 19% in sample AV3. The classes *Sordariomycetes*, *Leotiomycetes* and *Eurotiomycetes* were represented in all the samples with similar proportions, ranging from 7 to 17% of the total fungi, followed by the class *Pezizomycetes*, which contributed between 1 and 6% of the total fungal reads. The class *Lecanoromycetes* was well-represented in sample AV3 (21% of fungi) and in lower proportion in AV1 and EZ1 (1 and 7%, respectively). Finally, the class *Orbiliomycetes* was only significant in sample AV3 (0.6%).

The phylum *Basidiomycota* was mainly represented by the *Agaricomycetes* class, present with low relative abundance (~0.6% of fungi) in samples AV1, AV3 and EZ4.

Differently from the bacterial communities, which clearly showed the dominance of some genera, fungal communities revealed very high biodiversity but no dominant taxa, with most of the genera accounting for less than 1% of the total microbiome of each sample. Nevertheless, in samples AV3 and EZ1, a few genera contributing with higher proportion were detected (Figure 4). In sample AV3 the genus *Letharia* (class *Lecanoromycetes*) represented 6% of the total reads, while the genus *Tuber* (class *Pezizomycetes*) and the genus *Pseudogymnoascus* (class *Leotiomycetes*) accounted for 1.3% of the total microbiome. In sample EZ1, the genera *Alternaria* (class *Dothideomycetes*), *Lachnellula* (class *Leotiomycetes*), *Letharia* (class *Lecanoromycetes*) and *Metarhizium* (class *Sordariomycetes*) were present with a relative abundance of around 2.5%, while the genera *Diplodia* (class *Dothideomycetes*), *Talaromyces* (class *Eurotiomycetes*) and *Verruconis* (class *Dothideomycetes*) made up between 1 and 1.9% of the total reads.

3.2. Digital Optical Microscopy

The samples collected from the petroglyph sites were first studied under the microscope without any preparation to gather an overview of their physical characteristics (microscope images of the samples are reported in Supplementary Figure S4), while microscope observation of thin sections allowed us to evaluate different aspects such as the thickness of the varnish, the possible presence of weathered areas in the bedrock, and where visible, the penetration of microorganisms in the stone.

Six of the rock samples (AV1, AV2, AV3, EZ1, EZ2, EZ3) collected from the petroglyph sites are coated with a black crust (Supplementary Figure S4a), the so-called desert varnish. It has been observed that underneath the thin black outer layer, a bright reddish layer is present, which is intermixed with the stone grains (Supplementary Figure S4b). The black varnish has a compact, shiny appearance, but when observed under microscope, the surface is not homogeneous and presents reddish and white patches. Indeed, in the highest points, the black layer is missing and the orange layer or the white rock are exposed. In the lowest points (depressions) an accretion of brownish material is visible, with an accumulation of dust grains in the micro-basins. The appearance of the black crust can vary from sample to sample and be thinner and/or more intermixed with the reddish layer (Supplementary Figure S4c). In addition, the black crust does not occur homogeneously throughout the same sample, and different stages of weathering of the outer black layer can be observed. Sample EZ4 differs from the others and is covered by an orange crust (Supplementary Figure S4d). In some areas of sample EZ4, it appears that traces of the black varnish are present. Finally, the presence of cyanobacterial aggregates was extensively detected in the samples (Supplementary Figure S4e,f).

The thin sections showed that the limestone bedrock of all the samples mainly consists of microfossils of foraminifera, bryozoans, diatoms and turritella. Samples AV1, AV2, AV3, EZ1, EZ2 and EZ3, which are characterized by the black coating, presented similar features.

The black varnish has a variable width, ranging from about 20 μm in the thinnest areas to about 130 μm in the thickest parts (Figure 5a), but never exceeding 150 μm. The reddish layer is not homogeneously distributed and is present only in some areas. What is clearly visible in all samples is a weathered area underneath the varnish. In Figure 5b–e, examples of this phenomenon are shown. The zones below the black varnish appear to be moderately to heavily weathered, and grains of the bedrock have been leached, leaving behind a very porous area with an average depth of about 800 μm but that can also reach about 1.8 mm, as observed in sample EZ3 (Figure 5c), which shows different layers of weathering. In the porous weathered areas, it can be noticed that the cavities and pores close to the surface are coated by the dark crust (see also Supplementary Figure S5). In other areas, beneath the black coating, the bedrock is heavily weathered, and the stone appears completely fragmented into small particles with loose arrangement (Figure 5d). In these areas it can be observed that the shells of the fossils seem to be more resistant to the weathering processes and remain almost intact, while the softer stone matrix is leached. In sample AV1, mixed with the loose grains of the weathered stone, we observed a granular brown material.

Aggregates of cyanobacteria colonized this area, as is well-illustrated in Figure 5e. It can be noticed that in this zone, the black varnish appears completely fragmented and detached from the bedrock, and in some areas is no longer present.

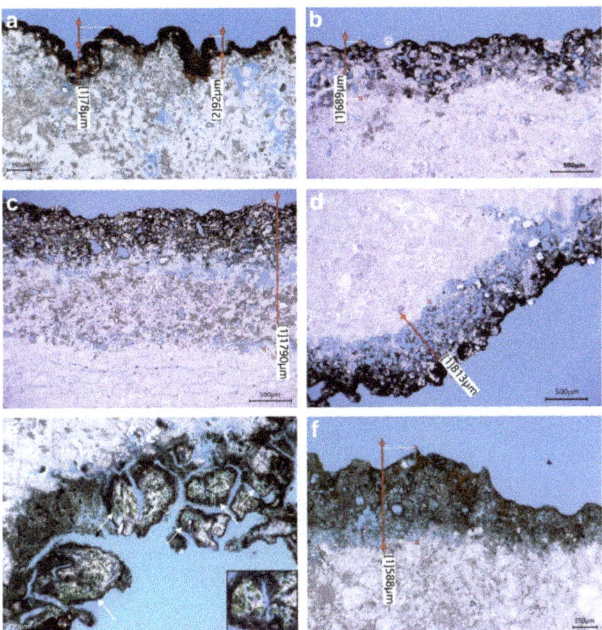

Figure 5. Pictures of the petrographic thin sections acquired with the Keyence digital microscope in transmitted light: (**a**) detail of the black varnish and of the reddish layer underneath it of sample AV1; (**b–d**) examples of the weathered area occurring beneath the black coating in samples EZ1, EZ3 and AV2; (**e**) weathered area colonized by cyanobacteria aggregates in sample AV1; (**f**) detail of the orange crust of sample EZ4.

A different situation was observed in the thin section of sample EZ4. The orange crust is very porous and reaches a thickness greater than 1 mm. Interestingly, underneath the orange layer, no weathered area was observed. The bedrock looks undamaged till the boundary with the crust. Here, the stone seems to gradually merge into the orange layer (Figure 5f).

3.3. X-ray Fluorescence Spectroscopy (XRF)

Preliminary analysis performed with the portable XRF showed that the dominant elements of the black varnish (samples AV1, AV2, AV3, EZ1, EZ2, EZ3) are Fe, Mn and Si, with lower counts of K, Ca and Ti, thus in agreement with the composition described in the literature, i.e., clay minerals mixed with iron and manganese oxides. In sample EZ4, which does not have the black crust but presents instead a thicker orange crust, Mn shows much lower counts or, depending on the spot of analysis, is not present at all. Fe presents high counts, while Ti, K, Ca and Si are detected with low counts. The bedrock is mainly composed of Ca and Si.

To better understand the composition of the different layers (white bedrock, black coating and orange layer beneath it) the powders obtained from each layer of sample AV1 were analyzed with WDXRF, and the results are reported in Table 2.

Table 2. XRF data (%) of the three layers of sample AV1.

Elements *	Black Varnish (%)	Reddish Layer (%)	Bedrock (%)
Mg	1.1	1.1	2.0
Al	8.4	7.0	3.2
Si	66.1	73.4	48.9
P	1.1	0.5	0
S	1.0	0.3	0.1
K	1.1	1.1	0.4
Ca	8.3	9.9	40.9
Ti	0.6	0.5	0.1
Mn	3.1	0.8	0.1
Fe	8.0	4.7	1.2
Ba	0.3	0	0

* Only the main elements are shown.

The results confirmed that the black varnish is rich in Fe and Mn with a high content of Si and lower of Ca and highlighted the presence of light elements (not detectable with the portable XRF) such as Mg, Al, P and S in the sample. Mg is present in higher amounts in the bedrock, while Al, S and P showed the highest percentages in the black varnish, followed by the reddish layer. Low amounts of Ti and Ba were also detected in the black and reddish layers; the latter being present only in the black varnish. While Mn is mostly present only in the black coating, Fe is present in higher amounts also in the reddish layer, which seems to be composed mainly of Si and lower percentages of Ca, Al, K and Mg. In the bedrock, in addition to the presence of Ca and Si, low percentages of Mg, Al and Fe were also detected.

Because of the difficulty in accurately separating the layers, some elements might show a higher percentage due to contamination coming from the adjacent layer.

The distribution of the elements in the samples is further discussed in the following section.

3.4. Scanning Electron Microscopy Coupled with EDX Microanalysis (SEM–EDX)

SEM–EDX was initially used for observation of the stone slabs without any preparation to detect the presence of microorganisms both in the crust and in the bedrock. On the black coating, no biological colonization was detected, but at the transversal section of the samples, putative bacteria (coccoid shape) were revealed (Figure 6a), occupying the microcavities close to the surface beneath the black layer. In other areas of the samples, cyanobacterial biofilms were detected (Figure 6b). In addition, on the surface of sample EZ4, fungal hyphae were observed (Figure 6c).

Observation of surface morphology was also performed by SEM–EDX, and punctual analyses were carried out to locally characterize the elemental composition of the black crust, which exhibits strongest maxima of Si and Al, followed by Fe, Mg and Mn and in lower proportions Ca, K and Ti, thus confirming the results of XRF analysis. The black coating observed from the external surface shows a structure with mounds and depressions, where the upper parts appear smoother and more compact while the lower parts appear more crumbled. EDX performed on different points revealed that Mn is present in higher amounts on the more compact parts, while in areas that appear more crumbled, the peak of Mn is lower or not detected at all.

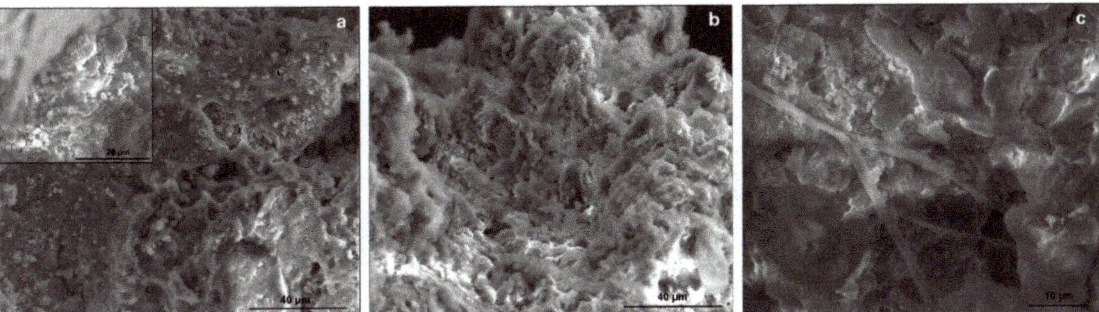

Figure 6. SEM images in secondary electron mode showing putative bacteria in the microcavities of the stone (**a**), a cyanobacterial biofilm (**b**) and fungal hyphae (**c**).

Examination of the black varnish in the petrographic thin sections revealed a clear micron-size laminated pattern (Figure 7a) with a curved shape and oriented parallel to the surface. This micro texture is defined as a stromatolite structure and is typical of desert varnish [17,19,27]. Mappings of the elemental composition of the black crust revealed that Mn and Fe are not homogeneously distributed, but their content varies in the different layers of the coating. Specifically, we observed that the upper part of the varnish is Mn-depleted, along with the part in contact with the bedrock, while an accumulation of Mn is present in the central area. Other elements such as Al, Mg and K are equally distributed throughout the crust. In addition, accumulation of detrital particles was observed in the depressions of the outer layer. These particles were mostly composed of clay minerals (Si, Al, K, Mg and Fe), while the Mn-rich material that characterizes the varnish is only present in the coating itself (Supplementary Figure S5). Surprisingly, no calcium was detected in the rock beneath the crust, but only silicon, which was in contrast with the results of XRF analyses. Therefore, a more thorough analysis of the bedrock was carried out through EDX punctual measurements, which revealed that the bedrock is composed of a calcitic matrix with lower content of Mg in which silicate grains and silicified microfossils are embedded. Interestingly, backscattered images of the thin sections (sample AV1 is reported as an example in Figure 7b) showed that the bedrock is not homogeneous and, in the areas corresponding to the porous weathered zones previously observed with the optical microscope, the calcitic matrix (light grey) is missing except for sporadic grains, and only the silicate component (dark grey) is left. This phenomenon is clearly illustrated in Figure 7c, where elemental mapping of the weathered area of sample EZ3 is reported. Mg is present only in the intact bedrock (and in the black varnish on the surface), while in the weathered layer in contact with it we observed an initial leaching of Ca, which then disappears completely in the upper layer, leaving a very porous zone. In this area we observed only the silicate component, which seems to be homogeneous throughout the sample. In the upper layer Al, K, Mg, Fe and Mn are distributed in the black crust, thus along the edges of the pores and on the surface of the sample (Supplementary Figure S5). In this area, barium sulphate was also detected (bright white grains in the backscattered image of Figure 7c) but mostly concentrated towards the surface of the sample (maximum of 400 µm depth).

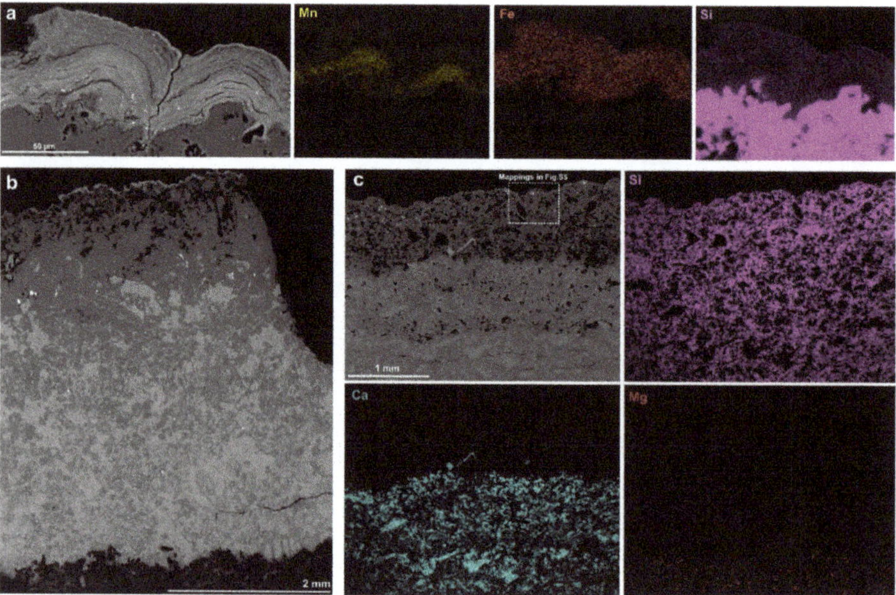

Figure 7. SEM–EDX analyses on petrographic thin sections: (**a**) SEM image in backscattered electron mode and EDX elemental mapping (Mn, Fe and Si) of the black varnish; (**b**) SEM image in backscattered electron mode of the thin section of sample AV1 showing the lack of Ca in the upper part of the sample and a heavily weathered area at the bottom; (**c**) SEM image in backscattered electron mode with EDX elemental mapping (Si, Ca and Mg) of the weathered area of sample EZ3 (additional elemental mappings of the area in the dashed box in Supplementary Figure S5).

The backscattered image of sample AV1 (Figure 7b) shows that in the heavily weathered area that lies at the bottom, where aggregates of cyanobacteria were extensively detected by optical microscopy, the Ca and Si components of the bedrock seem to be equally leached, and loosely arranged mineral particles are observed. Mapping of this area indeed revealed that the mineral fragments in the proximity of the cyanobacteria cells are mainly composed of Ca and Si, along with Mg-rich grains and other minerals such as Al, Fe and K (Supplementary Figure S6).

Punctual EDX analysis of the orange crust of sample EZ4 showed that it is mainly composed of Si, Fe, Al and K (as shown by XRF results), and a low proportion of Mn depending on the point of analysis. The elemental mapping carried out on this sample revealed some interesting results (Figure 8). We noticed that the distribution of the elements showed a pattern similar to the one displayed in the weathered areas of the other samples: Ca and Mg are detected only in the intact bedrock, while silicon is homogeneously distributed (in the bedrock and in the orange crust). In addition, the map of silicon showed the presence of silicified fossils in the orange layer. Aluminum is detected solely in the orange layer, with a distribution similar to that in the surface layer of sample EZ3, i.e., more concentrated along the edges of the pores, but in this case it appears more spread throughout the layer. Iron is mostly present in the orange part and seems to be homogenously distributed, while the mapping of Mn did not give a clear distribution.

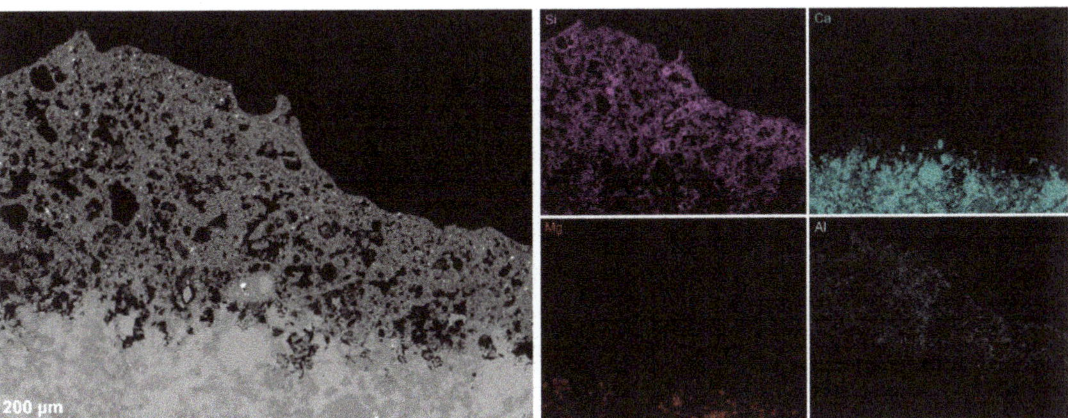

Figure 8. SEM image in backscattered electron mode and EDX elemental mapping (Si, Ca, Mg and Al) of sample EZ4.

3.5. Micro-Raman Spectroscopy

Raman spectra acquired on the black varnish were quite complex. Depending on the measurement spot, the spectra displayed different features confirming the heterogeneity of the varnish composition. In many cases, the overall spectral profile matches well the spectral features of the pigment burnt Sienna, which is a mixture of iron oxides, quartz and aluminum oxides. An example is reported in Figure 9a—spectrum A, in which, more specifically, the bands at 224, 291, 408 and 1322 cm^{-1} can be assigned to hematite (α-Fe_3O_4), while the band around 670 cm^{-1} is diagnostic for magnetite (Fe_2O_3) [40]. In the region between 500 and 700 cm^{-1}, the peaks of the spectra acquired on the black varnish are not well resolved and a broad band with shoulders is observed; therefore, clear assignment of signals is challenging. In this spectral region, the characteristic peaks of Mn oxides are found, so their presence could not be clearly identified. Nevertheless the peaks visible at around 400 cm^{-1}, at 502 cm^{-1} and at 574 cm^{-1} (Figure 9a—spectrum B) might be assigned to birnessite or birnessite-like minerals [41], a manganese oxide often reported in the literature as part of the desert varnish [19]. Quartz (sharp band at 465 cm^{-1}) was often identified in the spectra of the black varnish, and in some areas anatase (TiO_2) (band at 143 cm^{-1}) and pseudobrookite (Fe_2TiO_5) were also detected.

The spectra obtained on the orange layer emerging from the black varnish (Figure 9a—spectrum C) showed a spectral profile similar to the one of the orange crust of sample EZ4 (Figure 9a—spectrum D), showing characteristic features of iron oxides (bands around 220, 290 and 400 cm^{-1}) and quartz (bands at 207 and 465 cm^{-1}) at lower wavenumbers. In the range from 1000 to 1800 cm^{-1}, EZ4 presented a complex spectrum, with overlapping broad bands that did not allow clear interpretation. The peaks in this range might be related to organic pigments such as carotenoids combined with amorphous carbon (peaks at 1320 and 1600 cm^{-1}).

The presence of carotenoids was clearly confirmed in a reddish area beneath the black varnish (transversal section of the sample). Indeed, the Raman spectrum of the area analyzed (Figure 9B) showed three sharp bands at 1000 cm^{-1}, 1151 cm^{-1} and 1505 cm^{-1} typical of carotenoids, which are characterized by three main skeletal features: the stretching of C=C in the region around 1520 cm^{-1}, the stretching of C–C around 1150 cm^{-1} and the bending of C=CH around 1000 cm^{-1} [42].

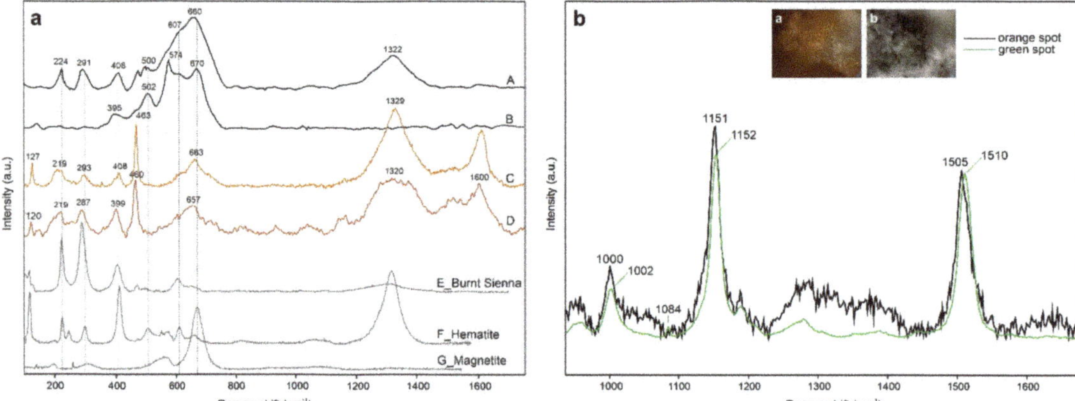

Figure 9. (**a**) Normalized Raman spectra collected from the black varnish on two different spots (A, B), on the orange layer emerging from the black varnish (C) and on the orange crust of sample EZ4 (D) compared with reference spectra of burnt Sienna (E) (Pigments Checker Raman Database [39]), hematite (F) and magnetite (G) (RRUFF [38]) in the spectral range between 150 and 1700 cm^{-1}. (**b**) Normalized Raman spectra collected on the orange layer beneath the black crust (a) and on a spot colonized by cyanobacteria (b) in the range from 900 cm^{-1} to 1700 cm^{-1}, showing the characteristic peaks of carotenoids.

On one sample, Raman spectra were acquired in a green area visibly colonized by cyanobacterial aggregates. The spectral features of carotenoids were again identified but with a shift in the wavenumbers (1002 cm^{-1}, 1152 cm^{-1} and 1510 cm^{-1}) compared to the spectrum obtained on the reddish area (Figure 9b). The shift might indicate the presence of a different type of carotenoid pigment in the green area. According to Jehlička et al. [42], the position of the stretching band of the C=C group is diagnostic for the discrimination of different carotenoids, and they assigned the band at 1506 cm^{-1} to the carotenoid pigment bacterioruberin and the C=C signal at higher wavenumbers (1510 cm^{-1}) to β-carotene. The positions of the two bands correspond very well to the one shown in the Raman spectra of carotenoids obtained in this study: 1505 cm^{-1} in the case of the orange spot and 1510 cm^{-1} in the case of the green spot. Despite this, attribution of the two spectra to one carotenoid pigment rather than the other cannot be certain solely by Raman spectroscopy.

In addition, the signal of carotenoids on the green spot was detected along with calcite (band at 1084 cm^{-1}), while on the orange area anatase (band at 141 cm^{-1}) was also identified.

4. Discussion

The multi-analytical approach used in this study allowed us to characterize the microbiome of stones associated with petroglyph sites and, at the same time, to characterize the lithic substrate, thus obtaining a deeper knowledge of possible mineral–microbial interactions.

The microbiomes of the samples were generated using nanopore sequencing technology, which has been employed in different studies in the cultural heritage field [37,43–45], showing some advantages compared to next-generation sequencing methods. In fact, this technology, which offers an easier workflow and much more affordable prices, when combined with whole genome amplification (WGA) protocol allows simultaneous obtainment of the real proportions of different groups of microorganisms (fungi, bacteria and archaea) as a whole in a given sample, which was one of the first aims of this study.

The microorganism communities associated with rock varnish have been previously studied to determine whether or not there is a connection between the varnish and the microorganisms inhabiting the stone. The results of our metagenomic study are consistent with the findings reported in the literature [25,33], revealing a microbiome dominated by bacteria, with lower proportions of eukaryotes, and archaea representing less than 1% of the total assigned reads. Within the bacteria, Actinobacteria and Cyanobacteria were the most abundant phyla, followed by Proteobacteria and Bacteroidetes, as found also by other authors in metagenomics studies on microbial communities of stones in desert environments [16,27,30]. Notable is that sample EZ4, which does not present the superficial black varnish, showed a different microbiome, in which the genus *Rubrobacter* (*Actinobacteria* phylum) together with the genus Acinetobacter (*Proteobacteria* phylum) and the genus *Pontibacter* (*Bacteroidetes* phylum) made up almost 90% of the total reads, while Cyanobacteria represented less than 1%.

In the other samples (AV1, AV3 and EZ1), Cyanobacteria represented between 26 and 39% of the bacterial communities, with a rather high biodiversity. Cyanobacteria are common photoautotroph bacteria inhabiting rocks in desert environments [29,46,47]; they are considered primary producers with the ability to withstand stressful environmental conditions due to their thick cell wall and the production of protective pigments such as scytonemin and carotenoids, which protect them against photooxidative damage [16,48]. Among the cyanobacteria, the order *Nostocales*, a nitrogen-fixing family that contains several pigment-secreting taxa, was dominant, with *Nostoc* spp. being the most abundant. The order *Nostocales* was previously reported to be desiccation resistant [49] and radiation tolerant and a dominant taxa in hypolith communities of the Atacama desert [50]. The genus *Crinalium* (order *Oscillatoriales*) and the genus *Gloeocapsa* (order *Chroococcales*) had relatively high abundance in our samples and are also known to be well adapted to harsh environments [51,52]. Within the order *Chroococcales*, the extremely desiccation and radiation resistant genus *Chroococcidiopsis* is often indicated as dominant in cyanobacterial communities on stone in deserts [27,53,54]; however, in our study it represented only around 1% of total reads.

Cyanobacteria, as many other microorganisms, can form biofilms, which have huge weathering potential [55] and can cause both physical and chemical degradation of the stone. The physical actions involve breaking of the surface and penetration of the microorganisms between the grains of the rock with consequent decreases in grain cohesion, while chemical actions include dissolution and chelating processes bringing chemical changes to the stone substrate [56]. In our study, through optical microscopy and the use of SEM, we could visualize areas of the samples with cyanobacterial biofilms that were heavily weathered. In fact, associated with cyanobacteria aggregates, we observed disintegration of the rock, with loose mineral fragments and consequent detachment of the black crust (Supplementary Figure S6). Species of the genus *Gloeocapsa*, which was relatively abundant in our samples, and of the genus *Leptolyngbya*, also detected in our metagenomic analysis but with lower proportions, were, for example, found responsible for the formation of biofilms in historic stone monuments, with resulting bioweathering of the lithic substrate [57].

Within the Actinobacteria, the genus *Rubrobacter* was very dominant in all the samples (as clearly visible in the heatmap of Figure 4). *Rubrobacter* spp. are often associated with desert varnish and in general as a bacterium inhabiting stones in desert environments [16,58]. This genus is known to be highly resistant to many types of environmental stress, such as exposure to gamma and UV radiation, desiccation, high temperature fluctuations, low water availability and high levels of salts [59,60]. In fact, *Rubrobacter* spp. are often reported to grow on walls and stone monuments rich in salts, causing the so-called rosy discoloration due to the release of carotenoid pigments such as Bacterioruberin [61–63].

The genera *Pontibacter*, *Hymenobacter* and *Adhaeribacter* are also highly radiation resistant bacteria [64] with the ability to synthesize carotenoids and indeed show pink to red shades [65,66].

The eukaryotic communities were affiliated with the fungi kingdom, which was mostly represented by the phylum Ascomycota. High biodiversity at the genus level was observed, but each genus showed low proportion and no dominant taxa, except for a few genera contributing with slightly higher proportions, such as *Letharia*, *Alternaria*, *Lachnellula*, *Metharizium* and *Talaromyces*. Most of the fungi detected are not stone-dwelling species, but they are found ubiquitously in the environment and are probably deposited on the stone as spores. In fact, we found little evidence that fungi are established on the rock surface. Within the fungi detected in this study, only the order *Chaetothyriales*, with *Exophiala* spp. and the genus *Coniosporium*, with the *Coniosporium apollinis* species, might actually be established on the rock. The aforementioned fungi are in fact part of the black fungi, or black meristematic fungi group, which have an extraordinary tolerance to extreme environmental conditions and are known to inhabit stone surfaces in hot deserts and moderate climates [67–69].

Previous studies mentioned the presence of fungi and lichens, such as *Caloplaca* sp., as a weathering agent of rock substrates due to the production of oxalic acid and consequent formation of calcium oxalates that can facilitate the disintegration of the calcareous matrix [33]. In this study, however, we could not detect the presence of calcium oxalates. Further culture-dependent and culture-independent microbiological studies are being carried out to have better insight of the fungal communities associated with the rocks of petroglyph sites.

The characterization of the black crust through XRF and SEM–EDX analysis confirmed that the varnish is composed of clay minerals and Fe and Mn oxides, as reported by many studies [18,23,24]. Specifically, the varnish is highly enriched in Mn compared to the windborne dust deposited in the micro-basins of the surface. This was proven both by punctual EDX analysis, which showed that the crumbled parts of the surface (loose material) have low Mn content compared to the compact surface of the varnish, and SEM–EDX elemental mapping, in which it is clearly visible that the dust deposited in the depressions of the crust is mainly composed of Al, K and Fe, while Mn is concentrated in the actual varnish layer. Enrichment of Mn in the varnish layer by a factor of ~100 [19] compared to its primary source of material, i.e., windborne dust, is one of the main points of discussion in the literature, especially related to whether or not Mn enrichment is due to abiotic [19,70,71] or biogenic factors [26,27,46].

Raman analysis allowed us to partially characterize the minerals present in the black varnish. Identification of iron oxides was quite straightforward, and hematite and magnetite were detected, consistent with the results of other Raman studies on rock varnish [72–74]. The identification of Mn oxides was more challenging, as also previously reported by Malherbe et al. [72], who were not able to identify Mn oxides by Raman spectroscopy on the rock varnish analyzed. Nevertheless, our results seem to indicate that birnessite-like Mn oxides might be present, which are addressed as the most common Mn compounds in desert varnish [19]. Interestingly, the Raman spectra of the orange layer of sample EZ4 and of the reddish layer emerging from the black crust of the other samples showed similar spectral features, indicating the presence of iron oxides, a possible presence of Mn oxides, and, at higher wavenumber, the peaks around 1320 and 1600 cm^{-1} suggested the presence of carbon. Although it is believed that Raman spectroscopy cannot distinguish between biogenic and abiotic sources of carbon, Malherbe et al. [72] address the possibility that the graphitic carbon found on the desert varnish in their study might be the result of "highly thermally process carbonaceous residues from living organisms".

Raman spectroscopy also pointed out the presence of carotenoid pigments in the samples, which are used by microorganism as protection against radiation, therefore indicative of microbial activity. A wide variety of bacteria detected in this study are known to synthesize carotenoids, for example the bacterioruberin from *Rubrobacter* spp., which might be the source of the carotenoid pigment we detected on the reddish layer beneath the black crust. We could also prove the presence of carotenoids directly on cyanobacteria aggregates, in which the most common pigment is β-carotene. The Raman measurements to detect biopigments were preliminary, and these results could be implemented in future studies.

The combination of microscope observations and SEM–EDX analysis on the petrographic thin sections showed very interesting results. Optical microscopy unveiled the presence of weathered areas beneath the black crust, and SEM–EDX elemental maps revealed that in these areas only the Si component of the bedrock is present while the Ca and Mg matrix is missing. The same elemental distribution was observed in sample EZ4, in which the orange layer was mainly composed of Si, also highlighting the presence of silicified fossils widely present in the bedrock and the lack of the Ca matrix. These observations brought us to the conclusion that the orange layer, rather than a different kind of superficial crust, might be a weathered area of the bedrock similar to the ones observed in the other samples, but in this case, the superficial black varnish is completely missing. The typical elements normally detected in it, such as Al, K and Fe, are distributed in the weathered area, including also a low proportion of Mn. These observations might suggest that elements present in the black crust had been mobilized by microorganisms, contributing to the slow degradation of the black varnish.

The lack of Ca in the superficial layers of rocks coated by the desert varnish was previously observed by other authors [33,46], who attribute the calcite matrix deficiency to dissolution actions by microorganisms. Biodeterioration of calcareous rocks due to microbial colonization is a well-studied phenomenon [56,75–77], and acidification due to the production of carbonic, sulfuric and nitric acids as byproducts of cell metabolisms has been proposed as a possible cause of calcium carbonate dissolution [29,78]. In particular Lian et al. [79] mention cyanobacteria genera such as *Gloeocapsa*, *Nostoc* and *Oscillatoria* as able to degrade calcareous rocks. These cyanobacteria genera were all detected as part of the bacterial communities by the metagenomic analysis carried out in this study. The presence of microorganisms in the microcavities below the crust was detected by SEM analysis; further, by means of Raman spectroscopy, we were able to identify carotenoid pigments in the reddish layer below the dark coating, indicating biological colonization. Moreover, extensive microbial colonization in the cavities and discontinuities of the rocks associated with petroglyph sites through SEM–BSE was previously broadly studied by Nir et al. [33], who identified a complex microbial community (cyanobacteria, heterotrophic bacteria, free-living fungi and lichens) inhabiting the superficial layers of the rocks in which the Ca matrix is no longer present.

Krumbein and Jens [46] state that cyanobacteria and fungi dissolve the carbonates in limestone rocks to reach optimal living conditions such as temperature, humidity and nutrient supply. Indeed, the interior of the rocks can be a suitable niche for microorganisms to survive harsh condition due to better nutrient and water availability [54,55]. In the case of stones covered with desert varnish, this might act as an extra shelter from the extreme environmental conditions [16]. Moreover, Lang-Yona et al. [25] hypothesized that some species might take advantage of the rich transition-metal coating, being able to utilize the varnish minerals; they give as an example the use of Mn by cyanobacteria in their photosynthetic apparatus. Moreover, the authors of this study report that they detect enrichment in the metabolic process genes of microorganisms that relay on varnish-enriched transition metals.

It is known that a wide variety of endolithic microorganism has the ability to mobilize metal cations [55] and in particular iron [72] and manganese, which are key metals used by microorganism for protection against UV radiation. Webb and Di Ruggiero [60] state that the cellular accumulation of Mn by microorganisms might be a widespread mechanism to survive oxidative stress, and they investigated Mn^{2+} function in *Rubrobacter* spp., suggesting that Mn–antioxidant complexes are involved in radiation resistance, while a recent study [80] on the radiation resistance of *Deinococcus Radiodurans* established the key role played by intracellular Fe and Mn in oxidative protection of this microorganism.

In this study, *Rubrobacter* spp. were dominant in all of our samples, and especially in sample EZ4, and might uptake both iron and manganese from the black crust. Moreover, the genus *Rubrobacter*, as mentioned previously, is known to grow on stone monuments with the ability to form biofilms and penetrate the porous rock, enhancing mechanical damage through the detachment of mineral grains [62].

These observations support the hypothesis that in sample EZ4, the black varnish might have been degraded by microorganism, including the genus Acinetobacter (18% of total reads of sample EZ4), which is reported to be able to solubilize Al, Si and Fe from rocks [58].

More and more recent studies seem to support the hypothesis that the black crust is of biological origin [25,27,81–83] and is possibly produced by microorganisms to protect themselves from the adverse conditions. In our study, patterns of weathering of the black crust are observed, and this might be explained as a succession of biogeochemical cycles. We hypothesize that what we observed in sample EZ4 (presence of a weathering layer and lack of the black crust) might be a later stage of weathering compared to the other samples. After an initial protective colonization, the microorganisms that are established in the superficial layer of the rock, shielded by the black crust, leach the Ca matrix to reach optimal living conditions. At this point, some species, for example *Rubrobacter* spp., might take advantage of the Mn and Fe and uptake the cations to survive oxidative stress, leading to a very slow deterioration of the crust, which is also enhanced by the mechanical detachment caused by weakening of the bedrock.

In support of this theory, we observed also a different microbial community in sample EZ4 that could be explained in terms of microbial succession. Cyanobacteria, which are the pioneering colonizers of the stone surface, although able to withstand a stressful environment, need the protection of the black crust to survive the harsh desert conditions. With the loss of the black crust, cyanobacterial communities are probably too exposed to the extreme environmental conditions (high T, low water availability, high irradiation) and thus gradually disappear, followed by the establishment of more resistant taxa, such as, in the case of sample EZ4, *Rubrobacter*, *Acinetobacter* and *Pontibacter*.

5. Conclusions

Our results pointed out that:

1. most of the microorganisms detected by metagenomic analysis are taxa adapted to live in harsh environmental condition such as the Negev desert, with a high resistance to radiation and desiccation;
2. weathering of the stone beneath the black crust was associated with the potential of the species detected to leach the calcareous matrix of the rock in order to find optimal conditions of growth while being sheltered by the black desert varnish;
3. in sample EZ4, the orange layer is probably not a different kind of coating, but it is actually a weathered area of the bedrock in which the black varnish is no longer present; therefore, we hypothesize that this sample is in a later stage of weathering compared to the others;
4. we acknowledge the possible microbial involvement in the origin of the crust, as previously reported by many authors in the literature, and, on the other hand, what we observe in our study are patterns of deterioration in the bedrock and in the black varnish, feasibly connected to microbial action. The two hypotheses can be viewed in terms of succession of biogeochemical processes and of microbial communities.

Further research is needed to better understand the mineral–microbial interaction; therefore, the goal of future research will be to further explore the genome of microbial communities inhabiting stones associated with petroglyph sites and get insight into functional genes.

Supplementary Materials: The following supporting information can be downloaded at: https://www.mdpi.com/article/10.3390/app12146936/s1, Table S1: Details of the sequencing runs. Figure S1: Charts of the meteorological measurements. Figure S2: Relative abundance of the bacterial communities at the class level. Figure S3: Relative abundance of the fungal communities at the class level. Figure S4: 3D pictures of the stone slabs acquired with the Keyence digital microscope. Figure S5: SEM–BSE image and digital OM picture of a detail of sample EZ3 with EDX elemental mappings. Figure S6: SEM–BSE image and digital OM picture of a cyanobacteria colonized area with EDX elemental mappings.

Author Contributions: Conceptualization, K.S., G.P. and A.K.; analytical measurements, L.R., I.N., E.E. and M.J.P.; molecular analysis, L.R. and G.P.; bioinformatic analysis, M.W. and A.G.; writing—original draft preparation, L.R., K.S. and G.P.; writing—review and editing, L.R., G.P., K.S., I.N. and A.K. All authors have read and agreed to the published version of the manuscript.

Funding: This research was funded by the Austrian Science Fund (FWF—Der Wissenschaftsfonds) project I 4748-B Stone cultural heritage in Austria and Israel.

Institutional Review Board Statement: Not applicable.

Informed Consent Statement: Not applicable.

Data Availability Statement: All data are available in the NCBI database.

Acknowledgments: The authors gratefully acknowledge the Austrian Science Fund (FWF, Der Wissenschaftsfonds) and the Ministry of Science and Technology (MOST), Israel for funding the project. We would also like to acknowledge Dubravka Jembrih-Simbürger, Academy of Fine Arts, Vienna, for help with initial XRF analysis, Mirsky Yelena, Ben-Gurion University, for WDXRF analysis, Chris Mckay, NASA-Aimes, U.S.A for climatic measurement support, and the Israel Nature and Parks Authorities (INPA) for sampling permission.

Conflicts of Interest: The authors declare no conflict of interest.

References

1. Aubert, M.; Lebe, R.; Oktaviana, A.A.; Tang, M.; Burhan, B.; Hamrullah; Jusdi, A.; Abdullah; Hakim, B.; Zhao, J.-X.; et al. Earliest hunting scene in prehistoric art. *Nature* **2019**, *576*, 442–445. [CrossRef] [PubMed]
2. Brumm, A.; Oktaviana, A.A.; Burhan, B.; Hakim, B.; Lebe, R.; Zhao, J.-X.; Sulistyarto, P.H.; Ririmasse, M.; Adhityatama, S.; Sumantri, I.; et al. Oldest cave art found in Sulawesi. *Sci. Adv.* **2021**, *7*, eabd4648. [CrossRef] [PubMed]
3. Francis, A.; Loendorf, L.L. *Ancient Visions: Petroglyphs and Pictographs of the Wind River and Bighorn Country, Wyoming and Montana*; University of Utah Press: Salt Lake City, UT, USA, 2004.
4. Bednarik, R.G. Pleistocene Palaeoart of Africa. *Arts* **2013**, *2*, 6–34. [CrossRef]
5. Dorn, R.I. Anthropogenic Interactions with Rock Varnish. In *Biogeochemical Cycles: Ecological Drivers and Environmental Impact*; John Wiley & Sons: New York, NY, USA, 2020; pp. 267–283.
6. Smith, B.W.; Black, J.L.; Hoerle, S.; Ferland, M.A.; Diffey, S.M.; Neumann, J.T.; Geisler, T. The Impact of Industrial Pollution on the Rock Art of Murujuga, Western Australia. *J. Aust. Rock Art Res. Assoc.* **2022**, *39*, 3–14.
7. Neumann, J.T.; Black, J.L.; Hœrlé, S.; Smith, B.W.; Watkins, R.; Lagos, M.; Ziegler, A.; Geisler, T. Artificial weathering of rock types bearing petroglyphs from Murujuga, Western Australia. *Herit. Sci.* **2022**, *10*, 77. [CrossRef]
8. Fernandes, A.P.B. The Conservation Programme of the Côa Valley Archaeological Park: Philosophy, Objectives and Action. *Conserv. Manag. Archaeol. Sites* **2007**, *9*, 71–96. [CrossRef]
9. Siegesmund, S.; Snethlage, R. *Stone in Architecture. Properties, Durability*, 5th ed.; Springer: Berlin/Heidelberg, Germany, 2014.
10. Gorbushina, A.A.; Krumbein, W. Role of Microorganisms in Wear Down of Rocks and Minerals. *Microorg. Soils Roles Genes. Funct.* **2005**, *3*, 59–84. [CrossRef]
11. Sterflinger, K.; Krumbein, W.E.; Lellau, T.; Rullkötter, J. Microbially mediated orange patination of rock surfaces. *Anc. Biomol.* **1999**, *3*, 51–65.
12. Steiger, M.; Charola, E.; Sterflinger, K. Weathering and Deterioration. In *Stone in Architecture: Properties, Durability, 5th ed*; Springer: Berlin/Heidelberg, Germany, 2014; pp. 225–316.
13. Louati, M.; Ennis, N.J.; Ghodhbane-Gtari, F.; Hezbri, K.; Sevigny, J.L.; Fahnestock, M.F.; Cherif-Silini, H.; Bryce, J.G.; Tisa, L.S.; Gtari, M. Elucidating the ecological networks in stone-dwelling microbiomes. *Environ. Microbiol.* **2020**, *22*, 1467–1480. [CrossRef]
14. Eisenberg-Degen, D.; Rosen, S.A. Chronological Trends in Negev Rock Art: The Har Michia Petroglyphs as a Test Case. *Arts* **2013**, *2*, 225–252. [CrossRef]
15. Eisenberg-Degen, D.; Nash, G. Hunting and Gender as Reflected in the Central Negev Rock Art, Israel. *Time Mind* **2014**, *7*, 259–277. [CrossRef]
16. Nir, I.; Barak, H.; Kramarsky-Winter, E.; Kushmaro, A. Seasonal diversity of the bacterial communities associated with petroglyphs sites from the Negev Desert, Israel. *Ann. Microbiol.* **2019**, *69*, 1079–1086. [CrossRef]
17. Perry, R.S.; Adams, J.B. Desert varnish: Evidence of cyclic deposition of manganese. *Nature* **1978**, *276*, 489–491. [CrossRef]
18. Martínez-Pabello, P.U.; Villalobos, C.; Sedov, S.; Solleiro-Rebolledo, E.; Solé, J.; Pi-Puig, T.; Chávez-Vergara, B.; Díaz-Ortega, J.; Gubin, A. Rock varnish as a natural canvas for rock art in La Proveedora, northwestern Sonoran Desert (Mexico): Integrating archaeological and geological evidences. *Quat. Int.* **2021**, *572*, 74–87. [CrossRef]
19. Goldsmith, Y.; Stein, M.; Enzel, Y. From dust to varnish: Geochemical constraints on rock varnish formation in the Negev Desert, Israel. *Geochim. Cosmochim. Acta* **2014**, *126*, 97–111. [CrossRef]

20. Andreae, M.O.; Al-Amri, A.; Andreae, C.M.; Guagnin, M.; Jochum, K.P.; Stoll, B.; Weis, U. Archaeometric studies on the petroglyphs and rock varnish at Kilwa and Sakaka, northern Saudi Arabia. *Arab. Archaeol. Epigr.* **2020**, *31*, 219–244. [CrossRef]
21. Harmon, R.S.; Khashchevskaya, D.; Morency, M.; Owen, L.A.; Jennings, M.; Knott, J.R.; Dortch, J.M. Analysis of Rock Varnish from the Mojave Desert by Handheld Laser-Induced Breakdown Spectroscopy. *Molecules* **2021**, *26*, 5200. [CrossRef]
22. Otter, L.M.; Macholdt, D.S.; Jochum, K.P.; Stoll, B.; Weis, U.; Weber, B.; Scholz, D.; Haug, G.H.; Al-Amri, A.M.; Andreae, M.O. Geochemical insights into the relationship of rock varnish and adjacent mineral dust fractions. *Chem. Geol.* **2020**, *551*, 119775. [CrossRef]
23. Andreae, M.O.; Al-Amri, A.; Andreae, T.W.; Garfinkel, A.; Haug, G.; Jochum, K.P.; Stoll, B.; Weis, U. Geochemical studies on rock varnish and petroglyphs in the Owens and Rose Valleys, California. *PLoS ONE* **2020**, *15*, e0235421. [CrossRef]
24. Macholdt, D.; Jochum, K.; Pöhlker, C.; Arangio, A.; Förster, J.-D.; Stoll, B.; Weis, U.; Weber, B.; Müller, M.; Kappl, M.; et al. Characterization and differentiation of rock varnish types from different environments by microanalytical techniques. *Chem. Geol.* **2017**, *459*, 91–118. [CrossRef]
25. Lang-Yona, N.; Maier, S.; Macholdt, D.S.; Müller-Germann, I.; Yordanova, P.; Rodriguez-Caballero, E.; Jochum, K.P.; Al-Amri, A.; Andreae, M.O.; Fröhlich-Nowoisky, J.; et al. Insights into microbial involvement in desert varnish formation retrieved from metagenomic analysis. *Environ. Microbiol. Rep.* **2018**, *10*, 264–271. [CrossRef] [PubMed]
26. Dorn, R.I.; Oberlander, T.M. Microbial Origin of Desert Varnish. *Science* **1981**, *213*, 1245–1247. [CrossRef] [PubMed]
27. Lingappa, U.F.; Yeager, C.M.; Sharma, A.; Lanza, N.L.; Morales, D.P.; Xie, G.; Atencio, A.D.; Chadwick, G.L.; Monteverde, D.R.; Magyar, J.S.; et al. An ecophysiological explanation for manganese enrichment in rock varnish. *Proc. Natl. Acad. Sci. USA* **2021**, *118*, e2025188118. [CrossRef] [PubMed]
28. Esposito, A.; Ahmed, E.; Ciccazzo, S.; Sikorski, J.; Overmann, J.; Holmström, S.J.M.; Brusetti, L. Comparison of Rock Varnish Bacterial Communities with Surrounding Non-Varnished Rock Surfaces: Taxon-Specific Analysis and Morphological Description. *Microb. Ecol.* **2015**, *70*, 741–750. [CrossRef] [PubMed]
29. Wierzchos, J.; de los Ríos, A.; Ascaso, C. Microorganisms in desert rocks: The edge of life on Earth. *Int. Microbiol.* **2012**, *15*, 173–183. [CrossRef]
30. Northup, D.E.; Snider, J.R.; Spilde, M.N.; Porter, M.L.; Van De Kamp, J.L.; Boston, P.J.; Nyberg, A.M.; Bargar, J.R. Diversity of rock varnish bacterial communities from Black Canyon, New Mexico. *J. Geophys. Res. Biogeosci.* **2010**, *115*, G2. [CrossRef]
31. Knight, K.; St. Clair, L.L.; Gardner, J. Lichen Biodeterioration at Inscription Rock, El Morro National Monument, Ramah, New Mexico, USA. In *Biodeterioration of Stone Surfaces*; St. Clair, L.L., Seaward, M.R.D., Eds.; Springer: Dordrecht, The Netherlands, 2004; pp. 129–163.
32. Dandridge, D.E. Lichens: The Challenge for Rock Art Conservation. Ph.D. Thesis, Texas A & M University, College Station, TX, USA, 2007.
33. Nir, I.; Barak, H.; Kramarsky-Winter, E.; Kushmaro, A.; Ríos, A. Microscopic and biomolecular complementary approaches to characterize bioweathering processes at petroglyph sites from the Negev Desert, Israel. *Environ. Microbiol.* **2022**, *24*, 967–980. [CrossRef]
34. Evenari, M.; Shanan, L.; Tadmor, N. *The Negev: The Challenge of a Desert*; Harvard University Press: Cambridge, MA, USA, 1982.
35. Hillel, D. *Negev Land, Water, and Life in a Desert Environment*; Praeger: New York, NY, USA, 1982.
36. Kidron, G.J. Altitude dependent dew and fog in the Negev Desert, Israel. *Agric. For. Meteorol.* **1999**, *96*, 1–8. [CrossRef]
37. Piñar, G.; Poyntner, C.; Lopandic, K.; Tafer, H.; Sterflinger, K. Rapid diagnosis of biological colonization in cultural artefacts using the MinION nanopore sequencing technology. *Int. Biodeterior. Biodegrad.* **2020**, *148*, 104908. [CrossRef]
38. RRUFF Database. Available online: https://rruff.info/ (accessed on 3 May 2022).
39. Caggiani, M.; Cosentino, A.; Mangone, A. Pigments Checker version 3.0, a handy set for conservation scientists: A free online Raman spectra database. *Microchem. J.* **2016**, *129*, 123–132. [CrossRef]
40. Hanesch, M. Raman spectroscopy of iron oxides and (oxy)hydroxides at low laser power and possible applications in environmental magnetic studies. *Geophys. J. Int.* **2009**, *177*, 941–948. [CrossRef]
41. Bruins, J.H.; Petrusevski, B.; Slokar, Y.M.; Kruithof, J.C.; Kennedy, M.D. Manganese removal from groundwater: Characterization of filter media coating. *Desalin. Water Treat.* **2015**, *55*, 1851–1863. [CrossRef]
42. Jehlička, J.; Edwards, H.; Oren, A. Bacterioruberin and salinixanthin carotenoids of extremely halophilic Archaea and Bacteria: A Raman spectroscopic study. *Spectrochim. Acta Part A Mol. Biomol. Spectrosc.* **2013**, *106*, 99–103. [CrossRef]
43. Piñar, G.; Sclocchi, M.C.; Pinzari, F.; Colaizzi, P.; Graf, A.; Sebastiani, M.L.; Sterflinger, K. The Microbiome of Leonardo da Vinci's Drawings: A Bio-Archive of Their History. *Front. Microbiol.* **2020**, *11*, 593401. [CrossRef]
44. Planý, M.; Pinzari, F.; Šoltys, K.; Kraková, L.; Cornish, L.; Pangallo, D.; Jungblut, A.D.; Little, B. Fungal-induced atmospheric iron corrosion in an indoor environment. *Int. Biodeterior. Biodegrad.* **2021**, *159*, 105204. [CrossRef]
45. Grottoli, A.; Beccaccioli, M.; Zoppis, E.; Fratini, R.S.; Schifano, E.; Santarelli, M.L.; Uccelletti, D.; Reverberi, M. Nanopore Sequencing and Bioinformatics for Rapidly Identifying Cultural Heritage Spoilage Microorganisms. *Front. Mater.* **2020**, *7*, 14. [CrossRef]
46. Krumbein, W.E.; Jens, K. Biogenic rock varnishes of the negev desert (Israel) an ecological study of iron and manganese transformation by cyanobacteria and fungi. *Oecologia* **1981**, *50*, 25–38. [CrossRef]
47. Vikram, S.; Guerrero, L.D.; Makhalanyane, T.P.; Le, P.; Seely, M.; Cowan, D.A. Metagenomic analysis provides insights into functional capacity in a hyperarid desert soil niche community. *Environ. Microbiol.* **2015**, *18*, 1875–1888. [CrossRef]

48. Hirschberg, J.; Chamovitz, D. *Carotenoids in Cyanobacteria BT-The Molecular Biology of Cyanobacteria*; Bryant, D.A., Ed.; Springer: Dordrecht, The Netherlands, 1994; pp. 559–579.
49. Perera, I.; Subashchandrabose, S.R.; Venkateswarlu, K.; Naidu, R.; Megharaj, M. Consortia of cyanobacteria/microalgae and bacteria in desert soils: An underexplored microbiota. *Appl. Microbiol. Biotechnol.* **2018**, *102*, 7351–7363. [CrossRef]
50. Lacap, D.C.; Warren-Rhodes, K.A.; McKay, C.P.; Pointing, S.B. Cyanobacteria and chloroflexi-dominated hypolithic colonization of quartz at the hyper-arid core of the Atacama Desert, Chile. *Extremophiles* **2010**, *15*, 31–38. [CrossRef]
51. Mishra, S. Chapter 1-Cyanobacterial imprints in diversity and phylogeny. In *Advances in Cyanobacterial Biology*; Singh, P.K., Kumar, A., Singh, V.K., Shrivastava, C.B., Eds.; Academic Press: Cambridge, MA, USA, 2020; pp. 1–15.
52. Abed, R.M.; Al Kharusi, S.; Schramm, A.; Robinson, M.D. Bacterial diversity, pigments and nitrogen fixation of biological desert crusts from the Sultanate of Oman. *FEMS Microbiol. Ecol.* **2010**, *72*, 418–428. [CrossRef] [PubMed]
53. Wierzchos, J.; Ascaso, C.; McKay, C.P. Endolithic Cyanobacteria in Halite Rocks from the Hyperarid Core of the Atacama Desert. *Astrobiology* **2006**, *6*, 415–422. [CrossRef] [PubMed]
54. Wierzchos, J.; DiRuggiero, J.; Vítek, P.; Artieda, O.; Souza-Egipsy, V.; Škaloud, P.; Tisza, M.; Davila, A.F.; Vílchez, C.; Garbayo, I.; et al. Adaptation strategies of endolithic chlorophototrophs to survive the hyperarid and extreme solar radiation environment of the Atacama Desert. *Front. Microbiol.* **2015**, *6*, 934. [CrossRef] [PubMed]
55. Mergelov, N.; Mueller, C.W.; Prater, I.; Shorkunov, I.; Dolgikh, A.; Zazovskaya, E.; Shishkov, V.; Krupskaya, V.; Abrosimov, K.; Cherkinsky, A.; et al. Alteration of rocks by endolithic organisms is one of the pathways for the beginning of soils on Earth. *Sci. Rep.* **2018**, *8*, 3367. [CrossRef] [PubMed]
56. Ríos, A.D.L.; Ascaso, C. Contributions of in situ microscopy to the current understanding of stone biodeterioration. *Int. Microbiol.* **2005**, *8*, 181–188.
57. Grbic, M.L.; Vukojevic, J.; Subakov-Simic, G.; Krizmanic, J.; Stupar, M. Biofilm forming cyanobacteria, algae and fungi on two historic monuments in Belgrade, Serbia. *Arch. Biol. Sci.* **2010**, *62*, 625–631. [CrossRef]
58. Genderjahn, S.; Lewin, S.; Horn, F.; Schleicher, A.; Mangelsdorf, K.; Wagner, D. Living Lithic and Sublithic Bacterial Communities in Namibian Drylands. *Microorganisms* **2021**, *9*, 235. [CrossRef]
59. Ferreira, A.C.; Nobre, M.F.; Moore, E.; Rainey, F.A.; Battista, J.R.; Da Costa, M.S. Characterization and radiation resistance of new isolates of Rubrobacter radiotolerans and Rubrobacter xylanophilus. *Extremophiles* **1999**, *3*, 235–238. [CrossRef]
60. Webb, K.M.; DiRuggiero, J. Role of Mn^{2+} and Compatible Solutes in the Radiation Resistance of Thermophilic Bacteria and Archaea. *Archaea* **2012**, *2012*, 845756. [CrossRef]
61. Piñar, G.; Ettenauer, J.; Sterflinger, K. 'La vie en ros': A review of the rosy discoloration of subsurface monuments. In *The Conservation of Subterranean Cultural Heritage*; Saiz-Jimenez, C., Ed.; CRC Press: Boca Raton, FL, USA, 2014; pp. 113–124. [CrossRef]
62. Laiz, L.; Miller, A.Z.; Jurado, V.; Akatova, E.V.; Sanchez-Moral, S.; Gonzalez, J.M.; Dionísio, A.; Macedo, M.F.; Sáiz-Jiménez, C. Isolation of five Rubrobacter strains from biodeteriorated monuments. *Naturwissenschaften* **2009**, *96*, 71–79. [CrossRef]
63. Tescari, M.; Visca, P.; Frangipani, E.; Bartoli, F.; Rainer, L.; Caneva, G. Celebrating centuries: Pink-pigmented bacteria from rosy patinas in the House of Bicentenary (Herculaneum, Italy). *J. Cult. Heritage* **2018**, *34*, 43–52. [CrossRef]
64. Maeng, S.; Park, Y.; Lee, S.E.; Han, J.H.; Cha, I.-T.; Lee, K.-E.; Lee, B.-H.; Kim, M.K. *Pontibacter pudoricolor* sp. nov., and *Pontibacter russatus* sp. nov. radiation-resistant bacteria isolated from soil. *Int. J. Gen. Mol. Microbiol.* **2020**, *113*, 1361–1369. [CrossRef]
65. Zhang, J.-Y.; Liu, X.-Y.; Liu, S.-J. *Adhaeribacter terreus* sp. nov., isolated from forest soil. *Int. J. Syst. Evol. Microbiol.* **2009**, *59*, 1595–1598. [CrossRef]
66. Klassen, J.L.; Foght, J.M. Differences in Carotenoid Composition among *Hymenobacter* and Related Strains Support a Tree-Like Model of Carotenoid Evolution. *Appl. Environ. Microbiol.* **2008**, *74*, 2016–2022. [CrossRef]
67. Sterflinger, K.; Tesei, D.; Zakharova, K. Fungi in hot and cold deserts with particular reference to microcolonial fungi. *Fungal Ecol.* **2012**, *5*, 453–462. [CrossRef]
68. Sterflinger, K.; De Baere, R.; De Hoog, G.; De Wachter, R.; Krumbein, W.E.; Haase, G. *Coniosporium perforans* and *C. apollinis*, two new rock-inhabiting fungi isolated from marble in the Sanctuary of Delos (Cyclades, Greece). *Int. J. Gen. Mol. Microbiol.* **1997**, *72*, 349–363. [CrossRef]
69. Coleine, C.; Stajich, J.E.; de Los Ríos, A.; Selbmann, L. Beyond the extremes: Rocks as ultimate refuge for fungi in drylands. *Mycologia* **2020**, *113*, 108–133. [CrossRef]
70. Engel, C.; Robert, S. Chemical data on desert varnish. *Bull. Geol. Soc. Am.* **1958**, *69*, 487–518. [CrossRef]
71. Liu, T.; Broecker, W.S. How fast does rock varnish grow? *Geology* **2000**, *28*, 183. [CrossRef]
72. Malherbe, C.; Ingley, R.; Hutchinson, I.; Edwards, H.; Carr, A.; Harris, L.; Boom, A. Biogeological Analysis of Desert Varnish Using Portable Raman Spectrometers. *Astrobiology* **2015**, *15*, 442–452. [CrossRef]
73. Ren, G.; Yan, Y.; Nie, Y.; Lu, A.; Wu, X.; Li, Y.; Wang, C.; Ding, H. Natural Extracellular Electron Transfer Between Semiconducting Minerals and Electroactive Bacterial Communities Occurred on the Rock Varnish. *Front. Microbiol.* **2019**, *10*, 293. [CrossRef]
74. Edwards, H.G.M.; Moody, C.A.; Villar, S.E.J.; Mancinelli, R. Raman spectroscopy of desert varnishes and their rock substrata. *J. Raman Spectrosc.* **2004**, *35*, 475–479. [CrossRef]
75. Skipper, P.J.A.; Skipper, L.K.; Dixon, R.A. A metagenomic analysis of the bacterial microbiome of limestone, and the role of associated biofilms in the biodeterioration of heritage stone surfaces. *Sci. Rep.* **2022**, *12*, 4877. [CrossRef]

76. Ríos, A.D.L.; Cámara, B.; del Cura, M.G.; Rico, V.J.; Galván, V.; Ascaso, C. Deteriorating effects of lichen and microbial colonization of carbonate building rocks in the Romanesque churches of Segovia (Spain). *Sci. Total Environ.* **2009**, *407*, 1123–1134. [CrossRef]
77. Subrahmanyam, G.; Vaghela, R.; Bhatt, N.P.; Archana, G. Carbonate-Dissolving Bacteria from 'Miliolite', a Bioclastic Limestone, from Gopnath, Gujarat, Western India. *Microbes Environ.* **2012**, *27*, 334–337. [CrossRef]
78. Uroz, S.; Calvaruso, C.; Turpault, M.; Sarniguet, A.; de Boer, W.; Leveau, J.; Frey-Klett, P. Efficient mineral weathering is a distinctive functional trait of the bacterial genus Collimonas. *Soil Biol. Biochem.* **2009**, *41*, 2178–2186. [CrossRef]
79. Lian, B.; Chen, Y.; Zhu, L.; Yang, R. Effect of Microbial Weathering on Carbonate Rocks. *Earth Sci. Front.* **2008**, *15*, 90–99. [CrossRef]
80. dos Santos, S.I.P.; Yang, Y.; Rosa, M.T.G.; Rodrigues, M.A.A.; De La Tour, C.B.; Sommer, S.; Teixeira, M.; Carrondo, M.A.; Cloetens, P.; Abreu, I.A.; et al. The interplay between Mn and Fe in *Deinococcus radiodurans* triggers cellular protection during paraquat-induced oxidative stress. *Sci. Rep.* **2019**, *9*, 17217. [CrossRef]
81. Bernardini, S.; Bellatreccia, F.; Columbu, A.; Vaccarelli, I.; Pellegrini, M.; Jurado, V.; Del Gallo, M.; Saiz-Jimenez, C.; Sodo, A.; Millo, C.; et al. Morpho-Mineralogical and Bio-Geochemical Description of Cave Manganese Stromatolite-Like Patinas (Grotta del Cervo, Central Italy) and Hints on Their Paleohydrological-Driven Genesis. *Front. Earth Sci.* **2021**, *9*, 642667. [CrossRef]
82. Saiz-Jimenez, C.; Miller, A.Z.; Martin-Sanchez, P.M.; Hernández-Mariné, M. Uncovering the origin of the black stains in Lascaux Cave in France. *Environ. Microbiol.* **2012**, *14*, 3220–3231. [CrossRef]
83. Barboza, N.R.; Amorim, S.S.; Santos, P.A.; Reis, F.D.; Cordeiro, M.M.; Guerra-Sá, R.; Leão, V.A. Indirect Manganese Removal by *Stenotrophomonas* sp. and *Lysinibacillus* sp. Isolated from Brazilian Mine Water. *BioMed Res. Int.* **2015**, *2015*, 925972. [CrossRef] [PubMed]

Article

Biodeterioration of Salón de Reinos, Museo Nacional del Prado, Madrid, Spain

Valme Jurado [1,†], José Luis Gonzalez-Pimentel [1,2,†], Bernardo Hermosin [1] and Cesareo Saiz-Jimenez [1,*]

[1] Instituto de Recursos Naturales y Agrobiologia, IRNAS-CSIC, 41012 Sevilla, Spain; vjurado@irnase.csic.es (V.J.); pimentel@irnas.csic.es (J.L.G.-P.); hermosin@irnase.csic.es (B.H.)
[2] Laboratorio Hercules, Universidade de Evora, 7000-809 Evora, Portugal
* Correspondence: saiz@irnase.csic.es
† These authors contributed equally to this study.

Abstract: The *Salón de Reinos*, a remnant of the 17th century *Palacio del Buen Retiro*, was built as a recreational residence under the reign of Felipe IV between 1632 and 1640 and was the main room for the monarch's receptions. This *Salón* owes its name to the fact that the coats of arms (shields) of the 24 kingdoms that formed Spain in Felipe IV's time were painted on the vault, above the windows. In addition, the ceiling shows an original decorative composition. The painted ceiling and window vaults showed deterioration evidenced by fissures, water filtration, detachments of the paint layer, and black stains denoting fungal colonization related to humidity. Ten strains of bacteria and 14 strains of fungi were isolated from the deteriorated paintings. Their biodeteriorative profiles were detected through plate assays. The most frequent metabolic functions were proteolytic and lipolytic activities. Other activities, such as the solubilization of gypsum and calcite and the production of acids, were infrequent among the isolates.

Keywords: biodeterioration; mural paintings; bacteria; fungi; biocide

Citation: Jurado, V.; Gonzalez-Pimentel, J.L.; Hermosin, B.; Saiz-Jimenez, C. Biodeterioration of Salón de Reinos, Museo Nacional del Prado, Madrid, Spain. *Appl. Sci.* **2021**, *11*, 8858. https://doi.org/10.3390/app11198858

Academic Editor: Daniela Isola

Received: 23 August 2021
Accepted: 19 September 2021
Published: 23 September 2021

Publisher's Note: MDPI stays neutral with regard to jurisdictional claims in published maps and institutional affiliations.

1. Introduction

The earliest known examples of wall paintings are found in Crete and in Egyptian tombs. The Etruscan necropolis of Tarquinia, included in the UNESCO World Heritage list, contains a large number of decorated tombs, the oldest of which date from the 7th century BC. Examples of extraordinary Roman wall paintings are found in the magnificent *Villa dei Misteri* (1st century BC) in Pompeii. The late medieval period and the Renaissance are examples of the noticeable use of wall paintings, particularly in Italy, where most churches display this type of decoration.

Wall paintings can be affected by various problems—environmental pollution, excessive humidity, water seepage from the ceilings, and poor quality of materials and construction methods—all contributing to the deterioration of the murals. In wall paintings, different phenomena can be observed; the most common include cracks, detachment of the painted surface, efflorescence formation, color changes and growth of films of microbial origin.

In the initial stages of colonization, the growth of microorganisms in a wall painting causes only aesthetic damage, as little or no alteration of the painted surface occurs. Subsequently, the cells and hyphae penetrate the paint layer, and the chemical attack results in pitting, peeling, cracking and loss of paint. To this damage is added that caused by microbial metabolites, which alter the original color and sometimes produce red, green or black pigmentations [1–5].

Deteriorated wall paintings contain a variety of hygroscopic salts, including carbonates, chlorides, nitrates and sulfates, which form efflorescences on their surfaces. This means that, in many cases, efflorescence deposits in murals mimic the conditions found in extreme habitats, favoring the growth of halophilic microorganisms. Due to its deteriorating effect, saline efflorescence in wall paintings is a common problem and can create micro-niches with a salt concentration high enough to allow the growth of halophilic *Archaea*, as demonstrated by Piñar et al. [6] in two deteriorated wall paintings subjected to different environmental and climatic conditions: the Chapel of Herberstein Castle (Austria) and the Tomb of Servilia in the Necropolis of Carmona (Spain).

Of particular importance is the colonization and growth of bacteria in wall paintings. Gorbushina et al. [7] studied the deterioration of the wall paintings in the church of St. Martin in Germany. In these paintings, the authors isolated numerous species of *Arthrobacter*, *Bacillus* and related genera (*Paenibacillus*, *Gracilibacillus*, *Salibacillus*), producers of spores, in addition to *Staphylococcus* and *Nocardioides*.

However, the most frequent and conspicuous biodeterioration phenomenon is the colonization of wall paintings by fungi, since they generally produce black or green colored patinas. In modern wall paintings from La Rábida Monastery, Spain, the most abundant fungus was *Cladosporium sphaerospermum*, followed by *Engyodontium album*, recently transferred to the novel genus *Parengyodontium album* [8] and *Aspergillus versicolor*. Less frequent were *Acremonium charticola*, *Alternaria alternata*, *Botrytis cinerea*, *Cunninghamella echinulata*, *Phoma glomerata*, etc. [9]. In addition, the mite *Tyrophagus palmarum* was found in the paintings, feeding on bacteria and fungi.

Karpovich-Tate and Rebrikova [10] investigated the biodeterioration of wall paintings at the Pafnutii-Borovskii Monastery, Russia, finding that *Parengyodontium album* comprised 90% of the fungi isolated from the deteriorated stone, while *Cladosporium sphaerospermum* accounted for 65% of the total isolates from the painted surfaces.

Gorbushina and Petersen [11] studied various wall paintings in Germany and found *Acremonium* spp., *Cladosporium* spp., *Verticillium* spp., *Aspergillus sydowii*, *Aureobasidium pullulans*, *Parengyodontium album*, *Scopulariosis brevicaulus*, *Beauveria* sp. and *Chrysosporium* sp. The walls commonly included springtails, mites, flies and spiders. Arthropod remains were frequently found in all investigated locations, and most of them were colonized by fungi. In fact, most of the available data suggest that fungi in wall paintings are related to arthropods [12]. Entomopathogenic fungi, in particular *Parengyodontium album*, are common on wall paintings from different countries [7,9–12].

In wall paintings, certain ecological conditions prevail, such as a low concentration of carbon sources, variable-to-high humidity levels and, generally low temperatures. Under these conditions the growth of microorganisms is possible, in most cases forming a biofilm or microbial film composed of many species of bacteria, archaea, fungi, etc. [13–16].

The possible carbon sources in wall paintings are protein-type binders, such as casein, animal glue, egg, etc., and some pigments, as well as synthetic polymers used for preservation. The destructive activity of microorganisms in the paint film is not limited to its surface, but penetrates deep into the layer of paint and plaster, and its destructive effect can be even more intense in the deeper layers than on the surface.

The use of binders by microorganisms as a nutritional source leads to deterioration and, consequently, to the detachment of pigment particles, devoid of any binding effect. This process was reported by Piñar et al. [6], who noticed the damage caused during restoration of the Chapel of Herberstein Castle, Austria, with 14th century frescoes. Around 1580, the frescoes were covered with plaster for a few hundred years. Restoration works were carried between 1942 and 1949, and the plaster layer was removed. In this restoration, the paints were fixed with casein-water (1:10), and casein was also added to the mortar for consolidation. Fungal growth started only five days after the application of casein, and the ceiling had to be cleaned again. However, after this restoration, the chapel was abandoned and used as a warehouse until the mid-1990s. At the beginning of the 21st century, the paintings showed considerable microbial growth, including loss of consistency in many areas as well as severe discoloration, with the formation of saline efflorescences and pink pigmentation due to microbial attack. This is an example of how erroneous restoration protocols can damage wall paintings. Therefore, it is of the utmost importance to know the biodeteriorative activity of the microorganisms isolated from mural paintings in order to decide which treatments can be safely used in restoration.

2. Materials and Methods

2.1. Historic Overview of Salón de Reinos

On 25 July 2021, UNESCO added the *Paseo del Prado* and *Buen Retiro*, a landscape of Arts and Sciences (Madrid, Spain) to the World Heritage List. This Paseo includes avenues, fountains, gardens and many buildings, a few of them dedicated to the arts and sciences: *Museo Nacional del Prado, Museo Nacional Centro de Arte Reina Sofía, Museo Thyssen-Bornemisza, Museo Nacional de Antropología, Museo Naval*, and *Museo Nacional de Artes Decorativas*.

The *Museo Nacional del Prado* is a museum campus composed of several buildings. Two of them, *Casón del Buen Retiro* and *Salón de Reinos*, are remnants of the 17th century *Palacio del Buen Retiro*. This palace was built as a recreational residence under the reign of Felipe IV, between 1632 and 1640. The palace was seriously damaged during the War of Independence (1808–1812), and part of the ruined palace was demolished in 1816–1819, under Fernando VII. Other palace buildings were demolished in 1868, except the *Casón del Buen Retiro*, an old ballroom and party hall, and the *Salón de Reinos*, the main room for the monarch's receptions. A program to expand the *Museo Nacional del Prado* integrated this *Salón*, used as an Army Museum from 1841 to 2005, into the museum properties.

In the 17th century, the *Salón de Reinos* housed famous paintings (Velázquez, Zurbarán, and others), now deposited in the *Museo Nacional del Prado*. The *Salón*, a room 34.6 m long, 10 m wide and 8 m high, owes its name to the fact that the coats of arms (shields) of the 24 kingdoms that formed Spain in Felipe IV's time were painted on the vault, above the windows (Figure 1).

Figure 1. *Salón de Reinos, Museo Nacional del Prado*, Madrid, Spain. (**Left**), as Army Museum. Photo J. Laurent y Cia. Archivo Ruiz Vernacci. IPCE, Ministerio de Cultura y Deporte. (**Right**), before restoration. Photo Juan Aguilar, Ágora S.L.

The conservation of *Salón de Reinos* was poor (Figure 2). The painted ceiling and window vaults showed deterioration evidenced by fissures, water filtration (Figure 2B), detachments of the paint layer (Figure 2C,D) and fungal colonization related to humidity (Figure 2D,E). Other issues include the urgency of the design and construction of the *Salón*, the use of low-quality materials and, above all, the different uses and adaptations of the building over 400 years, with rehabilitation interventions that caused scratching and hiding of the paintings in the walls and repainting of the ceiling (Figure 2A).

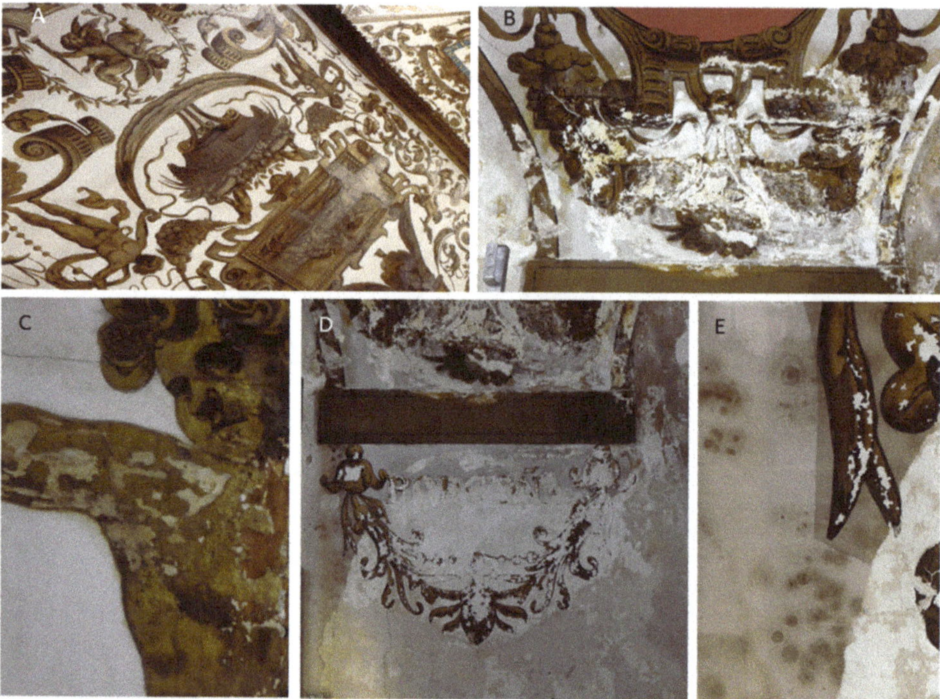

Figure 2. *Salón de Reinos, Museo Nacional del Prado*, Madrid, Spain. (**A**) Detail of the ceiling showing repaints. (**B**) Deterioration produced by a recent water leak. (**C**) Paint detachment in the ceiling. (**D**) Burgundy shield (sample M2) with fungal growth and painting detachments. (**E**) Detail of fungal growth in (**D**). Photos provided by Ágora S.L.

While the original decorative composition on the ceiling was preserved almost entirely, on the walls the decoration was covered by plastering. The scratching of all the wall surfaces to apply a new plaster caused damage to a large number of the compositional elements as well as physical deterioration of the pictorial layers and the gilding.

The ceiling was subject to repainting that covered almost its entire surface (Figure 2A). These patches, although they appeared as a homogeneous stratum, had different chronologies in the 19th and 20th centuries. At the beginning of the 20th century, a general repainting of the entire ceiling was carried out; tempera was used mainly for its application, and occasionally a greasy temper inside the shields was applied. The repaints were dated "14 April 1912".

Beneath the repainting, there was a deteriorated and detached original surface (pictorial layer and gilding) with an often significant degree of powder and detachment in the coats of arms or wide gaps due to different textures (Figure 2A). The natural aging of the pictorial layer was manifested mainly in the ceiling gilding, where the polymerization of the oils used in the glazes contributed to the generalized micro-exfoliation observed in large areas. In addition, on the ceiling paintings and on the repaintings, a thick dark layer of dust, a consequence of environmental pollution, appeared adhered to the surfaces.

Therefore, it was decided that *Salón de Reinos* needed rehabilitation and adaptation to meet modern museum requirements, and an international contest for project designs was launched in 2016. This was won by Foster + Partners and Rubio Arquitectura.

An unpublished study of Ágora Restauraciones de Arte, S.L., Madrid, for the *Museo Nacional del Prado*, provides information on *Salón de Reinos* mortars and paintings. The report stated the presence of a coarse mortar of gypsum and clays and fine-grained gypsum plasters, with binders of animal glue, linseed oil, pine resin, sandarac and turpentine of

Venice. The paintings were made of gold leaf (gilding), a mixture of cobalt and azurite glaze or natural lapis lazuli (blue color), and red lead (red color). The binders are excellent nutrient sources of carbon and nitrogen for microorganisms. This indicates the need to carry out a detailed study on the microbial populations and the possible use of the components of the mortars and paintings before carrying out a restoration.

This paper reports on the biodeterioration of *Salón de Reinos* before the restoration. The aims were focused on the isolated microorganisms and their ability to degrade cultural heritage materials and products commonly used in restoration works.

2.2. Sampling

The samples studied were located in the ceiling of *Salón de Reinos* (Table 1 and Figure 2). Sampling was restricted due to the protection of the monument, and only <500 mg from three samples were available for the study.

Table 1. Description and location of the samples studied.

Sample	Description	Location
M1	Surface dust	S-3. Mexico shield
M2	Black stains	S-4. Burgundy shield
M3	Black stains	S-02. Between Galicia and Murcia shields

2.3. Isolation of Culturable Microorganisms

The samples (<500 mg per sample) were homogenized in sterile saline solution (0.9% NaCl *w*/*v*) and inoculated into culture media such as Tryptone-Soy Agar (TSA) for bacteria and Malt Extract Agar (MEA) for fungi. The plates were incubated at 28 °C for 24–72 h to allow the growth of bacteria, and at 25 °C for 4–5 days for fungi. Individual colonies with different color, size, border shape, and texture were isolated and inoculated in TSA and MEA, respectively. Once the cultures were obtained from the colonies, the purity of each isolate was confirmed microscopically. The bacteria were conserved at −80 °C in cryovials (Microbank), and the fungi were preserved at 4 °C on MEA slants until studied.

2.4. DNA Extraction and Amplification

The extraction of DNA from the strains was carried out from bacterial biomass and fungal mycelia. The biomass was transferred to a 1.5 mL Eppendorf tube containing 500 µL of the TNE extraction buffer (10 mM Tris–HCl, 100 mM NaCl, 1 mM EDTA; pH 8) and 2 and 4 mm glass beads. The mixture was homogenized using a stirrer (Fast Prep–24), applying two cycles of 45 s at a speed of 4.5 m/s. The proteins were removed by adding 50 µL of sodium dodecyl sulfate (10% *w*/*v*) and 70 µL of proteinase K (10 mg/mL), and the mixture was incubated at 50 °C for one hour. DNA was purified with 750 µL phenol–chloroform–isoamyl alcohol (25:24:1). The mixture was centrifuged at 8000 rpm for 5 min. The supernatant obtained was subjected to further purification with 450 µL of chloroform–isoamyl alcohol (24:1). Centrifugation was repeated under identical conditions, and the supernatant was collected. DNA precipitation was performed by adding 50 µL of 3 M sodium acetate (pH 5.2) and 1 mL of absolute ethanol. To promote precipitation, the mixture was incubated at −20 °C for 30 min and centrifuged at 13,000 rpm for 10 min. The precipitate was washed with ethanol (70%) and allowed to dry at room temperature. Finally, the precipitated DNA was then dissolved in 100 µL of sterile ultra-pure water.

The resulting DNA concentration in the extraction products was quantified using a Qubit 2.0 fluorimeter (Invitrogen), following the manufacturer's instructions. The products were stored at −80 °C until use.

For the identification of bacteria, the 16S rRNA gene was amplified, and for fungi, the ITS region was between the small and large subunit of ribosomal RNA, 18S rRNA and 28S rRNA (ITS1 and ITS2, respectively). The primers 616F (5′–AGA GTT TGA TYM TGG CTC AG–3′) [17] and 1522R (5′–AAG GAG GTG ATC CAG CCG CA–3′) [18] were used for bacteria, and the primers ITS1 (5′–TCC GTA GGT CCT GAA GCG G–3′) and ITS4 (5′–TCC TCC GCT TAT TGA TAT GC–3′) described by White et al. [19] were used for fungi.

Each PCR reaction was run in a final volume of 50 µL. The reaction mixture consisted of 5 µL of 10X reaction buffer [750 mM Tris–HCl (pH 9), 500 mM KCl, 200 mM $(NH_4)_2SO_4$], 2 µL of 50 mM $MgCl_2$, 0.5 µL of each 50 µM primer (Invitrogen Life Technologies), 5 µL of deoxyribonucleotide mix (dNTPs) (Bioline GC Biotech, Alphen aan den Rijn, The Netherlands), 2 mM each, 0.25 µL of Taq polymerase enzyme (5 U/µL) (Biotools M&B Labs, Madrid, Spain), and 10–20 ng of template DNA; the reaction volume was completed with sterile ultra-pure water. All samples were run in duplicate, and two negative controls, containing sterile ultra-pure water replacing template DNA, were included for each batch of reactions. The PCR reactions were performed in an iCycler thermal cycler (Bio-Rad, CA, USA). The PCR protocol used for the 16S rRNA gene region consisted of the following steps: 2 min of initial denaturation at 94 °C; 35 cycles composed of 20 s at 94 °C, 20 s at 55 °C, and 2 min at 72 °C; followed by a 10 minute final extension at 72 °C. The protocol used for the amplification of the ITS regions consisted of 2 min of initial denaturation at 95 °C; 35 cycles consisting of 1 min at 95 °C, 1 min at 50 °C, and 1 min at 72 °C; followed by a 5 min final extension at 72 °C. To confirm the positive result of the PCR amplifications, horizontal SeaKem agarose gel electrophoresis (FMC Bioproducts, Rockland, ME, USA) was performed at 1% (*w*/*v*) with TAE 0.5X buffer (20 mM Tris, 10 mM glacial acetic acid, 0.5 mM EDTA).

2.5. DNA Sequencing

The PCR products corresponding to the bacterial and fungal strains were sequenced in duplicate at Secugen (CIB–CSIC, Madrid), with an ABI 3700 capillary sequencer (Applied Biosystems, Foster City, CA, USA) using the same primers as for PCR. The sequences obtained were edited with the BioEdit 7.2.5 program [20], establishing the consensus sequences for each strain analyzed. Homology searches of the sequences were performed using the BLASTn algorithm [21]. The fungal sequences were compared with the existing sequences from the National Center for Biotechnology Information (NCBI) databases, Standard Nucleotide and ITS Reference Sequence, and the fungi were identified according to the best matches. However, the ITS marker alone for identification does not discriminate at the species level in certain fungal genera such as *Aspergillus* and *Penicillium*, and it may be necessary to sequence one or more protein-coding genes to obtain a more precise identification at the species level [22]. The bacterial sequences were also compared using the NCBI Standard Nucleotide database and the EZBioCloud 16S database [23], which assigns the closest cultivated type species. All sequences were deposited in the NCBI GenBank database (http://www.ncbi.nlm.nih.gov/genbank/, accessed on 23 August 2021).

2.6. Biodeteriorative Activity of the Isolates

The strains were tested in different culture media to know their potential mortar and paint biodeterioration. Table 2 details the culture media tested for the analyses.

Mortar deterioration was tested for solubilization of gypsum and calcite (CaS and GyS media). Binder deterioration was tested by means of casein and gelatin hydrolysis (CH and GH media). The degradation of oils was tested with the lipolytic activity (LH medium). The production of acid or alkali (AAP medium) was also qualitatively measured. All tests were carried out in triplicate at 28 °C for bacteria and 25 °C for fungi for 1 week.

Table 2. Culture media and metabolic activities analyzed.

Medium	Composition *	Analysis	Reference
GH	TSA/MEA + 1% gelatin	Proteolytic activity	[24]
LH	1% Tween 80	Lipolytic activity	[25]
CH	Nutrient Casein Agar	Proteolytic activity	[26]
CaS	TSA/MEA + 1% $CaCO_3$	Calcite solubilization	[24]
GyS	TSA/MEA + 1% $CaSO_4$	Gypsum solubilization	[24]
AAP	Czapek-Dox broth	Acid/Alkali production	[26]

* TSA/MEA: TSA or MEA was used for testing bacterial or fungal strains, respectively.

3. Results and Discussion

A total of 10 strains of bacteria and 14 of fungi were isolated. Table 3 shows the identification of the bacteria isolated from the paintings and dust, and Table 4 shows the identification of fungi.

Table 3. Identification of bacteria isolated from *Salón de Reinos*.

Samples	Accession Number	Strains	Identification and Accession Number (% Identity)
M1	MZ827450	MP1B-1	*Priestia aryabhattai* EF114313 (99.7%)
	MZ827451	MP1B-2	*Bacillus tequilensis* AYTO01000043 (99.9%)
	MZ827452	MP1B-3	*Bacillus paralicheniformis* KY694465 (99.7%)
M2	MZ827453	MP2B-2	*Bacillus frigoritolerans* AM747813 (99.7%)
	MZ827454	MP2B-3	*Bacillus paralicheniformis* KY694465 (99.7%)
	MZ827455	MP2B-4	*Bacillus cereus* AE016877 (99.6%)
M3	MZ827456	MP3B-1	*Cytobacillus oceanisediminis* GQ292772 (99.4%)
	MZ827457	MP3B-2	*Streptomyces thinghirensis* FM202482 (99.6%)
	MZ827458	MP3B-3	*Priestia endophytica* AF295302 (99.4%)
	MZ827459	MP3B-6	*Bacillus frigoritolerans* AM747813 (99.7%)

Table 4. Identification of fungi isolated from *Salón de Reinos*.

Samples	Accession Number	Strains	Identification and Accession Number (% Identity)
M1	MZ827828	MP1B-4	*Penicillium chrysogenum* MF422150 (100%)
	MZ827829	MP1B-5	*Penicillium chrysogenum* MF422150 (99.7%)
	MZ827830	MP1H-1	*Alternaria angustiovoidea* MK910070 (100%)
	MZ827831	MP1H-2	*Cladosporium xylophilum* MH863875 (100%)
	MZ827832	MP1H-3	*Penicillium fuscoglaucum* NR163669 (100%)
	MZ827833	MP1H-4	*Mucor racemosus* NR126135 (99.0%)
	MZ827834	MP1H-5	*Cladosporium macrocarpum* NR119657 (100%)
	MZ827835	MP1H-6	*Penicillium chrysogenum* MF422150 (100%)
M2	MZ827836	MP2H-1	*Stagonosporopsis lupini* NR160205 (98.3%)
M3	MZ827837	MP3B-5	*Penicillium chrysogenum* MF422150 (99.7%)
	MZ827838	MP3H-1	*Mucor racemosus* NR126135 (99.0%)
	MZ827839	MP3H-2	*Penicillium chrysogenum* MF422150 (100%)
	MZ827840	MP3H-3	*Penicillium chrysogenum* MF422150 (100%)
	MZ827841	MP3H-5	*Botryotrichum domesticum* NR169944(98.6%)

3.1. Ecology of Bacterial Isolates

The abundance of strains of the genus *Bacillus* and bacteria phylogenetically related to that genus (*Cytobacillus, Peribacillus, Priestia*), which produce spores, as well as the species *Streptomyces thinghirensis*, was remarkable. Most of these species were not previously found on wall paintings (e.g., *Priestia aryabhattai, Bacillus tequilensis, Bacillus paralicheniformis, Priestia endophyticus*) but on plants and soils [27–31]. The habitat of *Streptomyces thinghirensis* is also the soil rhizosphere [32]. The possibility of the transport of spores by the air should not be excluded, given the proximity of *Salón de Reinos* to the Retiro Park and Botanical Garden.

Other bacilli (*Bacillus cereus, Bacillus frigoritolerans*) were commonly found in wall paintings [7,33–38]. *Cytobacillus oceanisediminis* was previously found in caves and mines [39,40].

The abundance of bacteria from the genus *Bacillus* and other related genera (*Paenibacillus, Gracilibacillus, Salibacillus*) on wall paintings is not new, since previous research reported their presence and isolation from the deteriorated wall paintings in the church of St. Martin in Germany [7]. The unique presence of spore-forming bacteria in the wall paintings was noticeable. The dominance of bacilli may be due to their ability to survive for a long time as spores, since in periods of favorable conditions (humidity), small populations of bacilli produce a high number of spores, using these favorable conditions for further growth. These spores tend to germinate rapidly in culture media.

Laiz et al. [38] isolated 11 species of *Bacillus*, four of *Paenibacillus*, three species of *Staphylococcus*, two of *Micrococcus* and *Cellulomonas*, and one of *Streptomyces* and *Arthrobacter* from the mural paintings located in Servilia tomb, Necropolis of Carmona, Spain. Most of these strains presented a facultative oligotrophic behavior coincident with the limitation of nutrients in the mural paintings.

Tomassetti et al. [35] investigated the *Tomba degli Scudi* in Tarquinia, Italy, after receiving a conventional treatment with Preventol. Subsequently, a number of bacteria and fungi were isolated from the walls, including *Bacillus frigoritolerans, Bacillus simplex, Fictibacillus barbaricus, Fictibacillus phosphorivorans,* and *Bacillus cereus*, as well as an unidentified fungus. Two of these bacteria match those found in the *Salón de Reinos*. The authors concluded that the treatment was not totally effective, since a few spores remained active and were able to grow in laboratory culture media.

3.2. Ecology of Fungal Isolates

The deterioration of wall paintings by fungi is a subject that has occupied the attention of restorers and microbiologists, even more than the biodeterioration induced by bacteria. For this reason, there is abundant literature on the topic, which reflects the harmful effects of the colonization of frescoes and wall paintings in churches, monasteries, tombs, and catacombs [6,7,9–11,34–37].

Penicillium chrysogenum is one of the fungi most frequently isolated in wall paintings, and obvious signs of biodeterioration have been found on the surface of paintings [9,41–43]. Therefore, their abundance in the *Salón de Reinos* paintings is not surprising.

The cosmopolitan genus *Alternaria* is made up of multiple saprophytic and pathogenic plant species. Woudenberg et al. [44] considered *Alternaria angustiovoidea* to be a synonym for *Alternaria alternata*, based on the study of the genome and comparison of the transcriptome and molecular phylogenies. *Alternaria alternata* is widespread on wall paintings [36,41,43].

The remaining fungi from Table 4 (*Cladosporium xylophilum, C. macrocarpum, Penicillium fuscoglaucum, Mucor racemosus, Stagonosporopsis lupini*) were commonly isolated from plants and foods [45–49], and *Botryotrichum domesticum* from a house [50]. As far as we know, no reports on their presence on wall paintings have been published, other than for *C. macrocarpum* [41,51].

3.3. Metabolic Activities of Bacterial and Fungal Isolates

Trovão and Portugal [5] described the importance of using plate assays for identifying the deteriorative abilities of microorganisms in order to evaluate the potential risks to mural paintings. The isolation of microorganisms and plate assays are widely used by microbiologists in deterioration studies to provide insight into their attack on different types of materials [5,14,26,52–55]. Table 5 presents the metabolic activities of the isolates and their role in the hydrolysis of gelatin, casein, and lipids, the solubilization of gypsum and calcite, and acid production (Figure 3).

Table 5. Microbial activities of bacteria and fungi isolated from *Salón de Reinos*.

Strain Reference	GH *	LH	CH	GyS	CaS	AAP
Priestia aryabhattai, strain MP1B-1	-	+	+	-	-	+
Bacillus tequilensis, strain MP1B-2	-	+	-	+	-	-
Bacillus paralicheniformis, strain MP1B-3	+	-	-	-	-	-
Penicillium chrysogenum, strain MP1B-4	-	-	-	-	-	-
Penicillium chrysogenum, strain MP1B-5	-	+	+	-	-	-
Alternaria angustiovoidea, strain MP1H-1	-	-	-	-	-	-
Cladosporium xylophilum, strain MP1H-2	-	+	-	-	-	-
Penicillium fuscoglaucum, strain MP1H-3	-	+	+	-	+	-
Mucor racemosus, strain MP1H-4	-	-	-	-	-	-
Cladosporium macrocarpum, strain MP1H-5	-	+	-	-	-	-
Penicillium chrysogenum, strain MP1H-6	-	-	+	-	+	-
Bacillus frigoritolerans, strain MP2B-2	-	+	+	+	-	+
Bacillus paralicheniformis, strain MP2B-3	+	-	-	-	-	-
Bacillus cereus, strain MP2B-4	-	-	-	-	-	-
Stagonosporopsis lupini, strain MP2H-1	-	-	+	-	-	-
Cytobacillus oceanisediminis, strain MP3B-1	-	-	+	-	-	-
Streptomyces thinghirensis, strain MP3B-2	-	+	+	-	-	-
Priestia endophytica, strain MP3B-3	-	+	+	+	-	+
Bacillus frigoritolerans, strain MP3B-6	-	+	+	-	-	+
Penicillium chrysogenum, strain MP3B-5	-	-	+	-	-	-
Mucor racemosus, strain MP3H-1	-	-	-	-	-	-
Penicillium chrysogenum, strain MP3H-2	-	+	+	-	-	-
Penicillium chrysogenum, strain MP3H-3	-	-	+	-	-	-
Botryotrichum domesticum, strain MP3H-5	-	+	-	-	-	-

* GH, gelatin hydrolysis; CH, casein hydrolysis; LH, lipolytic activity; GyS, gypsum solubilization; CaS, calcite solubilization; AAP, acid/alkali production.

The most conspicuous activities, with a greater number of positive strains, were the hydrolysis of casein and lipids, and to a lesser extent, the hydrolysis of gelatin and the production of acids and alkalis.

The activity of *Penicillium chrysogenum* stands out, with six out of seven strains hydrolyzing casein, in addition to the *Penicillium fuscoglaucum* strain. All the other fungal strains were negative for casein hydrolysis. Among the bacteria, five out of the nine *Bacillus* strains also hydrolyzed casein, as well as *Cytobacillus oceanisediminis*, *Streptomyces thinghirensis*, and *Stagonosporopsis lupini*.

The degradation of lipids is important in the *Bacillus* strains, of which six out of nine presented lipolytic activity, while only two out of the nine of *Penicillium chrysogenum* showed such activity, in addition to *Penicillium fuscoglaucum* and *Botryotrichum domesticum*. It should be noted that only the two strains of *Bacillus paralicheniformis* were capable of hydrolyzing gelatin.

Acid/alkali production was found in only five *Bacillus* strains. Calcite solubilization was very limited in the isolated strains and showed only in one strain of *Penicillium chrysogenum* and *Penicillium fuscoglaucum*. Gypsum solubilization was present in only three *Bacillus* strains.

Bacillus species generally produce enzymes such as lipases, proteases, xylanases, and others that allow for the breakdown of casein, gelatin, and lipids. Specifically, *Bacillus cereus* produces these types of enzymes. However, the strain isolated from *Salón de Reinos* does not possess these capacities, since it did not show hydrolytic activity against casein and lipids.

In light of the obtained data, the species with potential biodeterioration properties mainly belong to the genera *Bacillus, Priestia, Cladosporium,* and *Penicillium,* as described in other reports [2–5]. In addition, it should be noticed that airborne microorganisms, commonly isolated from plants, are able to use lipids and casein from mural paintings as nutrient sources and therefore represent a risk for mural paintings.

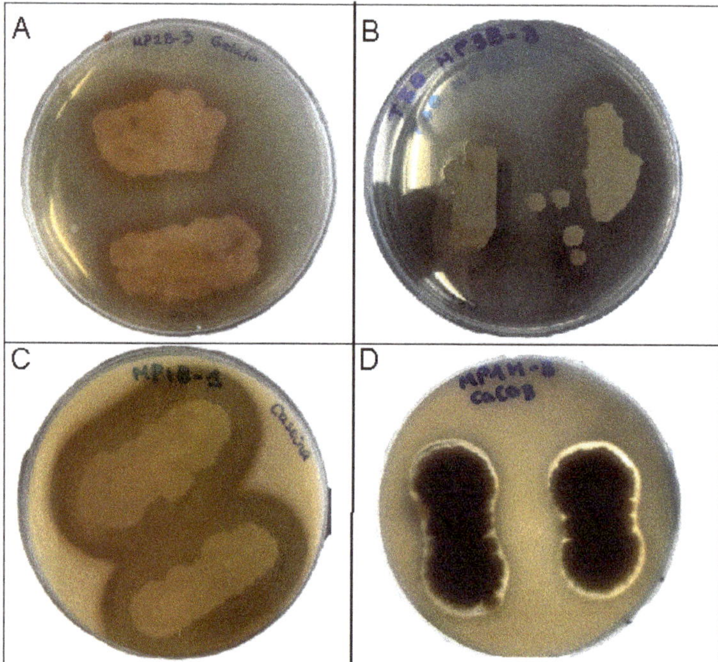

Figure 3. Metabolic activities of isolates. (**A**) Gelatin hydrolysis by *Bacillus paralicheniformis*. (**B**) Lipolytic activity by *Priestia endophytica*. (**C**) Casein hydrolysis by *Priestia aryabhattai*. (**D**) Calcite solubilization by *Penicillium fuscoglaucum*.

3.4. Cleaning and Biocide Treatment

When treating a biodeterioration, two different aspects must be taken into account: the elimination of the existing microorganisms and the prevention of new colonization.

The elimination of the effects of a biological attack is a delicate problem. If the growth is superficial, a mechanical cleaning and a soft biocide treatment should be effective. If the attack affected deep layers of the mural paintings, the fungal hyphae will be difficult to eliminate without affecting the painting, and for these situations an appropriate biocide is recommended.

The ideal biocide should have the following characteristics: broad spectrum of action at low concentrations, stability in solution, low vapor pressure to reduce evaporation, stability over time, and physical and chemical neutrality against the components of paints without action on the colors and texture of the paint.

One of the biocides most used by restorers and conservators in Italy [56–58] and also tested in Spain [59,60] is Preventol RI 80, based on a quaternary ammonium salt. It is active against a wide variety of bacteria, fungi, algae, and lichens on stones and building materials [57,59,61–63]. Nevertheless, quaternary ammonium salts have shown, in a few cases, to be unsuitable, such as in subterranean environments (caves with rock art paintings) with high and constant relative humidity (above 90%) [64,65]. However, these conditions are not present in the *Salón de Reinos*.

For the prevention of new biological attacks, an environmental control of the hall is needed. The main causes that favored the biodeterioration in the *Salón de Reinos* were the existence of fissures in the walls and water filtration that provoked the microbial growth. The restoration to be carried out will circumvent these problems.

Considering the variations due to the different compositions of the materials, the ideal conditions in which works of art (including wall paintings) should be preserved are a relative humidity of 40–65% and a temperature of 16–18 °C. These, or close, environmental conditions will prevent the growth and re-colonization of airborne microorganisms that can be deposited on the painted surfaces.

Bacteria need relative humidity >95% and fungi need 75–95%. In certain cases, some xerophilic microorganisms can grow at lower relative humidity. For this reason, the humidity limit in the *Salón* must never exceed values of 70% and the temperature must be maintained between 16 and 20 °C. If these conditions are maintained, the colonization and growth of microorganisms will be prevented. Once these conditions have been fixed, an additional precaution is the periodic cleaning of the painted surfaces (all this, of course, after the consolidation and restoration processes) to avoid the accumulation of dust particles and volatile and suspended organic matter as well as the deposition of airborne microorganisms.

As reported by Ágora, S.L., the microorganisms were eliminated using a protocol consisting of their mechanical removal with a natural bristle brush and the application of Preventol RI 80 at 0.5 % in demineralized water, which has a broad spectrum in low concentrations on bacteria, fungi, algae, and lichens. This biocide is effective because it interferes with the enzymatic processes of the cell and modifies the permeability of its membrane. However, the sensitivity to quaternary ammonium salts depends on the species tested [55]. The intervention at the *Salón de Reinos* was focused mainly on the areas of superficial fungal colonies, showing black stains, and in specific areas of the lintels of three windows that were blinded on the south side of the hall.

One of the treatments that wall paintings may require is fixing and consolidation. Regardless of the type of fixative product that is most suitable for the case, there are some whose use is not recommended at all, among them the traditional organic fixatives, made up of natural protein-based products such as albumin and casein. Restorations with casein tend to cause microbial growth in the short term, since a high number of bacteria and fungi use it as a nutrient source, as shown in Table 5. Therefore, an acrylic-based micro-emulsion was used [66–68].

4. Conclusions

Three samples were studied: M1, dust obtained from the paint surface, and M2 and M3, microbial colonizations subjected to the influence of humidity. The results obtained are consistent with the nature of the samples. The dust sample shows an exclusive preponderance of species of the genus *Bacillus* among the bacteria, and a diversity of fungi, composed of the genera *Penicillium, Cladosporium, Mucor,* and *Alternaria*, with distribution consistent with the diversity of airborne fungal spores generally found in the air.

The two samples of black biofilms show different profiles, compatible with microbial colonization, in an advanced status in sample M3, an area favored by humidity. In this sample, a greater diversity is observed, both of bacteria (*Bacillus, Cytobacillus, Streptomyces*) and fungi (*Penicillium, Mucor, Botryotrichum*), which indicates a better development of the colonization, or a more advanced stage of deterioration. Sample M2 has intermediate characteristics, with the presence of *Bacillus* and only one species of fungus (*Stagonosporopsis lupini*).

The nature of the isolated microorganisms points to a high influence of airborne spores. Most isolated bacteria and fungi have their origin in plants—saprophytes or pathogens—consistent with the abundant vegetation in parks and gardens near the *Museo Nacional del Prado* (Retiro Park and Botanical Garden). Likewise, the presence of a few bacteria and fungi with biodeteriorative activities, previously detected in other murals and frescoes, is also noteworthy.

Author Contributions: Conceptualization, C.S.-J.; investigation, V.J., J.L.G.-P., B.H.; writing—original draft preparation, C.S.-J.; writing—review and editing, C.S.-J. All authors have read and agreed to the published version of the manuscript.

Funding: This research was funded by Ágora Restauraciones de Arte, S.L., Madrid, Spain.

Institutional Review Board Statement: Not applicable.

Informed Consent Statement: Not applicable.

Data Availability Statement: Deposited in the NCBI GenBank database.

Acknowledgments: The authors wish to acknowledge the permission granted by *Museo Nacional del Prado* for this publication, the professional support of the CSIC Interdisciplinary Thematic Platform Open Heritage: Research and Society (PTI-PAIS), as well as the facilities of Ágora Restauraciones de Arte, S.L., and the help of Juan Aguilar.

Conflicts of Interest: The authors declare no conflict of interest.

References

1. Rosado, T.; Falé, A.; Gil, M.; Mirão, J.; Candeias, A.; Caldeira, A.T. Oxalate biofilm formation in mural paintings due to microorganisms—A comprehensive study. *Int. Biodeter. Biodegr.* **2013**, *85*, 1–7. [CrossRef]
2. Rosado, T.; Gil, M.; Caldeira, A.T.; Martins, M.R.; Dias, C.B.; Carvalho, L.; Mirão, J.; Candeias, A.E. Material characterization and biodegradation assessment of mural paintings: Renaissance frescoes from Santo Aleixo Church, Southern Portugal. *Int. J. Archit. Herit.* **2014**, *8*, 835–852. [CrossRef]
3. Rosado, T.; Falé, A.; Gil, M.; Mirão, J.; Candeias, A.; Caldeira, A.T. Understanding the influence of microbial contamination on colour alteration of pigments used in wall paintings—The case of red and yellow ochres and ultramarine blue. *Color Res. Appl.* **2019**, *44*, 783–789. [CrossRef]
4. Unković, N.; Grbić, M.L.; Stupar, M.; Savković, Ž.; Jelikić, A.; Stanojević, D.; Vukojević, J. Fungal-induced deterioration of mural paintings: In situ and mock-model microscopy analyses. *Microsc. Microanal.* **2016**, *22*, 410–421. [CrossRef] [PubMed]
5. Trovão, J.; Portugal, A. Current knowledge on the fungal degradation abilities profiled through biodeteriorative plate essays. *Appl. Sci.* **2021**, *11*, 4196. [CrossRef]
6. Piñar, G.; Saiz-Jimenez, C.; Schabereiter-Gurtner, C.; Blanco-Varela, M.T.; Lubitz, W.; Rölleke, S. Archaeal communities in two disparate deteriorated ancient wall paintings: Detection, identification and temporal monitoring by DGGE. *FEMS Microbiol. Ecol.* **2001**, *37*, 45–54. [CrossRef]
7. Gorbushina, A.A.; Heyrman, J.; Dornieden, T.; Gonzalez-Delvalle, M.; Krumbein, W.E.; Laiz, L.; Petersen, K.; Saiz-Jimenez, C.; Swings, J. Bacterial and fungal diversity and biodeterioration problems in mural painting environments of St. Martins church (Greene-Kreiensen, Germany). *Int. Biodeter. Biodegr.* **2004**, *53*, 13–24. [CrossRef]
8. Tsang, C.-C.; Chan, J.F.W.; Pong, W.-M.; Chen, J.H.K.; Ngan, A.H.Y.; Cheung, M.; Lai, C.K.C.; Tsang, D.N.C.; Lau, S.K.P.; Woo, P.C.Y. Cutaneous hyalohyphomycosis due to *Parengyodontium album* gen. et comb. nov. *Med. Mycol.* **2016**, *54*, 699–713. [CrossRef]
9. Saiz-Jimenez, C.; Samson, R.A. Microorganisms and environmental pollution as deteriorating agents of the frescoes of "Santa María de la Rábida," Huelva, Spain. In Proceedings of the 6th Triennial Meeting ICOM Committee for Conservation, Ottawa, Canada, 21–25 September 1981; p. 81/15/5.
10. Karpovich-Tate, N.; Rebrikova, N.L. Microbial communities on damaged frescoes and building materials in the cathedral of the Nativity of the Virgin in the Pafnutii-Borovskii monastery, Russia. *Int. Biodeter.* **1990**, *27*, 281–296. [CrossRef]
11. Gorbushina, A.A.; Petersen, K. Distribution of microorganisms on ancient wall paintings as related to associated faunal elements. *Int. Biodeter. Biodegr.* **2000**, *46*, 277–284. [CrossRef]
12. Jurado, V.; Sanchez-Moral, S.; Saiz-Jimenez, C. Entomogenous fungi and the conservation of the cultural heritage: A review. *Int. Biodeter. Biodegr.* **2008**, *62*, 325–330. [CrossRef]
13. Ma, Y.; Zhang, H.; Du, Y.; Tian, T.; Xiang, T.; Liu, X.; Wu, F.; An, L.; Wang, W.; Gu, J.-D.; et al. The community distribution of bacteria and fungi on ancient wall paintings of the Mogao Grottoes. *Sci. Rep.* **2015**, *5*, 7752. [CrossRef]
14. Ma, W.; Wu, F.; Tian, T.; He, D.; Zhang, Q.; Gu, J.-D.; Duan, Y.; Ma, D.; Wang, W.; Feng, H. Fungal diversity and its contribution to the biodeterioration of mural paintings in two 1700-year-old tombs of China. *Int. Biodeter. Biodegr.* **2020**, *152*, 104972. [CrossRef]
15. Duan, Y.; Wu, F.; Wang, W.; Gu, J.-D.; Li, Y.; Feng, H.; Chen, T.; Liu, G.; An, L. Differences of microbial community on the wall paintings preserved in situ and ex situ of the Tiantishan Grottoes, China. *Int. Biodeter. Biodegr.* **2018**, *132*, 102–113. [CrossRef]
16. He, D.; Wu, F.; Ma, W.; Zhang, Y.; Gu, J.-D.; Duan, Y.; Xu, R.; Feng, H.; Wang, W.; Li, S.-W. Insights into the bacterial and fungal communities and microbiome that causes a microbe outbreak on ancient wall paintings in the Maijishan Grottoes. *Int. Biodeter. Biodegr.* **2021**, *163*, 105250. [CrossRef]
17. Juretschko, S.; Timmermann, G.; Schmid, M.; Schleifer, K.-H.; Pommering-Röser, A.; Koops, H.-P.; Wagner, M. Combined molecular and conventional analyses of nitrifying bacterium diversity in activated sludge: *Nitrosococcus mobilis* and *Nitrospira*-like bacteria as dominant populations. *Appl. Environ. Microbiol.* **1998**, *64*, 3042–3051. [CrossRef] [PubMed]
18. Edwards, U.; Rogall, T.; Blöcker, H.; Emde, M.; Bottger, E.C. Isolation and direct complete nucleotide determination of entire genes. Characterization of a gene coding for 16S ribosomal RNA. *Nucleic Acids Res.* **1989**, *19*, 7843–7853. [CrossRef] [PubMed]
19. White, T.J.; Bruns, T.; Lee, S.; Taylor, J. Amplification and direct sequencing of fungal ribosomal RNA genes for phylogenetics. In *PCR, Protocols: A Guide to Methods and Applications*; Innis, M.A., Gelfand, D.H., Sninsky, J.J., White, T.J., Eds.; Academic Press: New York, NY, USA, 1990; pp. 315–322.
20. Hall, T.A. BioEdit: A user-friendly biological sequence alignment editor and analysis program for Windows 95/98/NT. *Nucleic Acids Symp. Ser.* **1999**, *41*, 95–98.

21. Altschul, S.F.; Gish, W.; Miller, W.; Myers, E.W.; Lipman, D.J. Basic local alignment search tool. *J. Mol. Biol.* **1990**, *215*, 403–410. [CrossRef]
22. Raja, H.A.; Miller, A.N.; Pearce, C.J.; Oberlies, N.H. Fungal identification using molecular tools: A primer for the natural products research community. *J. Nat. Prod.* **2017**, *80*, 756–770. [CrossRef] [PubMed]
23. Yoon, S.H.; Ha, S.M.; Kwon, S.; Lim, J.; Kim, Y.; Seo, H.; Chun, J. Introducing EzBioCloud: A taxonomically united database of 16S rRNA and whole genome assemblies. *Int. J. Syst. Evol. Microbiol.* **2017**, *67*, 1613–1617. [CrossRef] [PubMed]
24. Mazzoni, M.; Alisi, C.; Tasso, F.; Cecchini, A.; Marconi, P.; Sprocati, A.R. Laponite micro-packs for the selective celaning of multiple coherent deposits on wall paintings: The case study of Casina Farnese on the Palatine Hill (Rome-Italy). *Int. Biodeter. Biodegr.* **2014**, *94*, 1–11. [CrossRef]
25. Lanyi, B. Classical and rapid identification methods for medically important bacteria. In *Methods in Microbiology*; Colwell, R.R., Grigorova, R., Eds.; Academic Press: London, UK, 1987; pp. 1–67.
26. Unković, N.; Dimkić, I.; Supar, M.S.; Stanković, S.; Vukojević, J.; Grbić, M.L. Biodegradative potential of fungal isolates from sacral ambient: In vitro study as risk assessment implication for the conservation of wall paintings. *PLoS ONE* **2018**, *13*, e0190922. [CrossRef] [PubMed]
27. Bhattacharyya, C.; Bakshi, U.; Mallick, I.; Mukherji, S.; Bera, B.; Ghosh, A. Genome-guided insights into the plant growth promotion capabilities of the physiologically versatile *Bacillus aryabhattai* strain AB211. *Front. Microbiol.* **2017**, *8*, 411. [CrossRef]
28. Li, H.; Guan, Y.; Dong, Y.; Zhao, L.; Rong, S.; Chen, W.; Lv, M.; Xu, H.; Gao, X.; Chen, R.; et al. Isolation and evaluation of endophytic *Bacillus tequilensis* GYLH001 with potential application for biological control of *Magnaporthe oryzae*. *PLoS ONE* **2018**, *13*, e0203505. [CrossRef] [PubMed]
29. Liu, G.H.; Liu, B.; Wang, J.P.; Che, J.M.; Li, P.F. Reclassification of *Brevibacterium frigoritolerans* DSM 8801T as *Bacillus frigoritolerans* comb. nov. based on genome analysis. *Curr. Microbiol.* **2020**, *77*, 1916–1923. [CrossRef]
30. Dunlap, C.A.; Kwon, S.-W.; Rooney, A.P.; Kim, S.-J. *Bacillus paralicheniformis* sp. nov., isolated from fermented soybean paste. *Int. J. Syst. Evol. Microbiol.* **2015**, *65*, 3487–3492. [CrossRef] [PubMed]
31. Reva, O.N.; Smirnov, V.V.; Pettersson, B.; Priest, F.G. *Bacillus endophyticus* sp. nov., isolated from the inner tissues of cotton plants (*Gossypium* sp.). *Int. J. Syst. Evol. Microbiol.* **2002**, *52*, 101–107. [CrossRef]
32. Loqman, S.; Bouizgarne, B.; Barka, E.A.; Clément, C.; von Jan, M.; Spröer, C.; Klenk, H.-P.; Ouhdouch, Y. *Streptomyces thinghirensis* sp. nov., isolated from rhizosphere soil of *Vitis vinifera*. *Int. J. Syst. Evol. Microbiol.* **2009**, *59*, 3063–3067. [CrossRef] [PubMed]
33. Pepe, O.; Sannino, L.; Palomba, S.; Anastasio, M.; Blaiotta, G.; Villani, F.; Moschetti, G. Heterotrophic microorganisms in deteriorated medieval wall paintings in southern Italian churches. *Microbiol. Res.* **2010**, *165*, 21–32. [CrossRef] [PubMed]
34. Sugiyama, J.; Kiyuna, T.; Nishijima, M.; An, K.-D.; Nagatsuka, Y.; Tazato, N.; Handa, Y.; Hata-Yomita, J.; Sato, Y.; Kigawa, R.; et al. Polyphasic insights into the microbiomes of the Takamatsuzuka Tumulus and Kitora Tumulus. *J. Gen. Appl. Microbiol.* **2017**, *63*, 63–113. [CrossRef]
35. Tomassetti, M.C.; Cirigliano, A.; Arrighi, C.; Negri, R.; Mura, F.; Maneschi, M.L.; Gentili, M.D.; Stipe, M.; Mazzoni, C.; Rinaldi, T. A role for microbial selection in frescoes' deterioration in *Tomba degli Scudi* in Tarquinia, Italy. *Sci. Rep.* **2017**, *7*, 6027. [CrossRef] [PubMed]
36. Elhagrassy, A.F. Isolation and characterization of actinomycetes from mural paintings of Snu-Sert-Ankh tomb, their antimicrobial activity, and their biodeterioration. *Microbiol. Res.* **2018**, *216*, 47–55. [CrossRef] [PubMed]
37. Caneva, G.; Isola, D.; Lee, H.J.; Chung, Y.J. Biological risk for hypogea: Shared data from Etruscan tombs in Italy and ancient tombs of the Baekje dynasty in Republic of Korea. *Appl. Sci.* **2020**, *10*, 6104. [CrossRef]
38. Laiz, L.; Hermosin, B.; Caballero, B.; Saiz-Jimenez, C. Facultatively oligotrophic bacteria in Roman mural paintings. In *Protection and Conservation of the Cultural Heritage of the Mediterranean Cities*; Galan, E., Zezza, F., Eds.; Balkema: Lisse, The Netherlands, 2002; pp. 173–178.
39. Dominguez-Moñino, I.; Jurado, V.; Rogerio-Candelera, M.A.; Hermosin, B.; Saiz-Jimenez, C. Airborne bacteria in show caves from Southern Spain. *Microb. Cell* **2021**, in press.
40. Srinath, B.S.; Namratha, K.; Byrappa, K. Green synthesis of biocompatible gold nanoparticles from gold mine bacteria *Bacillus oceanisediminis* and their antileukemic activity. *Int. J. Pharm. Biol. Sci.* **2018**, *8*, 169–177.
41. Nugari, M.P.; Realini, M.; Roccardi, A. Contamination of mural paintings by indoor airborne fungal spores. *Aerobiologia* **1993**, *9*, 131–139. [CrossRef]
42. An, K.-D.; Kiyuna, T.; Kigawa, R.; Sano, C.; Miura, S.; Sugiyama, J. The identity of *Penicillium* sp. 1, a major contaminant of the stone chambers in the Takamatsuzuka and Kitora Tumuli in Japan, is *Penicillium paneum*. *Anton. Leeuw.* **2009**, *96*, 579–592. [CrossRef] [PubMed]
43. Ciferri, O. Microbial degradation of paintings. *Appl. Environ. Microbiol.* **1999**, *65*, 879–885. [CrossRef]
44. Woudenberg, J.H.C.; Seidl, M.F.; Groenewald, J.Z.; de Vries, M.; Thomma, B.P.H.J.; Crous, P.W. *Alternaria* section *Alternaria*: Species, *formae speciales* or pathotypes? *Stud. Mycol.* **2015**, *82*, 1–21. [CrossRef]
45. Bensch, K.; Groenewald, J.Z.; Dijksterhuis, J.; Starink-Willemse, M.; Andersen, B.; Summerell, B.A.; Shin, H.-D.; Dugan, F.M.; Schroers, H.-J.; Braun, U.; et al. Species and ecological diversity within the *Cladosporium cladosporioides* complex (*Davidiellaceae*, *Capnodiales*). *Stud. Mycol.* **2010**, *67*, 1–94. [CrossRef] [PubMed]
46. Bensch, K.; Groenewald, J.Z.; Meijer, M.; Dijksterhuis, J.; Jurjevic, Z.; Andersen, B.; Houbraken, J.; Crous, P.W.; Samson, R.A. *Cladosporium* species in indoor environments. *Stud. Mycol.* **2018**, *89*, 177–301. [CrossRef]

47. Vaghefi, N.; Pethybridge, S.J.; Ford, R.; Nicolas, M.E.; Crous, P.W.; Taylor, P.W.J. *Stagonosporopsis* spp. associated with ray blight disease of *Asteraceae*. *Australas. Plant Pathol.* **2012**, *41*, 675–686. [CrossRef]
48. Ropars, J.; Didiot, E.; Rodríguez de la Vega, R.C.; Bennetot, B.; Coton, M.; Poirier, E.; Coton, E.; Snirc, A.; Le Prieur, S.; Giraud, T. Domestication of the emblematic white cheese-making fungus *Penicillium camemberti* and its diversification into two varieties. *Curr. Biol.* **2020**, *30*, 4441–4453. [CrossRef] [PubMed]
49. Saito, S.; Michailides, T.J.; Xiao, C.L. Mucor rot–An emerging postharvest disease of mandarin fuit caused by *Mucor piriformis* and other *Mucor* spp. in California. *Plant Dis.* **2016**, *100*, 1054–1063. [CrossRef]
50. Schultes, N.P.; Strzalkowski, N.; Li, D.-W. *Botryotrichum domesticum* sp. nov., a new hyphomycete from an indoor environment. *Botany* **2019**, *87*, 311–319. [CrossRef]
51. Pangallo, D.; Kraková, L.; Chovanová, K.; Šimonovicová, A.; De Leo, F.; Urzì, C. Analysis and comparison of the microflora isolated from fresco surface and from surrounding air environment through molecular and biodegradative assays. *World J. Microbiol. Biotechnol.* **2012**, *28*, 2015–2027. [CrossRef] [PubMed]
52. Pangallo, D.P.; Chovanová, K.C.; Šimonovicová, A.Š.; Ferianc, P.F. Investigation of microbial community isolated from indoor artworks and air environment: Identification, biodegradative abilities, and DNA typing. *Can. J. Microbiol.* **2009**, *55*, 277–287. [CrossRef] [PubMed]
53. Savković, Ž.; Stupar, M.; Unković, N.; Ivanović, Ž.; Blagojević, J.; Vukojević, J.; Grbić, M.L. In vitro biodegradation potential of airborne *Aspergilli* and *Penicillia*. *Sci. Nat.* **2019**, *106*, 8. [CrossRef] [PubMed]
54. Trovão, J.; Tiago, I.; Catarino, L.; Gil, F.; Portugal, A. In vitro analyses of fungi and dolomitic limestone interactions: Bioreceptivity and biodeterioration assessment. *Int. Biodeter. Biodegr.* **2020**, *155*, 105107. [CrossRef]
55. Isola, D.; Zucconi, L.; Cecchini, A.; Caneva, G. Dark-pigmented biodeteriogenic fungi in etruscan hypogeal tombs: New data on their culture-dependent diversity, favouring conditions, and resistance to biocidal treatments. *Fungal Biol.* **2021**, *125*, 609–620. [CrossRef] [PubMed]
56. Bartolini, M.; Pietrini, A.M. La disinfezione delle patine biologiche sui manufatti lapidei: Biocidi chimici e naturali a confronto. *Boll. ICR* **2016**, *33*, 40–49.
57. Vannini, A.; Contardo, T.; Paoli, L.; Scattoni, M.; Favero-Longo, S.E.; Loppi, S. Application of commercial biocides to lichens: Does a physiological recovery occur over time? *Int. Biodeter. Biodegr.* **2018**, *129*, 189–194. [CrossRef]
58. Lo Schiavo, S.; De Leo, F.; Urzì, C. Present and future perspectives for biocides and antifouling products for stone-built cultural heritage: Ionic liquids as a challenging alternative. *Appl. Sci.* **2020**, *10*, 6568. [CrossRef]
59. Sanmartín, P.; Rodríguez, A.; Aguiar, U. Medium-term field evaluation of several widely used cleaning-restoration techniques applied to algal biofilm formed on a granite-built historical monument. *Int. Biodeter. Biodegr.* **2020**, *147*, 104870. [CrossRef]
60. Sanmartín, P.; Carballeira, R. Changes in heterotrophic microbial communities induced by biocidal treatments in the Monastery of San Martiño Pinario (Santiago de Compostela, NW Spain). *Int. Biodeter. Biodegr.* **2021**, *156*, 105130. [CrossRef]
61. Coutinho, M.L.; Miller, A.Z.; Martin-Sanchez, P.M.; Mirao, J.; Gomez-Bolea, A.; Machado-Moreira, B.; Cerqueira-Alves, L.; Jurado, V.; Saiz-Jimenez, C.; Lima, A.; et al. A multiproxy approach to evaluate biocidal treatments on biodeteriorated majolica glaze tiles. *Environ. Microbiol.* **2016**, *18*, 4794–4816. [CrossRef] [PubMed]
62. Romani, M.; Warscheid, T.; Nicole, L.; Marcon, L.; Di Martino, P.; Suzuki, M.T.; Lebaron, P.; Lami, R. Current and future chemical treatments to fight biodeterioration of outdoor building materials and associated biofilms: Moving away from ecotoxic and towards efficient, sustainable solutions. *Sci. Total. Environ.* **2021**, *802*, 149846. [CrossRef]
63. Presentato, A.; Armetta, F.; Spinella, A.; Chillura Martino, D.F.; Alduina, R.; Saladino, M.L. Formulation of mesoporous silica nanoparticles for controlled release of antimicrobials for stone preventive conservation. *Front. Chem.* **2020**, *8*, 699. [CrossRef]
64. Bastian, F.; Alabouvette, C.; Saiz-Jimenez, C. Impact of biocide treatments on the bacterial communities of the Lascaux Cave. *Naturwissenschaften* **2009**, *96*, 863–868. [CrossRef]
65. Martin-Sanchez, P.M.; Nováková, A.; Bastian, F.; Alabouvette, C.; Saiz-Jimenez, C. Use of biocides for the control of fungal outbreaks in subterranean environments: The case of the Lascaux Cave in France. *Environ. Sci. Technol.* **2012**, *46*, 3762–3770. [CrossRef] [PubMed]
66. Camaiti, M.; Borgioli, L.; Rosi, L. Photostability of innovative formulations for artworks restoration. *Chim. Ind.* **2011**, *9*, 100–195.
67. Gheno, G.; Badetti, E.; Brunelli, A.; Ganzerla, R.; Marcomini, A. Consolidation of Vicenza, Arenaria and Istria stones: A comparison between nano-based products and acrylate derivatives. *J. Cult. Herit.* **2018**, *32*, 44–52. [CrossRef]
68. Negri, A.; Nervo, M.; Di Marcello, S.; Castelli, D. Consolidation and adhesion of pictorial layers on a stone substrate. The study case of the Virgin with the Child from Palazzo Madama, in Turin. *Coatings* **2021**, *11*, 624. [CrossRef]

Article

Holistic Approach to the Restoration of a Vandalized Monument: The Cross of the Inquisition, Seville City Hall, Spain

Valme Jurado [1], Juan Carlos Cañaveras [2], Antonio Gomez-Bolea [3], Jose Luis Gonzalez-Pimentel [1], Sergio Sanchez-Moral [4], Carlos Costa [5] and Cesareo Saiz-Jimenez [1,*]

1 Instituto de Recursos Naturales y Agrobiologia, IRNAS-CSIC, 41012 Sevilla, Spain; vjurado@irnase.csic.es (V.J.); pimentel@irnas.csic.es (J.L.G.-P.)
2 Departamento de Ciencias de la Tierra, Universidad de Alicante, 03080 Alicante, Spain; jc.canaveras@ua.es
3 Departament de Biologia Evolutiva, Ecologia i Ciències Ambientals, Facultat de Biologia, Institut de Recerca de la Biodiversitat (IRBio), Universitat de Barcelona, 08028 Barcelona, Spain; agomez@ub.edu
4 Museo Nacional de Ciencias Naturales, MNCN-CSIC, 28006 Madrid, Spain; ssmilk@mncn.csic.es
5 Atelier Samthiago, 4900-374 Viana do Castelo, Portugal; ccosta@samthiago.com
* Correspondence: saiz@irnase.csic.es

Abstract: The Cross of the Inquisition, sculpted in 1903 and raised on a column with a fluted shaft and ornamented with vegetable garlands, is located in a corner of the Plateresque façade of the Seville City Hall. The Cross was vandalized in September 2019 and the restoration concluded in September 2021. A geological and microbiological study was carried out in a few small fragments. The data are consistent with the exposure of the Cross of the Inquisition to an urban environment for more than 100 years. During that time, a lichen community colonized the Cross and the nearby City Hall façades. The lichens, bryophytes and fungi colonizing the limestone surface composed an urban community, regenerated from the remains of the original communities, after superficial cleaning of the limestone between 2008 and 2010. This biological activity was detrimental to the integrity of the limestone, as showed by the pitting and channels, which evidence the lytic activity of organisms on the stone surface. Stone consolidation was achieved with Estel 1000. Preventol RI80, a biocide able to penetrate the porous limestone and active against bacteria, fungi, lichens, and bryophytes, was applied in the restoration.

Keywords: green algae; lichens; *Trebouxia aggregata*; black fungi; bryophytes; limestone; mineralogy; restoration; Seville City Hall

Citation: Jurado, V.; Cañaveras, J.C.; Gomez-Bolea, A.; Gonzalez-Pimentel, J.L.; Sanchez-Moral, S.; Costa, C.; Saiz-Jimenez, C. Holistic Approach to the Restoration of a Vandalized Monument: The Cross of the Inquisition, Seville City Hall, Spain. *Appl. Sci.* **2022**, *12*, 6222. https://doi.org/10.3390/app12126222

Academic Editor: Asterios Bakolas

Received: 12 May 2022
Accepted: 16 June 2022
Published: 18 June 2022

1. Introduction

The construction of the Seville City Hall building started in 1526 and the last works ended in 1928. The main façade, facing to Plaza Nueva was completed in 1867. The Plateresque façade from the 16th century, facing to Plaza de San Francisco, was continued in the 19th and reliefs were carved between the end of the 19th and the 20th centuries. Still today, part of this façade is unfinished. In a corner of the Plateresque façade is located the Cross of the Inquisition, sculpted in 1903, raised on a column with a fluted shaft and ornamented with vegetable garlands (Figure 1A). Previously, another Cross, dating from the end of the 18th century, devoid of ornamentation, was placed in the same position as a commemoration of the last case of death at the stake (year 1781), although a few historians believe that the primitive Cross was erected during a plague epidemic in the 16th century. The Cross and the City Hall façade were restored between 2008 and 2010, and were vandalized in the early morning of 10 September 2019 (Figure 1B). The Cross fragments were stored for two years until a restoration was decided, which should include the repair of damages caused by the act of vandalism, but also previous pathologies caused by environmental pollution and anthropic actions which resulted in cracks and losses of materials. The restoration concluded last September 2021 (Figure 1C), and included a geological and microbiological study that, unfortunately, was only possible on a few fragments. The limestone fragments showed dark stains and crusts, likely originated by lichen colonization. Lichen communities were found on the nearby City Hall limestone. Here, we present the data obtained in the study, which were to some extent limited by the scarce availability of samples, as all fragments should be used in the restoration, as well as the time that the fragments were kept in storage which prevented the study of fresh biological samples.

Figure 1. The Cross of the Inquisition before (**A**), after the vandalism (**B**), and restored (**C**).

2. Materials and Methods

Six small samples, three for geological and three for microbiological studies, were collected under the supervision of the restorers (Figure 2).

Mineralogy and textural properties of samples were studied on (30 μm-thick) polished thin sections on a transmitting light microscope. A staining procedure with combined alizarine red S and potassium ferricyanide solution was applied to distinguish ferroan and non-ferroan calcite and dolomite phases [1]. Two thin sections were made from the sample, one from the external surface of the sample and the other being transverse to that surface. Photomicrographs were performed by using a petrographic microscope: Zeiss Assioskop, a digital camera: USB UI-1490SE and an image capture software: uEye Cockpit (IDS). The sample was also studied under a scanning electron microscope (SEM) using BSE mode on a Hitachi S3000N SEM coupled with an X-ray detector, Bruker XFlash 3001, for microanalysis (EDS) and mapping. The samples' observation was carried out under variable pressure mode without coating with any conductive material. EDS analyses were performed at a 40 Pa chamber atmosphere, 20 kV accelerating voltage and 10–12 mm working distance.

The mineralogical composition was analyzed by powder X-ray diffraction in a Bruker D8 Discover A25 microdiffractometer. Microdiffraction X-ray equipment makes it possible to make precise measurements in very small areas without the need to prepare the sample. The analysis area was selected using an optical microscope and an X-ray micro-source with a copper anode, using its Kα spectral line. EVA and TOPAS computer programs were used for data processing and qualitative and quantitative identification, with the Rietveld method, of the crystalline phases present in the sample.

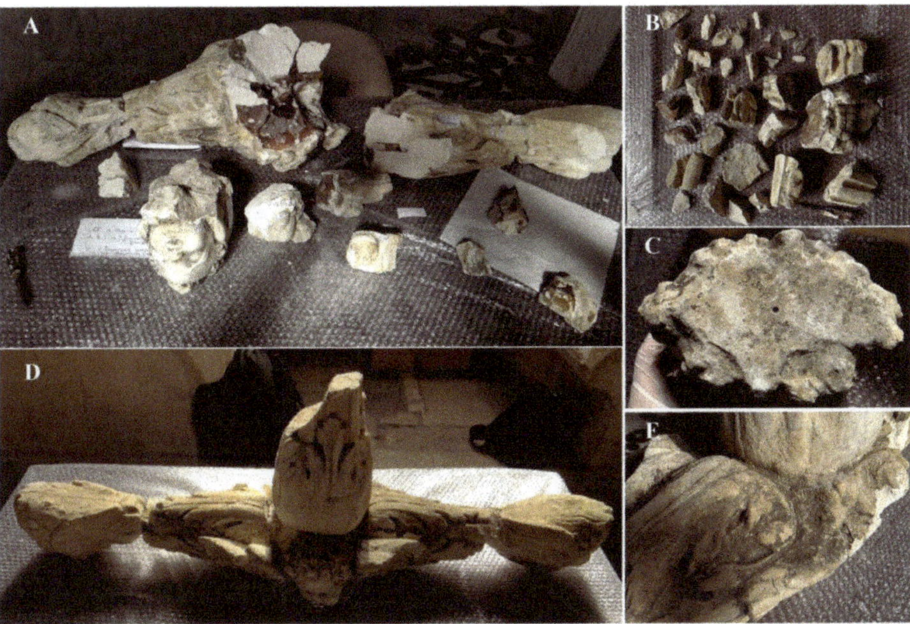

Figure 2. Vandalization of the Cross of the Inquisition. (**A**,**B**) Fragments of different sizes. (**C**,**E**) Fragments with black to brown crusts. (**D**) Reconstruction of some fragments.

For the identification of the organisms two approaches were adopted in this study. First, due to the fact that the vandalized fragments of the Cross were stored for two years, and no fresh samples from organisms could be obtained, we focused our attention on the identifications of DNA on the fragments available, in order to determine past biological colonizations. In addition, we sampled the lichen community colonizing the City Hall limestones in an area around 5 m from the Cross. It was expected that the community or at least the most abundant lichens in both the building façade and Cross would be the same. The protocols used were as follows:

Nucleic acids from each sample were extracted with commercial extraction kits, following the manufacturer's instructions. FastPrep matrix-E lysis tubes (Qbiogene, Carlsbad, CA, USA) with glass beads were used and physical disruption was performed using a shaker (Fast Prep-24, Solon, OH, USA) at a speed of stirring at 5.5 m/s for 2 × 30 s. The extracted nucleic acids (50 μL for each sample) were stored at −80 °C until shipment for sequencing. The concentration of the DNA extracted from the samples was measured by fluorometric quantification using a Qubit 2.0 fluorimeter (Invitrogen, Carlsbad, CA, USA). Eukaryotes were identified by sequencing the 18S ribosomal RNA gene. The genes were amplified by PCR using the primer pairs EUKA (5′-AAC CTG GTT GAT CCT GCC AGT-3′) and EUKB (5′-TGA TCC TTC TGC AGG TTC ACC TAC-3′) for eukaryotes.

Amplifications were carried out in an iCycler thermal cycler from BioRad (Hercules, CA, USA). Reactions were set up in 0.2 mL tubes (Greiner BioOne, Monroe, LA, USA). Each reaction contained: 5 μL of 10× BioTaq PCR buffer (Bioline, Randolph, MA, USA), 1.5 μL of MgCl2 (50 mM stock solution), 1 μL of a mixture of the four nucleotides (dNTP) (2 mM) (Invitrogen, Carlsbad, CA, USA), 0.5 μL of each of the primers with a concentration of 50 μM, 0.25 μL of the equivalent BioTaq DNA polymerase (Bioline, Randolph, MA, USA) at 2.5 units and template DNA. The final volume of the reaction was made up to 50 μL with ultrapure water. For each reaction, about 10–20 ng of DNA extract was added. The PCR protocol used was as follows: 5 min of initial denaturation at 94 °C; 30 cycles consisting of 2 min at 94 °C, 15 s at 55 °C, and 2 min at 72 °C; followed by a 10-min final extension at 72 °C. The results of PCR reactions were verified by horizontal electrophoresis in agarose gels (1% *w/v*) and the amplified products were purified using the JETquick Spin kit (Genomed, Bad Oeynhausen, Germany), and stored at −20 °C.

In order to identify the main organisms that were present in the limestone samples, libraries were constructed from the purified PCR products of the 18S rRNA genes. For this purpose, purified products were cloned using the commercial TOPO TA Cloning Kit for Sequencing kit (Invitrogen, Carlsbad, CA, USA) following manufacturer's instructions. The vector used was the plasmid pCR4-TOPO (Invitrogen, Carlsbad, CA, USA). The colonies obtained were picked at random using sterile toothpicks and grown in LB liquid medium with ampicillin. In order to verify that the collected clones had the insert (purified PCR product), a PCR was performed using the primers promoter T7 (5′-TAA TAC GAC TCA CTA TAG GG-3′) and M13 Reverse (5′-CAG GAA ACA GCT ATG AC-3′), which were specific to the vector sequence. This amplified the complete sequence of the insert and part of the vector at the ends. The clones were processed at the company Secugen (CIB-CSIC, Madrid, Spain), with an ABI 3700 capillary sequencer (Applied Biosystems, Foster City, CA, USA).

The sequences obtained were edited with the BioEdit 7.0.5.3 program. Sequences that were too short (<300 nucleotides) or whose chromatogram showed poor quality were directly removed. All non-redundant database sequences were compared with sequences deposited at the National Center for Biotechnology (NCBI) using the BLASTn algorithm [2]. We only considered sequences above 94.5%, which is the minimum threshold sequence identity for confirming a genus affiliation [3]. The nucleotide sequences generated in this study were deposited into the NCBI GenBank database under accession numbers ON479828–ON479853.

3. Results and Discussion

3.1. Mineralogical, Chemical and Petrophysical Characteristics of the Stone

The stone, being massive and granular in appearance, is not very compact, with a high porosity and a medium–low density, and exhibits a thin altered layer or surface crust. It is white to light beige (N9 to 10YR 8/2, Munsell rock color chart) with dark mottling in the altered surface area.

XRD analysis showed a predominant calcite mineral composition (90–95%), as well as dolomite (5–10%). Analysis of the surface of the sample by μXRD also revealed the presence of calcite as the main constituent, and other accessory minerals, such as thenardite ($NaSO_4$) and hematite-type iron oxides (Fe_2O_3).

The stone is a medium- to coarse-grained oolitic/peloidal limestone. According to the most used limestone classifications, it would be classified as oopelsparite [4] and oolitic/peloidal grainstone [5]. It is composed of medium- to coarse-sized (0.15–1 mm), rounded to sub-rounded grains that show poor sorting (Figure 3), developing a grain-supported fabric with point, long and concave–convex contacts between the grains (Figures 3 and 4A).

Highly micritized ooids and peloids (85–95%), ranging from 150 to 750 μm in size, (Figures 3 and 4) are the predominant framework constituents. Echinoderms, molluscs and foraminifera fragments of highly variable size were also observed, reaching 2 mm thick in some cases (Figure 3A–D). The bioclasts showed a good state of conservation in the case of echinoderm plates and are quite micritized in the rest of bioclasts (Figures 3 and 4). Non-carbonate grains are very scarce, mainly consisting of small grains of silicate (Figure 3F) or iron oxide minerals. A very scarce interstitial matrix (micrite) between framework grains was observed. The main cement types recognized were micro- and mesocrystalline calcite cements, showing both continuous circumgranular and occluding (syntaxial, mosaic, poikilotopic) geometries in intergranular positions, corresponding to different cementation phases (Figures 3C–F and 4A). Micro-mesocrystalline cements partially filling intraparticle pores (Figure 3C,D), and microcrystalline cements filling fractures or veins (10–25 μm thick) (Figure 4B) were also observed. Locally, some microcrystalline siliceous cement was recognized in an interparticle position (Figure 3F).

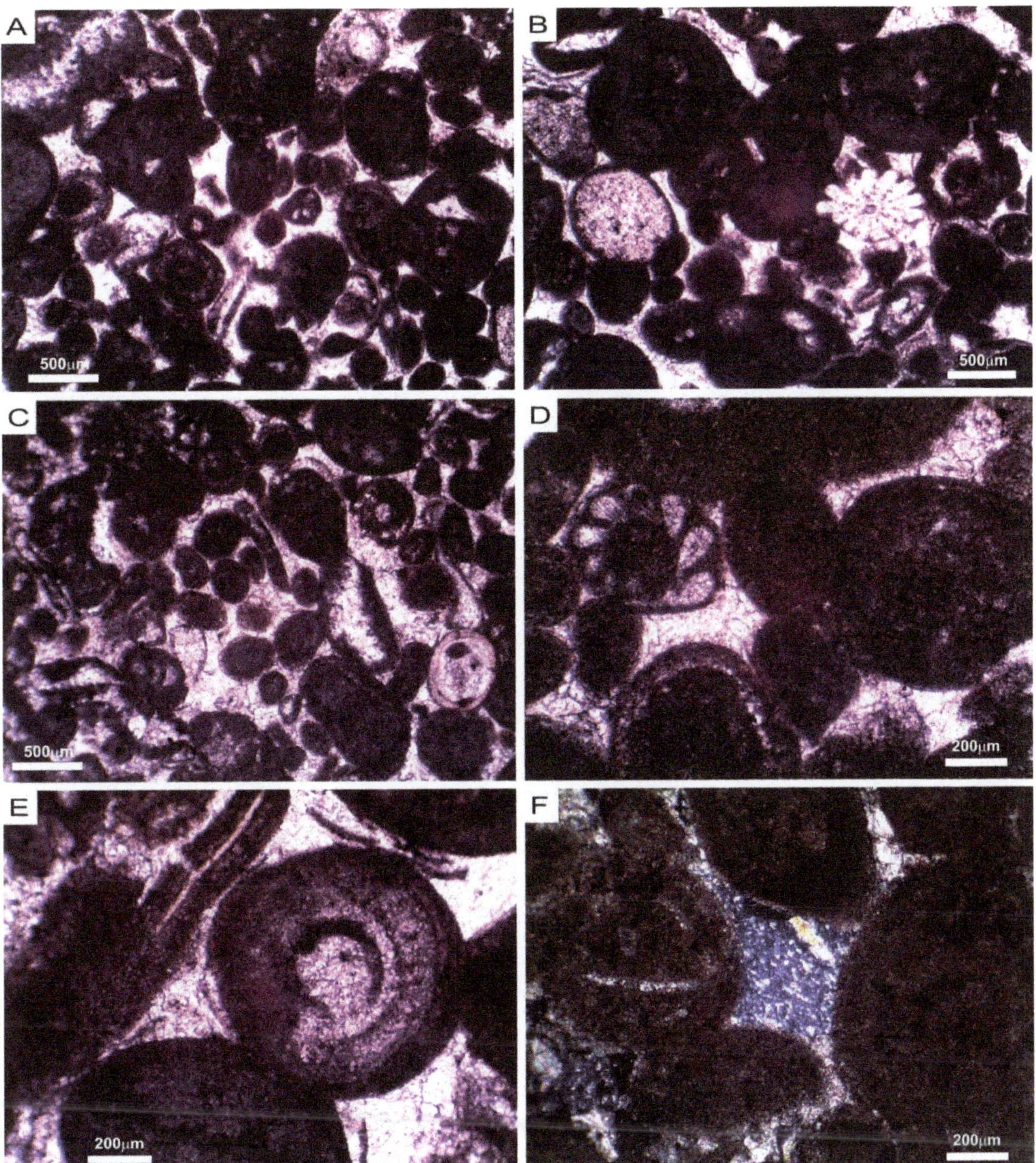

Figure 3. Optical photomicrographs. (**A–C**) General view of the sample showing a grain-supported fabric with framework composed of ooids, peloids and bioclast fragments, all of them bound by intercrystalline calcite cements. (**D,E**) Interparticle and intraparticle calcite cements between grains showing point, long and concave–convex contacts. (**F**) Detail of altered silicate grain and microcrystalline siliceous cement. (**A–E**) Plane polarized light. (**F**) Cross polarized light.

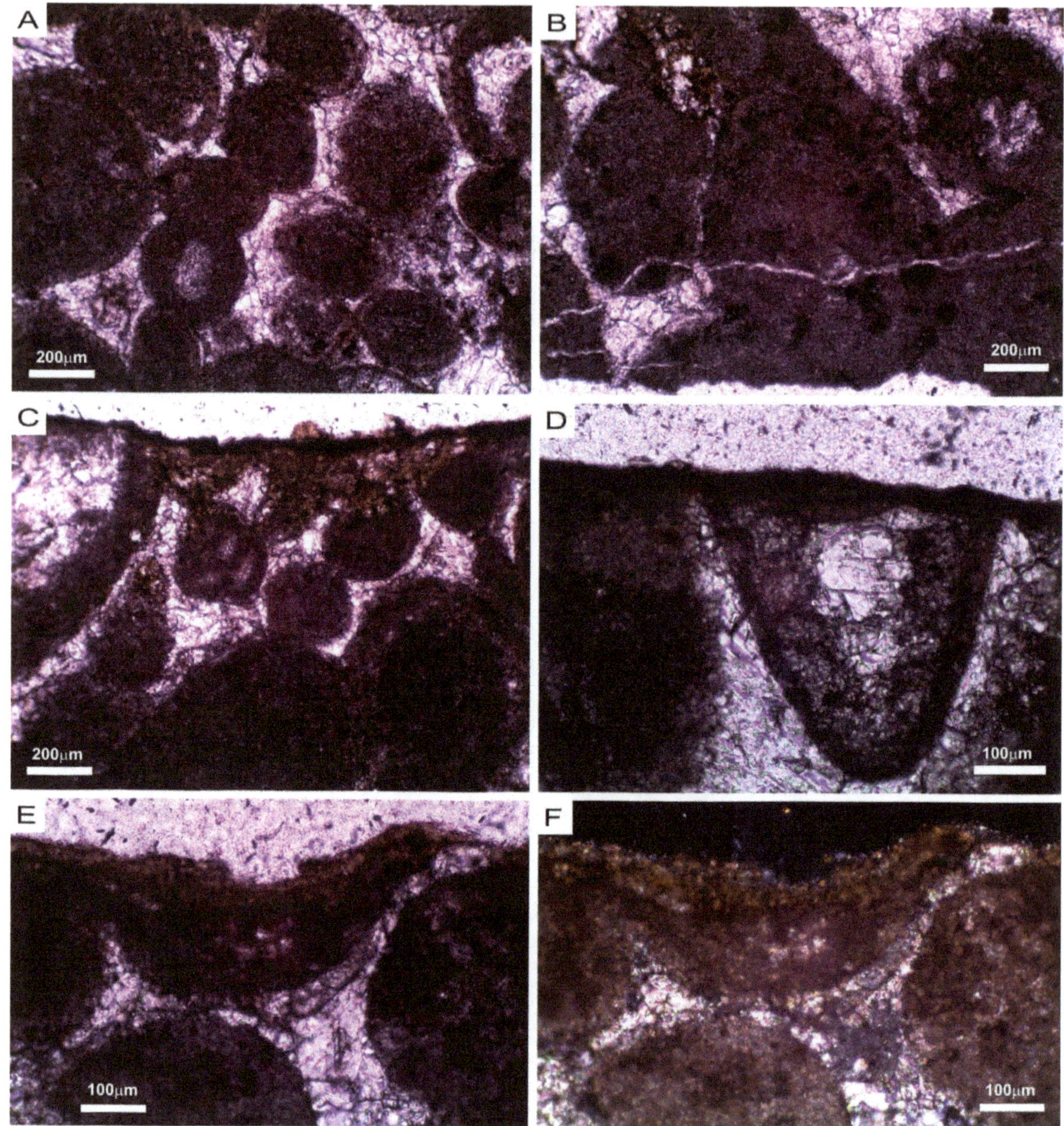

Figure 4. Optical photomicrographs. (**A**) Interparticle and intraparticle calcite cements between grains with point, long and concave–convex contacts. (**B**) Detail of a fracture filled by microcrystalline calcitic cement. (**C**) General view of the surface alteration crust. (**D–F**) Detail of the surface alteration crusts (see text). (**A–E**) Plane polarized light. (**F**) Cross polarized light.

The porosity of the sample is relatively high (5–15%) and fabric selective, mainly with interparticle and intercrystalline pore types. Framework grains (ooids, bioclasts) showed severe micritization processes (Figures 3 and 4). Most of the ooids appeared as rounded micritic grains that locally presented relics of primary microstructures (laminae, banded, internal chambers). The sample showed evidence of mechanical and/or chemical compaction (long and concave–convex contacts) (Figures 3D,E and 4A), as well as fracturing (Figure 4B).

The surface alteration crust showed a variable thickness (20–100 μm) (Figure 4C–F), a compact texture, a predominantly very small crystal/grain size (<25 μm), and a greater abundance of oxides.

3.2. Scanning Electron Microscopy and Microanalysis (SEM-EDS)

The surficial limestone appeared to be covered by a thin, darker layer with gray to black mottling, commonly discontinuous (Figure 5). The components (grains, crystals) that compose the surface layer include iron oxides, salts (sodium and calcium sulfates, sodium chloride) and carbonaceous particles, as well as organic components (biofilm, microorganisms). Salt crystals were also detected in more internal areas, in the porous system (Figure 5D), which contributes to the granular disintegration of the stone [6–9].

Figure 5. SEM micrographs showing the limestone surface of the Cross of the Inquisition. (**A**) General view of limestone surface showing zones with preferential Fe-staining (mottling). (**B**,**C**) Detail of the high porosity (interparticle and intercrystalline pores) at the sample's surface. (**D**) Detail of small idiomorphic halite (NaCl) crystals grown in interparticle porosity.

SEM-EDS analysis showed that the chemical composition of superficial layer is variable and includes (atom.%): O (66.3–64.93), C (11.64–17.44), S (1.38–5.42), Cl (0.16–0.26), Si (1.59–3.13), Al (0.61–0.92), Ca (5.56–8.63), Mg (1.17–1.53), Na (0.23), K (0.22–0.30), Ba (1.15–5.40), Fe (0.42–0.67) and Zn (0.14–0.41).

These data point to the existence of silicates (clays), soot and sulfated minerals (gypsum, thenardite) in addition to the observed halite, which seem to indicate a strong impact due to environmental contamination from vehicle exhaust pipes. The effects of this contamination were described in the Cathedral of Seville, near the City Hall [10–12]. In some stone surfaces, a deep deterioration caused by biological colonization, with abundant pitting and channels produced by organisms, was observed (Figure 6).

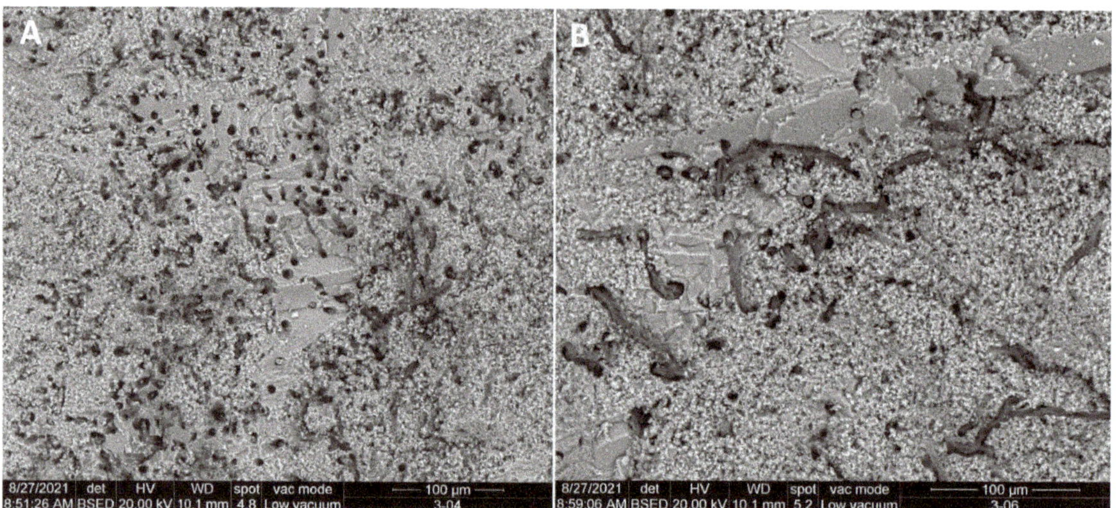

Figure 6. SEM micrographs showing biological weathering features on the limestone surface, Cross of the Inquisition. (**A**) Pits generally show a cylindrical morphology, a diameter between 2 and 10 µm and are not interconnected, although transitions patterns to interconnected pits can also be observed. (**B**) Channels produced by the action of filamentous microbes on the limestone surface.

Indeed, Figure 6A shows the abundance of pitting in the limestone. Pitting is generally produced by cyanobacteria, algae and fungi, and has been widely described as a product of the dissolution of limestone caused by the biological activity of colonizing organisms [13–15]. The existence of pitting, without biological structures inside, as observed in Figure 6A, denotes that the colonization was ancient, and the organisms have disappeared.

In addition to pitting, the dissolution of the limestone is produced by hyphae up to 3 microns in diameter, possibly from fungi, which are shown in Figure 6B, confirming that the colonization, in this case, was more recent.

3.3. *Organisms on the Cross of the Inquisition limestone*

Table 1 shows a summary of the eukaryotic organisms found in the limestone. Table 1 highlights the occurrence of *Trebouxia aggregata* in the black and brown crusts, which clone sequences show over 99% identity. *Trebouxia aggregata* is a unicellular green alga found in any habitat and is the most common photobiont of lichens. The green alga *Trebouxia* is the photobiont of many lichens such as *Lecanora, Caloplaca, Xanthoria, Ramalina, Buellia, Umbilicaria*, etc. [16–19].

The occurrence of *Trebouxia aggregata* suggests that the black crust observed on the limestone was actually lichens. The storage of the Cross fragments for two years prevented the collection of fresh material for an accurate taxonomical identification of the lichen community.

Table 1. Description of samples from the Cross of the Inquisition, Seville City Council, and list of eukaryotic organisms.

Sample	Species	Accession Number	Group *
Brown crust (CD1)	*Trebouxia aggregata*	ON479828	*Chlorophyta*
	Tortula truncata	ON479827	*Bryophyta*
	Pseudostichococcus monallantoides	ON479830	*Chlorophyta*
	Phoma sp.	ON479831	*Fungi*
	Naganishia albida	ON479825	*Fungi*
	Myrmecia sp.	ON479826	*Chlorophyta*
	Alternaria alternata	ON479829	*Fungi*
Black crust (CD3)	*Scleroconidioma sphagnicola*	ON479833	*Fungi*
	Pseudostichococcus monallantoides	ON479834	*Chlorophyta*
	Neofusicoccum parvum	ON479835	*Fungi*
	Lichinella cribellifera	ON479836	*Lichen*
	Aureobasidium pullulans	ON479837	*Fungi*
	Phoma sp.	ON479832	*Fungi*
Black crust (LQ1)	*Trebouxia aggregata*	ON479838	*Chlorophyta*
	Pseudostichococcus monallantoides	ON479839	*Chlorophyta*
	Dothidea berberidis	ON479840	*Fungi*

* Lichens are taxonomically classified by their fungal partner, so all lichens belong to the *Fungi* kingdom. However, in this work we distinguish between fungi and lichens for convenience.

The green alga *Pseudostichococcus monallantoides* appeared in the three crusts (Table 1). This microalga is halotolerant and salt tolerance mechanisms are associated with its resistance to desiccation, a necessary behavior for microalgae in terrestrial environments [20]. *Myrmecia* sp., a green alga in the brown crust, is frequent in desert biological crusts [21], and is a lichen photobiont [22].

Together with the green algae, the bryophyte *Tortula truncata*, synonymy of *Pottia truncata*, was identified in the brown crust. Casas et al. [23] considered that this species is rarely distributed in Spain, although specimens collected in the provinces of Seville and Badajoz are deposited in the Herbarium of the Faculty of Sciences, University of Oviedo.

Tortula species are usually common on the walls of buildings in urban environments [24], where they tend to resist air pollution and periods of desiccation. Isermann [25] found *Tortula truncata, Tortula muralis, Syntrichia ruralis* and many other species of bryophytes in the campus of Bremen University. Ekwealor and Fisher [26] found *Syntrichia caninervis* and *Tortula inermis*, with a high tolerance to desiccation, under the rocks of the Mojave Desert, demonstrating their resistance in hypolithic habitats.

Table 1 comprises the fungi *Phoma* sp., *Naganishia albida, Alternaria alternata, Scleroconidioma sphagnicola, Neofusicoccum parvum, Aureobasidium pullulans*, and *Dothidea berberidis*. Most of these fungi form the group of dematiaceous fungi, which are characterized by their black color due to the synthesis of melanin and by inhabiting rocks in arid environments [27,28]. The high identity values of the clone sequences allow for the identification at the genus and species level.

The genus *Phoma* comprises saprophytic and many other lichenicolous species [29,30]. *Phoma* spp. have been isolated from the lichens *Caloplaca, Cladonia, Ramalina*, etc. [31]. *Phoma* is frequent on rock surfaces in urban environments, with *Phoma glomerata* being the most abundant species.

Naganishia albida, a saprophytic yeast found in the Cross, can be isolated from different environments such as air, soils, bryophytes and plants [32,33]. *Alternaria alternata*, a cosmopolitan species, is one of the most frequent fungi in the air, stones and plants [34,35].

Scleroconidioma sphagnicola is a necrotrophic parasite of bryophytes [36,37]. *Neofusicoccum parvum*, a common and cosmopolitan plant pathogen species, has been isolated from a wide variety of hosts [38,39].

Aureobasidium pullulans is a ubiquitous species of black yeast, saprophytic in plants and frequent in the air of urban environments [40], but also found on monument surfaces [41–43].

Dothidea berberidis is included in the order *Dothideales*, comprising melanized fungi from rocks [44]. *Aureobasidium pullulans* also belong to this order.

Table 1 also comprises *Lichinella cribellifera*, a lichen growing on dry rock surfaces and representative of the xerophilous lichen flora of Southern Spain [45,46].

Rock-inhabiting fungi usually colonize the surfaces, forming a dark patina adhered to the substratum [47–50]. Electron microscope observations demonstrated the presence of pitting, whose shape and size were compatible with the endolithic action of filamentous microorganisms, observed in the Cross limestone (Figure 6A). These fungi are characterized by enduring extreme changes in humidity and long periods of desiccation, which undoubtedly is an advantage in the colonization of limestone in the climate of Seville.

Sterflinger and Prillinger [51] studied the buildings of Vienna, finding a diversity of fungi greater than in the same types of rocks in rural environments. Among the fungi, three species of *Phoma* were isolated: *Phoma exigua* var. *foveata*, *Phoma glomerata*, *Phoma macrostoma* and *Aureobasidium pullulans*, among others. *Aureobasidium pullulans* and *Phoma* sp. were identified in the Cross, which suggests a similar ecology of these species in limestone monuments in urban environments.

To discern the origin of the photo- and mycobionts listed in Table 1 we surveyed the lichens colonizing the City Hall façade, close to the Cross location. Three main lichen species were abundant on the façade, with white, yellow and blackish brown thalli (Figure 7).

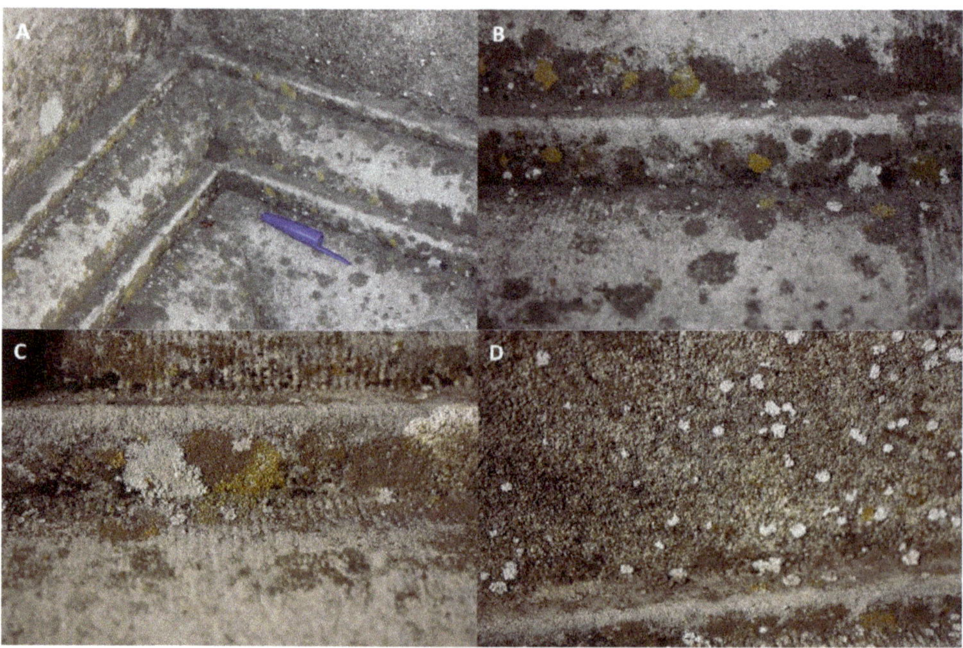

Figure 7. Lichen communities on the City Hall façade, close to the Cross of the Inquisition. (**A**) Community dominated by *Pyrenodesmia variabilis*. (**B**) Same community with *Caloplaca citrina* group species. (**C**) Community of *Pyrenodesmia variabilis*, *Caloplaca citrina* group species and *Kuettlingeria teicholyta*. (**D**) Community with *Kuettlingeria teicholyta*.

The white lichen (Figure 7D) was identified as being *Kuettlingeria teicholyta* (Ach.) Trevis. (=*Caloplaca teicholyta* (Ach.) J. Steiner). The yellow lichen (Figure 7B,C) was included in the *Caloplaca citrina* group [52]. This species complex, which still needs a thorough revision in Spain, occurs in a wide variety of substrata, from asbestos–cement, concrete and mortar, to basic siliceous rocks or even wood, and is very tolerant to, and even favored by eutrophication [53]. The blackish brown lichen (Figure 7A,B) was affiliated with *Pyrenodesmia variabilis* (Pers.) A. Massal. (=*Caloplaca variabilis* (Pers.) Th. Fr.). Taxonomical reviews on the family *Telochistaceae* transferred species of the genus *Caloplaca* to the new genera *Kuettlingeria* and *Pyrenodesmia* [54,55]. Muggia et al. [56] revised the black endolithic *Caloplaca* and reported that almost all the species shared *Trebouxia* as a photobiont. *Caloplaca* is a cosmopolitan lichen genus found in most xeric and mesic habitats [57].

The clones retrieved from lichen samples are shown in Table 2. The three lichens showed *Trebouxia aggregata* as the photobiont, which confirmed that the finding of this green alga in the Cross fragments was associated to the occurrence of a lichenic crust. Muggia et al. [56] stated that the photobiont of *Caloplaca variabilis* and other closely related *Caloplaca* spp. (e.g., *Caloplaca chalybaea*) was an unknown taxon of *Trebouxia*, typical of the Mediterranean area. These three lichens (among others) identified in the City Hall were also found on natural limestone outcrops, buildings and monuments [58–62].

Table 2. Identification of eukaryotic organisms associated with the three main lichens colonizing the limestones from Seville City Council.

Lichen	Associated Species	Accession Number	Group *
Kuettlingeria teicholyta	*Trebouxia aggregata*	ON479841	*Chlorophyta*
	Pyrenodesmia chalybaea	ON479847	*Lichen*
	Xanthoria parietina	ON479846	*Lichen*
	Xanthoria sp.	ON479845	*Lichen*
	Phoma sp.	ON479844	*Fungi*
	Ceratobasidium sp.	ON479843	*Fungi*
	Pyrenidium cf. *actinellum*	ON479842	*Fungi*
	Chaetothyriales sp.	ON479848	*Fungi*
	Rhinocladiella sp.	ON479849	*Fungi*
Caloplaca citrina group	*Trebouxia aggregata*	ON479852	*Chlorophyta*
	Xanthoria parietina	ON479850	*Lichen*
	Xanthoria sp.	ON479851	*Lichen*
Pyrenodesmia variabilis	*Trebouxia aggregata*	ON479854	*Chlorophyta*
	Capnobotriella sp.	ON479853	*Fungi*

* Lichens are taxonomically classified by their fungal partner, so all lichens belong to the *Fungi* kingdom. However, in this work we distinguish between fungi and lichens for convenience.

Clones of other lichens (*Pyrenodesmia chalybaea, Xanthoria parietina* and *Xanthoria* sp.) were present in the sample of *Kuettlingeria teicholyta*, and in the *Caloplaca citrina* group lichen were identified as being *Xanthoria parietina* and *Xanthoria* sp.

In addition, a few fungi were retrieved from *Kuettlingeria teicholyta* (*Pyrenidium* cf. *actinellum*, *Phoma* sp., *Rhinocladiella* sp., *Ceratobasidium* sp., and *Chaetothyriales* sp.). Most of these fungi are lichenicolous from arid habitats (e.g., *Phoma*, *Pyrenidium*, *Rhinocladiella*), as well as *Capnobotriella* sp., associated with *Pyrenodesmia variabilis* [49–51,63,64].

Table 2 shows that a complex crustose lichen community colonizes the building façade of the Seville City Hall. This community is the result of the biological regeneration after the restoration of the Cross and façade, carried out between 2008 and 2010. The protocol consisted in mechanical cleaning of the limestone surfaces with water and natural soap and the joints were filled with lime mortar [65].

Previous studies on the efficacy of mechanical procedures and laser cleaning for removal of crustose lichens from granite showed that these methods were unable to eliminate the thalli into fissures [66,67]. In our case, Figures 5 and 6 showed that different organisms penetrate the limestone through pitting and channels. Likely, the 2008–2010 cleaning did not affect the endolithic hyphae, which grow from the crustose lichen thalli and penetrate the porous limestone. These endolithic hyphae facilitate the subsequent recovery of the lichen community, in the same way that the roots in the soil allow the regeneration of the aerial part of the plants after harvesting.

Favero-Longo and Viles [68] recognized that rock-dwelling organisms are agents of the deterioration of cultural heritage stone surfaces, but their removal is usually followed by a rapid recolonization. Ariño et al. [59] reported that epilithic lichens can provide a protective cover for stones because biodeterioration is a slower process than physical and chemical weathering produced by environmental factors. This fact was confirmed in a study on colonized and uncolonized flagstones in an archaeological site. Concha-Lozano et al. [69] found that limestone secondary porosity was filled by lichen hyphae, and the thalli gave a waterproofing effect that conferred preservation to ancient monuments. Pinna [70] stated that a lichen crust protects porous stones from weathering by stabilizing the surfaces, and although the removal of lichens from statues and monuments is widely practiced, it can damage the stone due to the intimate association of biological structures with the stone components.

Salvadori and Municchia [71] in a review on the role of fungi and lichens on monuments stated that biodeterioration and bioprotection are not mutually exclusive and can occur simultaneously in a lichen community.

Indeed, Nascimbene et al. [72] studied the re-colonization of a limestone statue by lichens 12 years after cleaning and restoration. The lichen flora was dominated by nitrophilic species of *Caloplaca* and *Verrucaria*, which gave an orange–grey–black color to the community. The restoration protocol of the statue included the application of a biocide (Metatin N-58-10/101), mechanical removal of the biomass, a new application onto the cleaned surfaces and consolidation with Akeogard CO. The authors reported that the long-term effectiveness of the restoration was low because the total number of species (25) before restoration was similar to that found years after (20), indicating a progressive colonization along the time. This colonization was favored by the location of the statue surrounded by trees in a park [73]. In the case of the restored Cross of the Inquisition, located in an urban environment with scarce trees in the surroundings, it was expected that recolonization by ascospores from epiphytic lichens will take place in the very long-term.

A few reports suggested that the application of biocides, water-repellent products and consolidants was an effective treatment for preventing biological growth on stones [74–78], as performed in this restoration. However, due to the short distance of the lichen community colonizing the façade of the City Hall (Figure 7) to the Cross of Inquisition, the dispersal and settlement of ascospores on the restored limestone cannot be ruled out.

3.4. *Restoration of the Cross of the Inquisition*

The restoration of the Cross was carried out by Atelier Samthiago, Conservaçao e Restauro. The result of the final restoration is shown in Figure 1C. The main steps included:

Photographic survey: mapping of pathologies, registration of existing pathologies, three-dimensional record.

Dry cleaning of superficial dirtiness: removal of superficial dirt deposited on the monument's surfaces, using soft brushes and spatulas.

Biocide treatment: Preventol RI80 was applied with a soft paintbrush and by spraying on the entire area including the pedestal.

Physico-chemical cleaning of surfaces: after biocide treatment, a cleaning with nylon brushes and water was applied. For the homogenization of the surface layer with chromatic alteration, a poultice soaked with an aqueous solution of ammonium bicarbonate, EDTA and drops of neutral Teepol detergent was applied. In the cruise and the pedestal, poultices with different products and aqueous solutions were applied.

Consolidation of the stone surface: to reinforce the inter-granular cohesion of the stone, which was very weakened in some areas, all the fragments and the pedestal of the cross were subjected to a general consolidation with Estel 1000, applied with a brush. The consolidation was reinforced in areas of cracks, applying the product with a syringe.

Elimination and replacement of corroded internal metallic structures: removal of resin remains from previous restorations and metal elements that lost their structural function.

Volumetric reconstitution: adhesion of the fragments allowing volumetric reconstitution and reintegration of structural gaps with Sikadur 31EF epoxy resin adhesive. In some joints, the adhesion was reinforced with small fiberglass or stainless steel spigots.

Coating and micro-coating for the reintegration of surface discontinuities: application of lime putty with chromatic correction for coating and micro-coating of surface gaps, fissures and joints. Lime putty was used in a 1:1 ratio between inert hydraulic lime binder and lime Glue Bio Pasta Gordillos, adding natural pigments.

Opening, cleaning and sealing of joints, with volumetric correction: the cement filling the joints was mechanically removed using a chisel, then cleaned with nylon brushes and water in a controlled mode. The joints were cleaned and reintegrated with the same lime putty.

Assembling the cross on the cruise: fixation of the cross on the cruise with epoxy resin (Sikadur 31EF) by means of a metallic spigot treated and reinforced with two more fiberglass spikes of 8 mm diameter and about 90 mm long, distributed between the two elements.

Final protection: the water-repellent and anti-graffiti product Aguasil ST was applied with a brush to the entire monument surface, paying special attention to the carving, the cruise and the pedestal.

4. Conclusions

The data obtained are consistent with the exposure of the Cross of the Inquisition to an urban environment, where it has remained for more than 100 years. During that time, a lichen community colonized the limestone, associated with bacteria, fungi and bryophytes. Evidences of ancient (pitting) and recent (fungal hyphae) colonizations have been found. The occurrence of lichens, as denoted also by the green algae photobiont, lichenicolous and black fungi, and other microorganisms in the limestone have composed a mature and well-developed urban community on the limestone over the years, even after the 2008–2010 restoration, due to its regeneration capacity. Likewise, the species found in the Cross limestone were also identified in other urban environments and monuments, located in the Mediterranean Basin.

In arid and semi-arid environments of the Mediterranean Basin the climatic conditions are too extreme for most fungi, so the communities move towards the so-called black yeasts and other rock-inhabiting fungi, which form small black colonies on the surface and within the limestone and often occur in close association with lichens.

The microbial activity has been detrimental for the integrity of the limestone. The activity of rhizines (hyphae that keep the lichen fixed to the substratum), the rhizoids of bryophytes, and endolithic microorganisms has produced channels and pitting, which have evidenced the lytic activity of microorganisms on the stone surface. Therefore, stone consolidation was achieved with Estel 1000. In addition, a biocide (Preventol RI80), able to penetrate the porous limestone and active against bacteria, fungi, lichens and bryophytes, was used.

Author Contributions: Conceptualization, C.S.-J.; methodology, C.S.-J. and C.C.; investigation, V.J.; J.L.G.-P., A.G.-B., J.C.C. and S.S.-M.; restoration, C.C.; writing—original draft preparation, C.S.-J. and J.C.C.; writing—review and editing, C.S.-J. All authors have read and agreed to the published version of the manuscript.

Funding: The research was funded by Atelier Samthiago and the restoration of the Cross of the Inquisition was supported by the Seville City Hall.

Institutional Review Board Statement: Not applicable.

Informed Consent Statement: Not applicable.

Data Availability Statement: The nucleotide sequences were deposited into the NCBI GenBank database under accession numbers ON479828–ON479853.

Acknowledgments: The authors wish to acknowledge the professional support of the CSIC Interdisciplinary Thematic Platform Open Heritage: Research and Society (PTI-PAIS).

Conflicts of Interest: The authors declare no conflict of interest.

References

1. Lindholm, R.C.V.; Finkelman, R.B. Calcite staining: Semiquantitative determination of ferrous iron. *J. Sediments Res.* **1972**, *42*, 239–242. [CrossRef]
2. Altschul, S.F.; Gish, W.; Miller, W.; Myers, E.W.; Lipman, D.J. Basic local alignment search tool. *J. Mol. Biol.* **1990**, *215*, 403–410. [CrossRef]
3. Yarza, P.; Yilmaz, P.; Pruesse, E.; Glöckner, F.O.; Ludwig, W.; Schleifer, K.-H.; Whitman, W.B.; Euzéby, J.; Amann, R.; Rosselló-Móra, R. Uniting the classification of cultured and uncultured bacteria and archaea using 16S rRNA gene sequences. *Nat. Rev.* **2014**, *12*, 635–645. [CrossRef] [PubMed]
4. Folk, R.L. Spectral subdivision of limestone types. In *Classification of Carbonates Rocks—A Syposium*; Ham, W.E., Ed.; American Association of Petroleum Geologist: Tulsa, OK, USA, 1962; pp. 62–84.
5. Dunham, R.J. Classification of carbonate rocks according to depositional texture. In *Classification of Carbonates Rocks—A Syposium*; Ham, W.E., Ed.; American Association of Petroleum Geologist: Tulsa, OK, USA, 1962; pp. 108–121.
6. Rodriguez-Navarro, C.; Doehne, E. Salt weathering: Influence of evaporation rate, supersaturation and crystallization pattern. *Earth Surf. Process. Landf. J. Br. Geomorphol. Res. Group* **1999**, *24*, 191–209. [CrossRef]
7. Sabbioni, C. Mechanisms of air pollution damage of stone. In *The Effects of Air Pollution on the Built Environment*; Brimblecombe, P., Ed.; Air Pollution Reviews; Imperial College Press: London, UK, 2003; Volume 2, pp. 63–106.
8. Benavente, D.; Cueto, N.; Martinez-Martinez, J.; Garcia-del-Cura, M.A.; Cañaveras, J.C. The influence of petrophysical properties on the salt weathering of porous building rocks. *Environ. Geol.* **2007**, *52*, 197–206. [CrossRef]
9. Benavente, D.; Sanchez-Moral, S.; Fernandez-Cortes, A.; Cañaveras, J.C.; Elez, J.; Saiz-Jimenez, C. Salt damage and microclimate in the Postumius Tomb, Roman Necropolis of Carmona, Spain. *Environ. Earth Sci.* **2011**, *63*, 1529–1543. [CrossRef]
10. Cachier, H.; Sarda-Esteve, R.; Oikonomou, K.; Sciare, J.; Bonazza, A.; Sabbioni, C.; Greco, M.; Reyes, J.; Hermosin, B.; Saiz-Jimenez, C. Aerosol characterization and sources in different European atmospheres: Paris, Seville, Florence and Milan. In *Air Pollution and Cultural Heritage*; Saiz-Jimenez, C., Ed.; Balkema: Leiden, The Netherlands, 2004; pp. 3–14.
11. Reyes, J.; Hermosin, B.; Saiz-Jimenez, C. Organic analysis of aerosols in Seville atmosphere. In *Air Pollution and Cultural Heritage*; Saiz-Jimenez, C., Ed.; Balkema: Leiden, The Netherlands, 2004; pp. 15–20.
12. Reyes, J.; Hermosin, B.; Saiz-Jimenez, C. Organic composition of Seville aerosols. *Org. Geochem.* **2006**, *37*, 2019–2025. [CrossRef]
13. Caneva, G.; Nugari, M.P.; Ricci, S.; Salvadori, O. Pitting of marble Roman monuments and the related microflora. In Proceedings of the 7th International Congress on Deterioration and Conservation of Stone, Lisbon, Portugal, 15–18 June 1992; Delgado Rodrigues, J., Henriques, F., Telmo Jeremias, F., Eds.; NLCE: Lisbon, Portugal, 1992; pp. 521–529.
14. Sterflinger, K.; Krumbein, W.E. Dematiaceous fungi as a major agent of biopitting for Mediterranean marbles and limestones. *Geomicrobiol. J.* **1997**, *14*, 219–230. [CrossRef]
15. De Leo, F.; Antonelli, F.; Pietrini, A.M.; Ricci, S.; Urzì, C. Study of the euendolithic activity of black meristematic fungi isolated from a marble statue in the Quirinale Palace's Gardens in Rome, Italy. *Facies* **2019**, *65*, 18. [CrossRef]
16. Ahmadjian, V. *The Lichen Symbiosis*; John Wiley: New York, NY, USA, 1993.
17. Cordeiro, L.M.C.; Reis, R.A.; Cruz, L.M.; Stocker-Wörgotter, E.; Grube, M.; Iacomini, M. Molecular studies of photobionts of selected lichens from the coastal vegetation of Brazil. *FEMS Microbiol. Ecol.* **2005**, *54*, 381–390. [CrossRef]
18. Petrzik, K.; Vondrák, J.; Kvíderová, J.; Lukavský, J. Platinum anniversary: Virus and lichen alga together more than 70 years. *PLoS ONE* **2015**, *10*, e0120768. [CrossRef] [PubMed]
19. Sanders, W.B.; Masumoto, H. Lichen algae: The photosynthetic partners in lichen symbioses. *Lichenologist* **2021**, *53*, 347–393. [CrossRef]
20. Sommer, V.; Mikhailyuk, T.; Glaser, K.; Karsten, U. Uncovering unique green algae and cyanobacteria isolated from biocrusts in highly saline potash tailing pile habitats, using an integrative approach. *Microorganisms* **2020**, *8*, 1667. [CrossRef]
21. Lewis, L.A.; Flechtner, V.R. Green algae (Chlorophyta) of desert microbiotic crusts: Diversity of North American taxa. *Taxon* **2002**, *51*, 443–451. [CrossRef]
22. Voytsekhovich, A.; Beck, A. Lichen photobionts of the rocky outcrops of Karadag massif (Crimean Peninsula). *Symbiosis* **2016**, *68*, 9–24. [CrossRef]
23. Casas, C.; Brugués, M.; Cros, R.M.; Sérgio, C. *Handbook of Mosses of the Iberian Peninsula and the Balearic Islands: Illustrated Keys to Genera and Species*; Institut d'Estudis Catalans: Barcelona, Spain, 2020.
24. Kosior, G.; Prell, M.; Samecka-Cymerman, A.; Stankiewicz, A.; Kolon, K.; Kryza, R.; Brudzinska-Kosior, A.; Frontasyeva, M.; Kempers, A.J. Metals in *Tortula muralis* from sandstone buildings in an urban agglomeration. *Ecol. Indic.* **2015**, *58*, 122–131. [CrossRef]
25. Isermann, M. Diversity of bryophytes in an urban area of NW Germany. *Lindbergia* **2007**, *32*, 75–81.
26. Ekwealor, J.T.B.; Fisher, K.M. Life under quartz: Hypolithic mosses in the Mojave Desert. *PLoS ONE* **2020**, *15*, e0235928. [CrossRef]
27. Gorbushina, A.A. Life on the rocks. *Environ. Microbiol.* **2007**, *9*, 1613–1631. [CrossRef]
28. Sterflinger, K. Fungi: Their role in deterioration of cultural heritage. *Fungal Biol. Rev.* **2010**, *24*, 47–55. [CrossRef]
29. Diederich, P.; Kocourkova, J.; Etayo, J.; Zhurbenko, M. The lichenicolous *Phoma* species (coelomycetes) on *Cladonia*. *Lichenologist* **2007**, *39*, 153–163. [CrossRef]

30. Aveskamp, M.M.; de Gruyter, J.; Woudenberg, J.H.C.; Verkley, G.J.M.; Crous, P.W. Highlights of the *Didymellaceae*: A polyphasic approach to characterise *Phoma* and related pleosporalean genera. *Stud. Mycol.* **2010**, *65*, 1–60. [CrossRef] [PubMed]
31. Lawrey, J.D.; Diederich, P.; Nelsen, M.P.; Freebury, C.; Van den Broeck, D.; Sikaroodi, M.; Ertz, D. Phylogenetic placement of lichenicolous *Phoma* species in the Phaeosphaeriaceae (Pleosporales, Dothideomycetes). *Fungal Divers.* **2012**, *55*, 195–213. [CrossRef]
32. Tosi, S.; Casado, B.; Gerdol, R.; Caretta, G. Fungi isolated from Antarctic mosses. *Polar Biol.* **2002**, *25*, 262–268. [CrossRef]
33. Buzzini, P.; Lachance, M.-A.; Yurkov, A. *Yeasts in Natural Ecosystems: Diversity*; Springer: Cham, Switzerland, 2017.
34. Woudenberg, J.H.C.; Seidl, M.F.; Groenewald, J.Z.; de Vries, M.; Stielow, J.B.; Thomma, B.P.H.J.; Crous, P.W. *Alternaria* section *Alternaria*: Species, *formae speciales* or pathotypes? *Stud. Mycol.* **2015**, *82*, 1–21. [CrossRef] [PubMed]
35. Jurado, V.; Gonzalez-Pimentel, J.L.; Hermosin, B.; Saiz-Jimenez, C. Biodeterioration of Salón de Reinos, Museo Nacional del Prado, Madrid, Spain. *Appl. Sci.* **2021**, *11*, 8858. [CrossRef]
36. Tsuneda, A.; Thormann, M.N.; Currah, R.S. Characteristics of a disease of *Sphagnum fuscum* caused by *Scleroconidioma sphagnicola*. *Can. J. Bot.* **2001**, *78*, 1294–1298.
37. Davey, M.L.; Currah, R.S. Interactions between mosses (Bryophyta) and fungi. *Can. J. Bot.* **2006**, *84*, 1509–1519. [CrossRef]
38. Phillips, A.J.L.; Alves, A.; Abdollahzadeh, J.; Slippers, B.; Wingfield, M.J.; Groenewald, J.Z.; Crous, P.W. The *Botryosphaeriaceae*: Genera and species known from culture. *Stud. Mycol.* **2013**, *76*, 51–167. [CrossRef]
39. Dissanayake, A.J.; Phillips, A.J.L.; Li, X.H.; Hyde, K.D. *Botryosphaeriaceae*: Current status of genera and species. *Mycosphere* **2016**, *7*, 1001–1073. [CrossRef]
40. Lugauskas, A.; Sveistyte, L.; Ulevicius, V. Concentration and species diversity of airborne fungi near busy streets in Lithuanian urban areas. *Ann. Agric. Environ. Med.* **2003**, *10*, 233–239. [PubMed]
41. Urzì, C.; De Leo, F.; Lo Passo, C.; Criseo, G. Intra-specific diversity of *Aureobasidium pullulans* strains isolated from rocks and other habitats assessed by physiological methods and by random amplified polymorphic DNA (RAPD). *J. Microbiol. Methods* **1999**, *36*, 95–105. [CrossRef]
42. Simonovicová, A.; Gödyová, M.; Sevc, J. Airborne and soil microfungi as contaminants of stone in a hypogean cemetery. *Int. Biodeter. Biodegr.* **2004**, *54*, 7–11. [CrossRef]
43. Mansour, M. Effects of the halophilic fungi *Cladosporium sphaerospermum*, *Wallemia sebi*, *Aureobasidium pullulans* and *Aspergillus nidulans* on halite formed on sandstone surface. *Int. Biodeter. Biodegr.* **2017**, *117*, 289–298. [CrossRef]
44. Ruibal, C.; Platas, G.; Bills, G.F. High diversity and morphological convergence among melanised fungi from rock formations in the Central Mountain System of Spain. *Persoonia* **2008**, *21*, 93–110. [CrossRef]
45. Moreno, P.P.; Egea, J.M. El género *Lichinella* Nyl. en el sureste de España y norte de Africa. *Cryptogam. Bryol. Lichenol.* **1992**, *13*, 237–259.
46. Egea, J.M.; Alonso, F.L. Patrones de distribución en la flora liquénica xerófila del sureste de España. *Acta Bot. Malacit.* **1996**, *21*, 35–47. [CrossRef]
47. Liu, B.; Fu, R.; Wu, B.; Liu, X.; Xiang, M. Rock inhabiting fungi: Terminology, diversity, evolution and adaptation mechanisms. *Mycology* **2022**, *13*, 1–31. [CrossRef]
48. Harutyunyan, S.; Muggia, L.; Grube, M. Black fungi in lichens from seasonally arid habitats. *Stud. Mycol.* **2008**, *61*, 83–90. [CrossRef]
49. Gorbushina, A.A.; Beck, A.; Schulte, A. Microcolonial rock inhabiting fungi and lichen photobionts: Evidence for mutualistic interactions. *Mycol. Res.* **2005**, *109*, 1288–1296. [CrossRef]
50. Zakharova, K.; Tesei, D.; Marzban, G.; Dijksterhuis, J.; Wyatt, T.; Sterflinger, K. Microcolonial fungi on rocks: A life in constant drought? *Mycopathologia* **2013**, *175*, 537–547. [CrossRef] [PubMed]
51. Sterflinger, K.; Prillinger, H. Molecular taxonomy and biodiversity of rock fungal communities in an urban environment (Vienna, Austria). *Antonie van Leeuwenhoek* **2001**, *80*, 275–286. [CrossRef] [PubMed]
52. Vondrák, J.; Říha, P.; Arup, U.; Søchting, U. The taxonomy of the *Caloplaca citrina* group (*Teloschistaceae*) in the Black Sea region; with contributions to the cryptic species concept in lichenology. *Lichenologist* **2009**, *41*, 571–604. [CrossRef]
53. Nimis, P.L. *ITALIC—The Information System on Italian Lichens*, Version 5.0; Department of Biology, University of Trieste: Trieste, Italy, 2016. Available online: https://italic.units.it(accessed on 16 February 2022).
54. Arup, U.; Sochting, U.; Frödén, P. A new taxonomy of the family Teloschistaceae. *Nord. J. Bot.* **2013**, *31*, 16–83. [CrossRef]
55. Frolov, I.; Vondrák, J.; Kosnar, J.; Arup, U. Phylogenetic relationships within *Pyrenodesmia* sensu lato and the role of pigments in its taxonomic interpretation. *J. Syst. Evol.* **2021**, *59*, 454–474. [CrossRef]
56. Muggia, L.; Grube, M.; Tretiach, M. A combined molecular and morphological approach to species delimitation in black-fruited, endolithic *Caloplaca*: High genetic and low morphological diversity. *Mycol. Res.* **2008**, *112*, 36–49. [CrossRef]
57. Gaya, E.; Navarro-Rosinés, P.; Llimona, X.; Hladun, N.; Lutzoni, F. Phylogenetic reassessment of the *Teloschistaceae* (lichen-forming *Ascomycota*, *Lecanoromycetes*). *Mycol. Res.* **2008**, *112*, 528–546. [CrossRef]
58. Barreno, E.; Merino, A. Catálogo liquénico de las calizas de Madrid (España). *Lazaroa* **1981**, *3*, 247–268.
59. Ariño, X.; Ortega-Calvo, J.J.; Gomez-Bolea, A.; Saiz-Jimenez, C. Lichen colonization of the Roman pavement of Baelo Claudia (Cádiz, Spain): Biodeterioration vs. bioprotection. *Sci. Total Environ.* **1995**, *167*, 353–363. [CrossRef]
60. Marcos Laso, B. Biodiversidad y colonización liquénica de algunos monumentos en la ciudad de Salamanca (España). *Bot. Complut.* **2001**, *25*, 93–102.

61. Wilk, K. Calcicolous species of the genus *Caloplaca* in the Polish Western Carpathians. *Pol. Bot. Stud.* **2012**, *29*, 1–91.
62. Caneva, G.; Fidanza, M.R.; Tonon, C.; Favero-Longo, S.E. Biodeterioration patterns and their interpretation for potential applications to stone conservation: A hypothesis from allelopathic inhibitory effects of lichens on the Caestia Pyramid (Rome). *Sustainability* **2020**, *12*, 1132. [CrossRef]
63. Hawksworth, D.L. The identity of *Pyrenidium actinellum* Nyl. *Trans. Br. Mycol. Soc.* **1983**, *80*, 547–549. [CrossRef]
64. Etayo, J.; López de Silanes, M.E. Hongos liquenícolas del norte de Portugal, especialmente del Parque Natural Montesinho. *Nova Acta Cient. Compost. Biol.* **2020**, *27*, 35–50.
65. Robador, M.D.; Arroyo, F.; Perez-Rodriguez, J.L. Study and restoration of the Seville City Hall façade. *Constr. Build. Mater.* **2014**, *53*, 370–380. [CrossRef]
66. Rivas, T.; Pozo-Antonio, J.S.; López de Silanes, M.E.; Ramil, A.; López, A.J. Laser versus scalpel cleaning of crustose lichens on granite. *Appl. Surf. Sci.* **2018**, *440*, 467–476. [CrossRef]
67. Pozo-Antonio, J.S.; Rivas, T.; López de Silanes, M.E.; Ramil, A.; López, A.J. Dual combination of cleaning methods (scalpel, biocide, laser) to enhance lichen removal from granite. *Int. Biodeterior. Biodegrad.* **2022**, *168*, 105373. [CrossRef]
68. Favero-Longo, S.E.; Viles, H.A. A review of the nature, role and control of lithobionts on stone cultural heritage: Weighing-up and managing biodeterioration and bioprotection. *World J. Microbiol. Biotechnol.* **2020**, *36*, 100. [CrossRef]
69. Concha-Lozano, N.; Gaudon, P.; Pages, J.; de Billerbeck, G.; Lafon, D.; Eterradossi, O. Protective effect of endolithic fungal hyphae on oolitic limestone buildings. *J. Cult. Herit.* **2012**, *13*, 120–127. [CrossRef]
70. Pinna, D. Biofilms and lichens on stone monuments: Do they damage or protect? *Front. Microbiol.* **2014**, *5*, 133. [CrossRef]
71. Salvatori, O.; Municchia, A.C. The role of fungi and lichens in the biodeterioration of stone monuments. *Open Conf. Proc. J.* **2016**, *7*, 39–54. [CrossRef]
72. Nascimbene, J.; Salvadori, O.; Nimis, P.L. Monitoring lichen recolonization on a restored calcareous statue. *Sci. Total Environ.* **2009**, *407*, 2420–2426. [CrossRef] [PubMed]
73. Nascimbene, J.; Salvadori, O. Lichen recolonization on restored calcareous statues of three Venetian villas. *Int. Biodeterior. Biodegrad.* **2008**, *62*, 313–318. [CrossRef]
74. Ariño, X.; Canals, A.; Gomez-Bolea, A.; Saiz-Jimenez, C. Assessment of the performance of a water-repellent/biocide treatment after 8 years. In *Protection and Conservation of the Cultural Heritage of the Mediterranean Cities*; Galán, E., Zezza, F., Eds.; Balkema: Lisse, The Netherlands, 2002; pp. 121–125.
75. Nugari, M.P.; Salvadori, O. Biocides and treatment of stone: Limitations and future prospects. In *Art, Biology and Conservation: Biodeterioration of Works of Art*; Koestler, R.J., Koestler, V.H., Charola, A.E., Nieto-Fernandez, F.E., Eds.; Metropolitan Museum of Art: New York, NY, USA, 2003; pp. 519–535.
76. Urzì, C.; De Leo, F. Evaluation of the efficiency of water-repellent and biocide compounds against microbial colonization of mortars. *Int. Biodeterior. Biodegrad.* **2007**, *60*, 25–34. [CrossRef]
77. Pinna, D.; Salvadori, B.; Galeotti, M. Monitoring the performance of innovative and traditional biocides mixed with consolidants and water-repellents for the prevention of biological growth on stone. *Sci. Total Environ.* **2012**, *423*, 132–141. [CrossRef]
78. Pinna, D.; Galeotti, M.; Perito, B.; Daly, G.; Salvadori, B. In situ long-term monitoring of recolonization by fungi and lichens after innovative and traditional conservative treatments of archaeological stones in Fiesole (Italy). *Int. Biodeterior. Biodegrad.* **2018**, *132*, 49–58. [CrossRef]

Article

Spectroscopic and Microscopic Characterization of Flashed Glasses from Stained Glass Windows

Teresa Palomar [1,*], Marina Martínez-Weinbaum [2], Mario Aparicio [1], Laura Maestro-Guijarro [2], Marta Castillejo [2] and Mohamed Oujja [2,*]

1 Instituto de Cerámica y Vidrio (ICV), CSIC, 28049 Madrid, Spain; maparicio@icv.csic.es
2 Instituto de Quimica Fisica Rocasolano (IQFR), CSIC, 28006 Madrid, Spain; mgmartinez@iqfr.csic.es (M.M.-W.); lmaestro@iqfr.csic.es (L.M.-G.); marta.castillejo@iqfr.csic.es (M.C.)
* Correspondence: t.palomar@csic.es (T.P.); m.oujja@iqfr.csic.es (M.O.)

Abstract: Flashed glasses are composed of a base glass and a thin colored layer and have been used since medieval times in stained glass windows. Their study can be challenging because of their complex composition and multilayer structure. In the present work, a set of optical and spectroscopic techniques have been used for the characterization of a representative set of flashed glasses commonly used in the manufacture of stained glass windows. The structural and chemical composition of the pieces were investigated by optical microscopy, field emission scanning electron microscopy-energy dispersive X-ray spectrometry (FESEM-EDS), UV-Vis-IR spectroscopy, laser-induced breakdown spectroscopy (LIBS), and laser-induced fluorescence (LIF). Optical microscopy and FESEM-EDS allowed the determination of the thicknesses of the colored layers, while LIBS, EDS, UV-Vis-IR, and LIF spectroscopies served for elemental, molecular, and chromophores characterization of the base glasses and colored layers. Results obtained using the micro-invasive LIBS technique were compared with those retrieved by the cross-sectional technique FESEM-EDS, which requires sample taking, and showed significant consistency and agreement. In addition, LIBS results revealed the presence of additional elements in the composition of flashed glasses that could not be detected by FESEM-EDS. The combination of UV-Vis-IR and LIF results allowed precise chemical identification of chromophores responsible for the flashed glass coloration.

Keywords: flashed glass; multianalytical characterization; chemical composition; chromophores; laser-induced breakdown spectroscopy; thickness measurements

Citation: Palomar, T.; Martínez-Weinbaum, M.; Aparicio, M.; Maestro-Guijarro, L.; Castillejo, M.; Oujja, M. Spectroscopic and Microscopic Characterization of Flashed Glasses from Stained Glass Windows. *Appl. Sci.* **2022**, *12*, 5760. https://doi.org/10.3390/app12115760

Academic Editor: Asterios Bakolas

Received: 19 May 2022
Accepted: 1 June 2022
Published: 6 June 2022

Publisher's Note: MDPI stays neutral with regard to jurisdictional claims in published maps and institutional affiliations.

1. Introduction

Flashed glasses are constituted by two layers of different thicknesses; a thick one of clear or light glass, and a thin colored layer applied on top of the first one. The production of flashed glasses usually involves dipping a colorless base glass into a colored bath, producing a multi-layered glass pane in which the thickness of the colored layers controls the color tone [1]. By sandblasting or etching the flashed (thinner, brighter) layer, a striking two-tone effect can be achieved [2].

Flashed glasses have been used since medieval times in stained glass windows, especially for obtaining those of ruby-red color. The procedure to obtain this type of flashed glass is very complex as it involves the nucleation of copper nanoparticles of a suitable size in the glass matrix. If the nanoparticles are too small, light could pass through the material without interaction, while if they are too large, the glass would appear dark brown or black [3,4]. Additionally, the thickness of the colored glass layer also affects the color hue, being more intense when the glass is thicker [1]. This complexity, added to the expensive raw materials, has favored the development of these two-layer glasses, with the suitable color and the appropriate thickness to be installed in stained glass windows.

Although red-flashed glass is the most common [5–10], other colors have also been used in historical stained glass windows, including pink-purple [7,11,12], green [1,7], or blue [7]. Nowadays, flashed glasses are elaborated in almost all chromatic palettes and are commonly used for the creation and restoration of historical stained glass windows.

The characterization of glasses from stained glass windows has quickly improved in the last years. Laboratory-based techniques, such as field emission scanning electron microscopy/energy-dispersive X-ray spectroscopy (FESEM-EDS) and X-ray fluorescence spectrometry (XRF), are usually destructive, as sampling is usually necessary [13–16]. On the other hand, portable XRF instruments are becoming more and more available for in situ measurements. This technique allows one to distinguish different glass types within the same panel, identifying chromophores, enamels, grisailles, and other decoration layers; however, it is also sensible for unpolished and weathered glass surfaces because the analysis can only be undertaken very near to the glass surface due to low penetration. Therefore, the real bulk glass composition cannot be properly determined. Additionally, low Z elements, such as magnesium and sodium, elements normally used for glass type differentiation, are not detected by XRF. Another analytical technique frequently used for the study of stained glass windows is UV-Vis-IR spectroscopy, which working in transmission or absorption modes allows us to identify the glass chromophores [5,17,18]. UV-Vis-IR spectroscopy is a point analysis technique, and recent developments in in situ visible hyperspectral imaging offer alternatives for the overall analysis of whole glass windows in a few hours using sunlight as the illumination source [19].

Nowadays, valuable glass objects benefit from the availability of non-invasive or micro-invasive techniques, which allow distinguishing different glass types within the same glass panel and identifying chromophores, enamels, grisailles, and other decoration layers [20–23]. Laser spectroscopic techniques, such as laser-induced breakdown spectroscopy (LIBS) and laser-induced fluorescence (LIF), have been combined in several studies to provide complementary elemental and molecular composition in a non- or micro-invasive way [24–32]. LIBS is a micro-invasive technique based on the spectral analysis of the luminous plume generated by the pulsed laser ablation of a small amount of material from the surface of the sample and has the capacity for quantitative determination. LIBS has been shown to be an effective technique for the characterization of glasses from a wide variety of perspectives [30,33–39]. Previous works, using LIBS, have focused on finding the optimal parameters for the analysis of model soda-lime silicate [33] and historical lead silicate glasses [34], on the characterization of chromophores and opacifiers of ancient glasses, and on characterizing degradation pathologies [35,37,39,40]. From a complementary perspective, LIF spectroscopy provides valuable molecular information and facilitates the detection of trace elements and/or chromophores responsible for the glass coloration in a totally non-invasive way [30,41–44].

In this work, we present a multianalytical approach based on optical microscopy (OM) and FESEM-EDS for thickness determination of a multi-layered set of modern flashed glasses commonly used in stained glass windows. Besides, FESEM-EDS, LIBS, UV-Vis-IR spectroscopy, and LIF techniques served for their chemical characterization. FESEM-EDS and LIBS allowed the identification of most chemical elements present in both the base glasses and in their corresponding-colored layers, while UV-Vis-IR and LIF spectroscopies served for glass chromophore characterization.

2. Materials and Methods

The study was conducted in a set of 11 flashed glasses, provided by LambertsGlas®, consisting of a colorless base glass covered by layers of different colors and thicknesses (Figure 1).

The samples were compositionally and structurally characterized by OM, FESEM-EDS, UV-Vis-IR spectroscopy, LIBS, and LIF.

OM observations were carried out with a reflected light microscope Nikon SMZ1000 coupled to an Axiocam 105 color controlled with the software Zen2012 (blue edition) from Carl Zeiss Microscopy GmbH, 2011. Cross-section examinations and chemical linear analyses were performed by a field emission scanning electron microscope HITACHI S-

4700. For elemental chemical linear analysis by EDS, a NORAN system six was connected to the FESEM.

Figure 1. Images of the flashed glasses considered in the present study.

Transmission spectra of original glasses were recorded with a Perkin Elmer Lambda 950 UV-Vis-IR spectrophotometer. The illumination was provided by a tungsten-halogen and deuterium light source in a double optical path covering the 200–2500 nm range with a resolution of 10 nm.

LIBS analyses were performed upon laser excitation at 266 nm (4th harmonic of a Q-switched Nd:YAG laser, 15 ns pulses, 1 Hz repetition rate) and a 0.2 m spectrograph (Andor, Shamrock Kymera-193i-A) equipped with a grating of 1200 grooves/mm, blazed at 500 nm. The output of the spectrograph was coupled to an intensified charge-coupled device (ICCD) camera from Andor Technology, Belfast, Ireland (iStar CCD 334, 1024 × 1024 active pixels, 13 µm × 13 µm pixel size). The laser beam was steered to the surface of the samples at an angle of 45° by using different mirrors. A cut-off filter of 300 nm was placed at the entrance of the spectrograph to reduce the scattered laser light and avoid second-order diffraction. The shot-to-shot fluctuation of laser pulse energy was less than 10%. The laser beam was focused on the surface of the sample with a 10 cm focal length lens achieving a fluence of 8.3 J cm^{-2} (2.6 mJ per pulse in a 200 µm laser spot diameter). The spectra were recorded at 50 nm intervals in the 230–300 nm wavelength range and step and glue spectrograph mode in the range of 300–600 nm with 0.2 nm resolution and with a gate delay and width of 200 ns and 3 µs, respectively, with respect to the arrival of the laser pulse to the sample surface. The spectra resulted from summing up the emissions of the ablation products after ten successive laser pulses, a number that provided good signal-to-noise ratios.

For the LIF measurements, we used the same laser source as for LIBS analysis and a 0.30 m spectrograph (TMc300 Bentham) blazed at 500 nm with a 300 grooves/mm grating coupled to an ICCD, 2151 Andor Technology device, Belfast, Ireland. In this case, the time gate was operated with a zero-time delay with respect to the arrival of the laser pulse to the sample surface and with a width of 3 µs. The sample was illuminated at an incidence angle of 45° through a pinhole, to select the central part of the unfocused laser beam, giving rise to a spot on the sample of elliptical shape with axes dimensions between 1 and 2 mm and with fluences of around 6 × 10^{-3} J/cm^2. LIF spectra were recorded at 300 nm intervals in the wavelength range of 300–700 nm with 5 nm resolution. A cut-off filter at 300 nm was also used for the LIBS measurements. Each spectrum resulted from the accumulation of 10 individual ones.

3. Results and Discussion

3.1. Thickness Measurements

The different colored layers on the base glasses from the considered flashed glass samples were measured by OM and FESEM-EDS.

Analysis of the flashed glasses by OM (Figure 2 and Table 1) showed a structure of parallel layers with a large range of thicknesses for the colored layers. Following the Lambert–Beer law, the thicker the colored layer, the higher the intensity of the color. Nevertheless, some of the pieces with thin layers showed dark colorations due to a high concentration of chromophores.

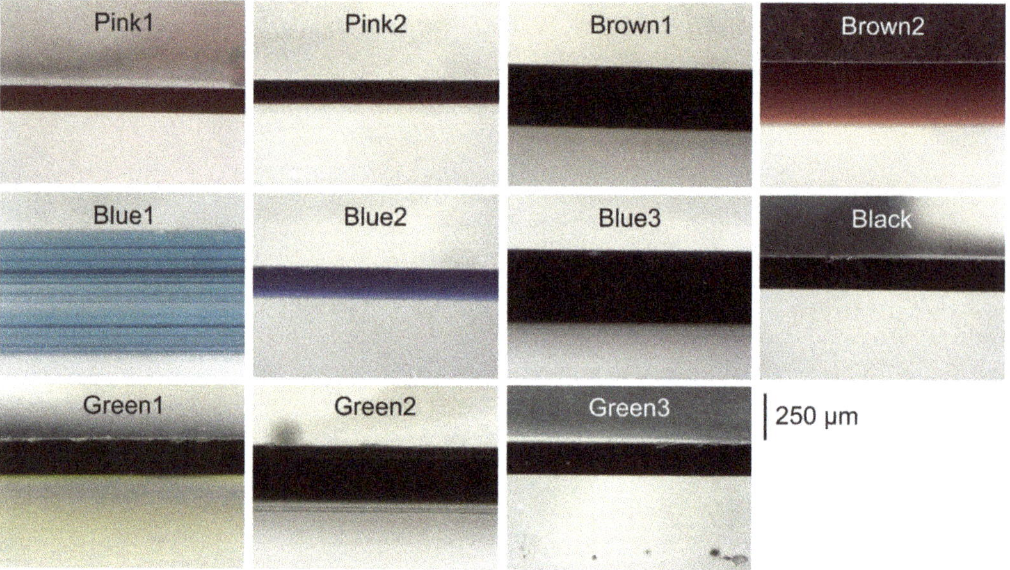

Figure 2. OM images of the colored layers on the flashed glasses in cross-section. Colored layer (**top**) and base glass (**bottom**). All the samples have the same scale.

Table 1. Thickness of the colored layers measured in the cross-section of the glasses using OM, FESEM, and chemical linear analyses with FESEM-EDS.

Sample	Thickness (μm)		
	OM	**FESEM**	**Linear FESEM-EDS Analyses**
Pink1	137 ± 1	147 ± 1	-
Pink2	127 ± 1	-	106–206
Brown1	331 ± 2	334 ± 2	-
Brown2	351 ± 1	-	306–403
Blue1	675 ± 1		652–747
Blue2	173 ± 2	-	176–226

Table 1. *Cont.*

Sample	Thickness (μm)		
	OM	FESEM	Linear FESEM-EDS Analyses
Blue3	386 ± 2	-	373–391
Black	175 ± 1	177 ± 2	-
Green1	193 ± 2	195 ± 1	-
Green2	352 ± 1	-	235–314
Green3	175 ± 1	180 ± 2	-

The pinkish and brownish glasses showed similar thicknesses; however, the bluish and greenish ones displayed completely different thicknesses. It is also interesting to note that the Blue1 and Green2 samples present a multilayer structure that could be related to their fabrication processes. The Brown2 and Blue3 glasses showed a non-homogeneous coloration in cross-section that can be related to the diffusion of the colored layer into the base colorless glass or that the fragment was not completely perpendicular in cross-section, inducing an optical effect of color fading.

The same set of glass pieces was inspected by FESEM-EDS and two different behaviors were detected. Some of the pieces (Pink1, Brown1, Black, Green1, and Green3) showed a clear surface layer that corresponds to the colored one, but the other samples behave differently (Figure 3). The differences in the hue shown in the micrographs are related to the dissimilar chemical composition (Section 3.2.) and not to the contribution of chromophores in the layer, as these could not be detected by FESEM-EDS due to their low concentrations.

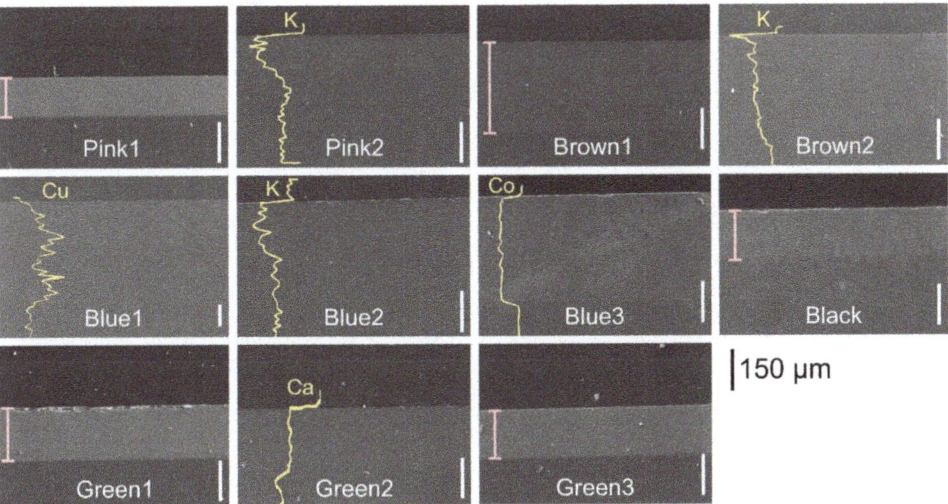

Figure 3. FESEM images of the colored layers in the flashed glasses in cross-section. The scale is the same for all the images.

In the samples with a clear layer, FESEM-EDS thickness measurements were comparable to those quantified by OM (Table 1); however, when this layer was not observable, the thickness was estimated through FESEM-EDS linear analyses by determining the profile of the chemical concentration of different elements (Figure 3). In the Pink2, Brown2, Blue2, and Blue3 samples, it was observed that the surface layer has a higher concentration of potassium or cobalt than the base glass; however, the thickness of the surface layer on the Green2 sample was estimated using calcium element, which presents a lower concentration in the colored layer in comparison with the base glass (Figure 3). The thickness was estimated as a range between the following two values: the depth at which the composition of the colored layer loses its homogeneity and the depth when the homogeneous material of the base glass appears. In this range, the chemical composition varies as a result of the ion exchange produced between the two glasses during manufacturing. This measurement method is not as accurate as that provided by OM, although the values obtained by the two methods are in good agreement (Table 1). Green2 was the only sample in which the thickness measurements did not match because OM revealed the presence of some colored lines in the bottom of the colored layer that was too thin to be detected by EDS linear analysis.

3.2. Chemical Characterization

The chemical composition of the flashed glasses was determined using FESEM-EDS, UV-Vis-IR spectroscopy, LIBS, and LIF.

3.2.1. Field Emission Scanning Electron Microscopy/Energy-Dispersive X-ray Spectroscopy and UV-Vis-IR Spectroscopy

The FESEM-EDS analysis of the cross-sections revealed that all the base glasses are soda-lime silicate glasses with similar chemical compositions (Table 2). The content of silica varied between 63 and 72 wt.%, sodium oxide between 14 and 20 wt.%, and calcium between 9 and 14 wt.%. The base glass of the Green2 sample was slightly different, with an extra amount of around 9 wt.% of BaO. This oxide is sometimes used in substitution of PbO as it increases the refraction index of the glass and, therefore, its shine, and additionally avoids unwanted devitrification [45].

Table 2. Chemical composition (wt.%) of the surface-colored layer and of the base glass analyzed by FESEM-EDS. CL: Colored layer, BG: Base glass.

		Na_2O	MgO	SiO_2	K_2O	CaO	TiO_2	Cr_2O_3	MnO	Fe_2O_3	CoO	NiO	CuO	ZnO	BaO	PbO
Pink1	CL	0.9	-	29.1	5.5	-	-	-	-	-	-	-	-	-	3.7	60.8
	BG	14.5	-	71.3	1.1	13.1	-	-	-	-	-	-	-	-	-	-
Pink2	CL	10.7	-	71.6	5.1	11.5	-	-	-	-	-	-	1.0	-	-	-
	BG	15.0	-	69.6	1.4	14.0	-	-	-	-	-	-	-	-	-	-
Brown1	CL	8.7	-	48.9	2.3	9.3	-	16.7	-	14.1	-	-	-	-	-	-
	BG	15.3	-	70.4	0.8	13.5	-	-	-	-	-	-	-	-	-	-
Brown2	CL	11.8	-	72.4	5.4	9.7	-	-	-	-	-	-	0.7	-	-	-
	BG	15.3	0.8	70.4	0.7	12.6	-	-	-	-	-	-	-	-	-	-
Blue1	CL	14.9	-	70.7	1.6	10.9	-	-	-	-	-	-	1.9	-	-	-
	BG	17.0	-	70.3	1.2	11.5	-	-	-	-	-	-	-	-	-	-
Blue2	CL	15.0	0.6	69.8	2.1	10.6	-	-	-	-	1.8	-	-	-	-	-
	BG	17.6	0.6	70.4	0.6	10.8	-	-	-	-	-	-	-	-	-	-
Blue3	CL	16.6	1.2	64.3	1.1	8.9	-	-	-	-	6.7	-	-	1.1	-	-
	BG	19.0	1.6	69.0	0.4	9.9	-	-	-	-	-	-	-	-	-	-
Black	CL	12.9	0.6	60.0	1.6	8.9	-	-	12.8	-	1.2	1.9	-	-	-	-
	BG	17.1	-	71.9	0.5	10.5	-	-	-	-	-	-	-	-	-	-
Green1	CL	2.5	-	46.6	9.4	-	4.4	1.5	-	-	-	-	-	-	-	35.6
	BG	18.3	1.8	70.4	-	9.5	-	-	-	-	-	-	-	-	-	-
Green2	CL	11.7	1.2	55.5	0.4	6.6	-	-	-	6.3	-	-	6.7	-	11.5	-
	BG	16.7	1.4	63.1	0.4	9.1	-	-	-	-	-	-	-	-	9.3	-
Green3	CL	2.3	-	44.4	10.5	-	-	1.4	-	-	-	-	3.9	-	-	37.4
	BG	19.3	1.9	68.2	0.4	10.2	-	-	-	-	-	-	-	-	-	-

Regarding the colored layers, three samples (Pink1, Green1, and Green3) display a high content of lead. This element is commonly used to reduce the melting point of the surface layer, favoring its adherence to the base glass. These three samples, together with the Black one, are those in which the colored layer showed a clear hue under FESEM observation (Figure 3), an effect attributed to the high content of heavy elements such as lead.

About the colorant agents, most of the surface layers showed a relatively high concentration of chromophores. Pink1 is the only sample in which the chromophores were not detected by FESEM-EDS; however, in the UV-Vis-IR spectrum (Figure 4a), the band of Mn^{3+} was detected with a slight shift to higher wavelengths as a result of the high concentration of lead in the colored layer [19]. A low concentration of manganese may color the surface layer with a pink hue, although its concentration could not be high enough to be detected by FESEM-EDS.

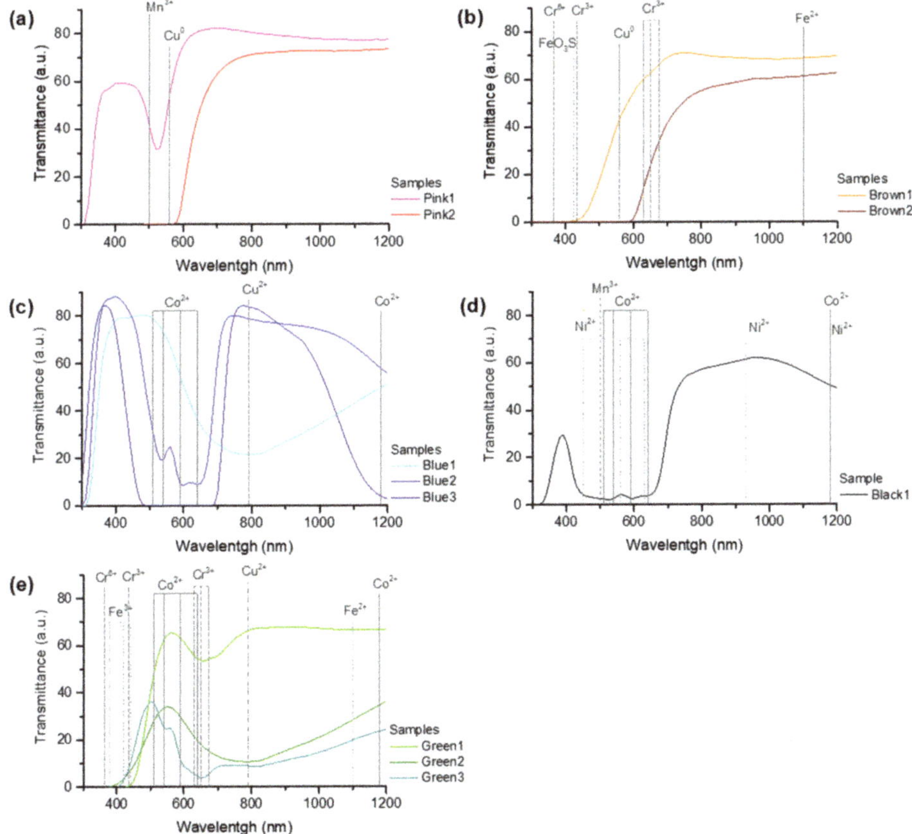

Figure 4. Transmittance spectra of the different colored flashed glasses: (**a**) pink, (**b**) brown, (**c**) blue, (**d**) black, and (**e**) green.

In the colored layer of the Pink2 piece, CuO was detected by FESEM-EDS. However, no clear band of copper was observed in the UV-Vis-IR spectrum, except for a shift of the absorption front to ~560 nm (Figure 4a), a behavior that is related to the plasmon resonance of copper nanoparticles [19]. The chemical composition of the Brown2 sample is similar to that of Pink2 (Table 2), but the color was more intense due to the presence of larger copper nanoparticles and/or a thicker colored layer.

In the Brown1 sample, which displays a yellowish hue, high contents of Cr_2O_3 and Fe_2O_3 (Table 2) were detected. Some of the bands of Cr^{3+} at 630, 650, and 675 nm appeared in the visible spectrum; whereas, the bands at 435 nm (Cr^{3+}) and 365 nm (Cr^{6+}) are overlapped with the shift of the absorption front to ~430 nm (Figure 4b). This shift, shorter than in the Brown2 spectrum, can be attributed to an iron-amber glass [45]. In this case, the chromophore is formed by mixed tetrahedral coordination, in which one Fe^{3+} ion is surrounded by three oxygen ions (bonded to silicon) and one sulfide anion (bonded to alkali ions for preserving electro-neutrality) (FeO_3S). To obtain this color, reductive conditions are necessary during glass fabrication. For this reason, the band at 1100 nm is assigned to Fe^{2+} (Figure 4b). The color observed in the surface layer is produced by the overlap of the bands of the different chromophores.

Regarding the blue samples, Blue1 has a relatively high content of CuO (~2 wt.%), which is also observed in the visible spectrum by virtue of the broad band of Cu^{2+} (Figure 4c). On the contrary, in Blue2 and Blue3, cobalt instead of copper was detected in the EDS analyses. The bands of Co^{2+} are clearly observed in the visible spectrum, being saturated in the darker glass (Figure 4c). Blue3 also contains ~1 wt.% of ZnO. The presence of this oxide decreases the dilatation coefficient of glasses, favoring their resistance to thermal shock that can be generated in bluish glasses in direct contact with the colorless ones due to their highest NIR absorption [46].

In the Black glass, a high concentration of MnO, CoO, and NiO was detected in the FESEM-EDS analyses. Mn^{3+} and Co^{2+} ions in high concentration give intense purple and blue colorations, respectively, due to their high molar extinction coefficient. NiO was also detected by FESEM-EDS and by virtue of the bands of Ni^{2+} in the visible spectrum (Figure 4d). The color associated with this ion depends on its type of coordination and on the nature of the glass. In silicate glasses, Ni^{2+} in tetrahedral coordination produces a violet color, while in octahedral coordination gives a yellow hue. If both types of coordination are present, it is common to find grey glasses. The intense coloration of the three chromophores yields the black color to the colored layer.

Finally, in the green glasses, common greenish chromophores such as Cr_2O_3, Fe_2O_3, or CuO were detected. Green1 has a high content of PbO (~36 wt.%) together with 4.4 wt.% of TiO_2 and 1.5 wt.% Cr_2O_3. PbO favors the decrease in the melting temperature of the colored layer, but it also gives a yellowish hue due to the shift of the absorption front into the visible range. The triplet band of chromium is clearly observed in the visible spectrum, which is not the case with the band of Ti^{3+} at 570 nm (Figure 4e). Nevertheless, this ion may produce an amber coloration when reacting with impurities of iron oxide, giving rise to the formation of ilmenite ($FeTiO_3$) [45]. The Green3 sample spectrum also showed the overlap of bands from chromium and copper, as well as a shift of the absorption front produced by lead. Finally, the visible spectrum of Green2, with ~6 wt.% of Fe_2O_3 and CuO, predominantly showed the band of Cu^{2+} at 790 nm (Figure 4e). This glass also has ~11 wt.% BaO, a compound that helps to stabilize the glass and increase its shine [45].

3.2.2. Laser-Induced Breakdown Spectroscopy

The LIB spectra from flashed glasses, displayed in Figure 5 were recorded on colorless base glass (lower black spectra) and on their corresponding colored layers (upper red spectra). The assignment of the spectral lines was based on the NIST database [47]. Table 3 summarizes the main elemental components of the base glasses and of the corresponding-colored layers of the considered samples, determined by the assignment of the spectral emission lines.

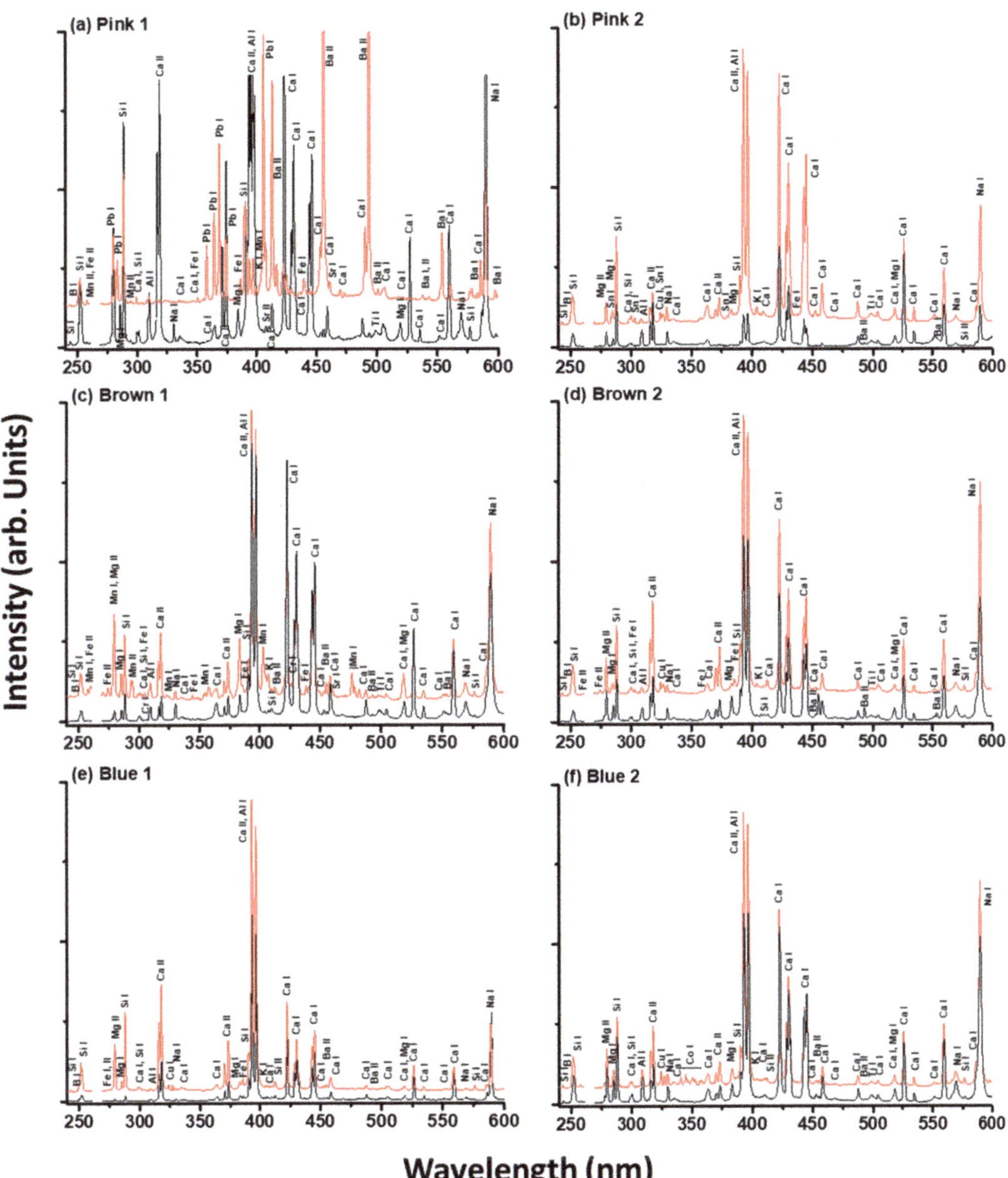

Figure 5. *Cont.*

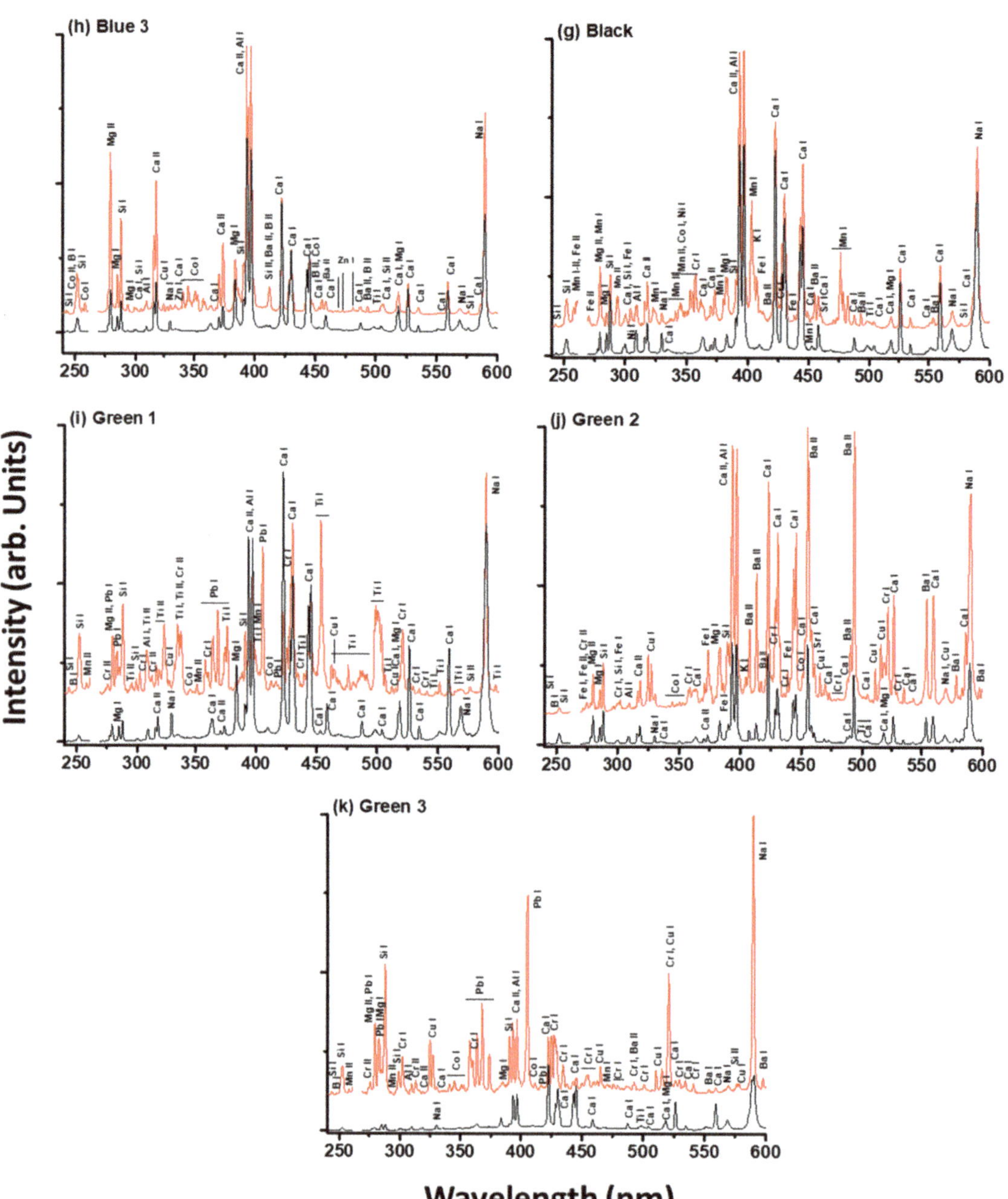

Figure 5. LIB spectra of base glass (lower black spectra) and colored layers (upper red spectra) from flashed glasses. The assigned neutral (I) and ionic (II) atomic lines are indicated. The gap at wavelengths between 260 and 270 nm has been intentionally left to avoid the scattering light from the excitation laser at 266 nm. (**a–k**) correspond to LIBS spectra of samples Pink1, Pink2, Brown1, Brown2, Blue1, Blue2, Blue3, Black, Green1, Green2 and Green3, respectively.

Table 3. Elemental composition of base glass and colored layers of flashed glasses as determined by LIBS. The main components and possible chromophores are indicated in bold black and bold red, respectively. BS: Base glass, CL: Colored layer.

Samples		Elemental Composition as Determined by LIBS
Pink1	CL	**Si**, B, Mn, **Fe**, **Pb**, Al, Ca, K, **Ba**, Sr, Ti, **Na**
	BG	**Si**, Mg, Al, **Ca**, **Na**, K, Ba, Sr, Ti
Pink2	CL	**Si**, B, Mg, Sn, Al, **Ca**, **Cu**, **Na**, **K**, Fe, Ti
	BG	**Si**, Mg, Al, **Ca**, **Na**, K, Ba, Sr, Ti
Brown1	CL	**Si**, B, **Fe**, **Mn**, **Mg**, **Ca**, Al, Cr, K, Ba, Sr, Ti, **Na**
	BG	**Si**, Mg, Al, **Ca**, **Na**, K, Ba, Sr, Ti
Brown2	CL	**Si**, B, Fe, Mg, **Ca**, Al, **Cu**, **Na**, K, Ti
	BG	**Si**, Mg, Al, **Ca**, **Na**, K, **Ba**, Ti
Blue1	CL	**Si**, B, Mg, **Ca**, Al, **Cu**, **Na**, Fe, K, Ba, Ti
	BG	**Si**, Mg, Al, **Ca**, **Na**, Fe, K, Ba, Ti
Blue2	CL	**Si**, B, **Mg**, **Ca**, **Al**, **Cu**, **Na**, **Co**, K, Ba, Ti
	BG	**Si**, **Mg**, Al, **Ca**, **Na**, K, Ba, Ti
Blue3	CL	**Si**, B, **Co**, **Mg**, **Ca**, Al, **Cu**, **Na**, Ba, Ti, Zn
	BG	**Si**, **Mg**, Al, **Ca**, **Na**, Ba, Ti
Black	CL	**Si**, Mg, **Fe**, **Mn**, **Ca**, Al, **Ni**, **Na**, **Co**, K, Ba, **Cr**, Sr, Ti
	BG	**Si**, Mg, Al, **Ca**, **Na**, K, Ti
Green1	CL	**Si**, B, Mg, Mn, **Ca**, Al, **Pb**, **Cr**, **Ti**, **Cu**, Co, K, **Na**
	BG	**Si**, **Mg**, Al, **Ca**, **Na**, K, Ti
Green2	CL	**Si**, B, **Fe**, **Cr**, Mg, Al, **Ca**, **Cu**, Co, **K**, **Ba**, Sr, **Na**
	BG	**Si**, **Mg**, Al, Ca, **Na**, **Ba**, Sr, Ti
Green3	CL	**Si**, B, Mn, **Cr**, **Mg**, **Pb**, Al, Ca, **Cu**, **Co**, Ba, **Na**
	BG	**Si**, **Mg**, Al, **Ca**, **Na**, Ti

The spectra reveal the elemental composition of the glasses by virtue of the emission lines of the main and minor components. These are silicon, calcium, aluminum, potassium, sodium, magnesium, barium, strontium, titanium, and iron. The elements Si, Ca, Al, K, and Na correspond to the main glass components. Other components (Mg, Ba, Sr, Fe, and Ti) can be attributed to impurities from the raw materials employed in the manufacturing processes and to stabilizing agents. Sr, Fe, Ti, with Al, which could not be detected by FESEM-EDS due to their low concentrations below the detection limit of the technique, were observed by LIBS. Besides, the Brown2 and Green2 glasses showed a high content of barium in comparison to the rest of the glasses, in good agreement with FESEM-EDS results. The high content of Na observed in all base glasses indicates that these are mostly based on soda-lime silicate glass. This corroborates the EDS results (Table 2).

The LIB spectra acquired on the different colored layers are richer in line emissions than those corresponding to the colorless base glass (upper red and lower black lines, respectively, in Figure 5) due to additional elements responsible for their coloration. Besides, in most of the spectra of coloration layers, boron lines are present, testifying to the intentional use of this element alone or in combination with lead to lower its melting temperature to obtain better adhesion. Figure 5a shows the LIB spectrum of the colored layer of the Pink1 sample, revealing the presence of Si, B, Mn, Fe, Pb, Al, Ca, K, Ba, Sr, Ti, and Na emissions. The high intensity of Na lines indicates that the material is a soda-lime silicate glass. Additionally, the high-intensity Pb lines (an element also detected by FESEM-EDS) with B lines are obtain lower melting temperatures and to avoid thermal bending or other deformations of the base glass substrate during the production of the flashed glass. The presence of Fe lines, and the minor contribution of Mn lines, indicate the presence of the chromophores [45] responsible for the slight pink coloration of these colored layers. The rest of the components (Si, Al, Ca, K, Ba, Sr, and Ti) are attributed to impurities, stabilizer agents, and elements used in the fabrication of the glass.

Figure 5b shows the LIB spectrum derived from the Pink2 sample revealing a qualitatively similar composition than that of the Pink1 sample, except for the additional presence of Cu chromophore, as determined also by FESEM-EDS. The contribution of Fe to the dark pink coloration of the Pink2 sample is revealed to be minor. Emission lines assigned to Si, Ca, Al, K, and Na correspond to the main glass components, while other emissions due to Ti, Mg, and Sn are attributed to stabilizing agents and impurities of the raw materials.

The LIB spectrum from the colored layer of the Brown1 sample (Figure 5c) displays emission lines of Fe, Mn, and Cr corresponding to the chromophores. Other lines in the spectrum are attributed to Si, B, Mg, Ca, Al, K, Ba, Sr, Ti, and Na. The composition of the colored layer from sample Brown2, displayed in Figure 5d, differs from that of sample Brown1, as in the latter, Cu is the main chromophore with little Fe contribution. Sr and Ba lines are not observed for the Brown2 sample.

The LIB spectra acquired on the colored layers of Blue1, Blue2, and Blue3 samples (Figure 5e,f,h) are qualitatively similar with the exception of the lines associated with their chromophores. Cu and Fe (with less contribution) are responsible for the coloration of sample Blue1, while Cu and Co are associated with the coloration of Blue2 and Blue3 samples. With respect to the minor components, the presence of potassium is observed in samples Blue1 and Blue2 and titanium in sample Blue3, in addition to Si, B, Mg, Ca, Al, Na, and Ba.

The LIB spectrum of the colored layer of the Black sample (Figure 5g) shows the presence of a higher number of elements related to chromophores when compared to the rest of the studied samples (Fe, Mn, Ni, Co, and Cr). The additional elements are Si, Mg, Ca, Al, Na, K, Ba, Sr, and Ti.

LIB spectra of Green1, Green2, and Green3 samples are shown in Figure 5i–k. The different shades of green color of these samples (from light green for Green1 to dark green for Green3) are due to quantitative differences in the main chromophores, Cr and Cu for Green1, Fe, Cr, and Cu for Green2, and Cr, Cu, and Co for Green3.

A comparison of the results of the elemental composition of the samples retrieved from FESEM-EDS and LIBS analyses (Tables 2 and 3, respectively) reveals that LIBS is able to detect the presence of all elements identified by FESEM-EDS in addition to other elements that could not be observed by this technique, such as light and trace elements.

3.2.3. Laser-Induced Fluorescence

The LIF spectra collected on the flashed glasses of this study are displayed in Figure 6. All spectra consist of a broad feature in the 300–700 nm wavelength range, although each sample displays different characteristics that are associated with its chemical composition, including chromophores.

The LIF spectra corresponding to the base glass of the different samples (black lines in Figure 6) are rather similar and can be described as the sum of emissions from different glass components. The emission in the 300–500 nm range encompasses the contribution of oxygen deficiency centers (ODC) from the glass network [42,44,48]. Besides, an additional shoulder with different relative intensities in each sample is noticed around 520 nm and assigned to Ca^{2+} of the glass lattice [39,49].

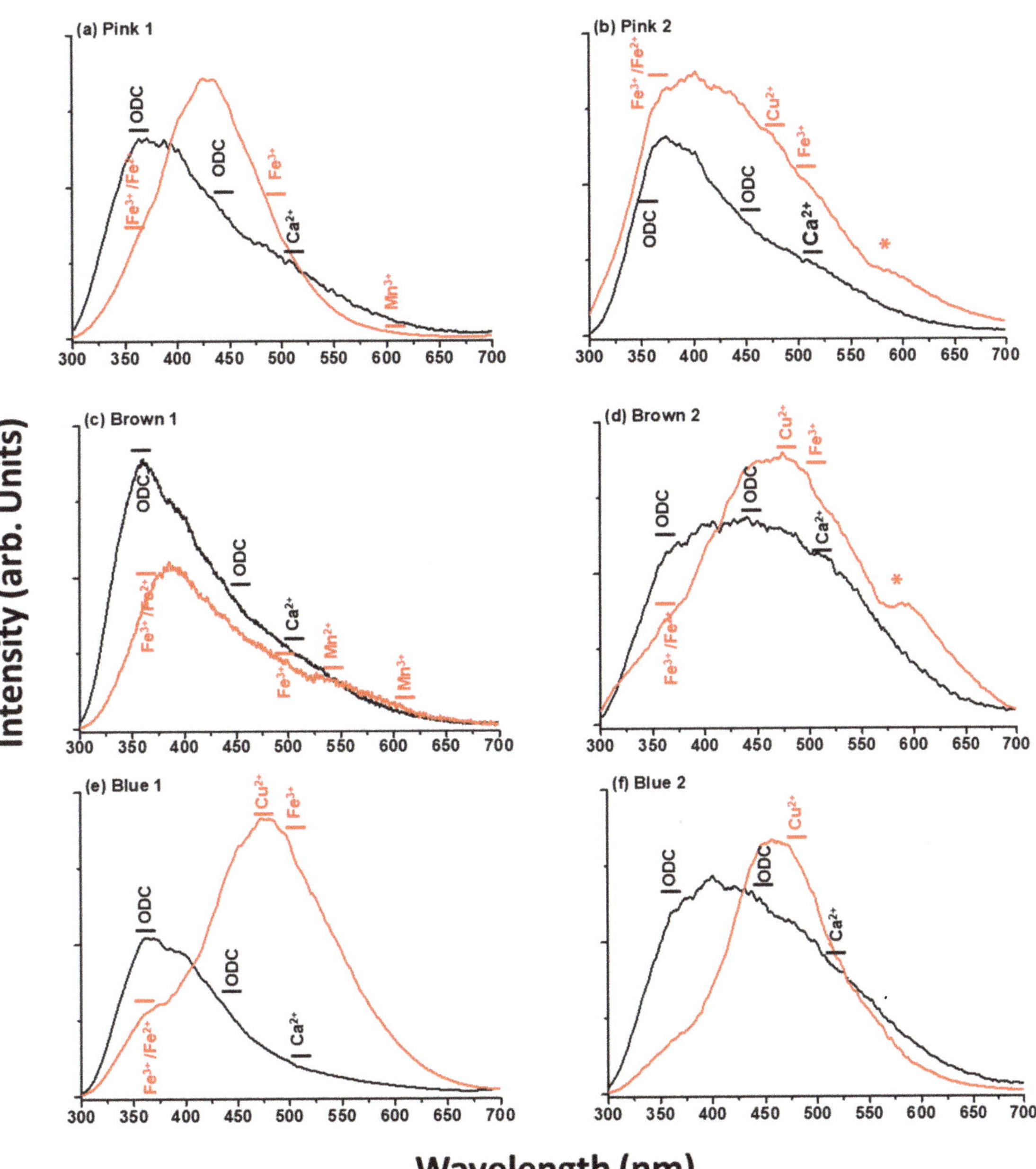

Figure 6. Cont.

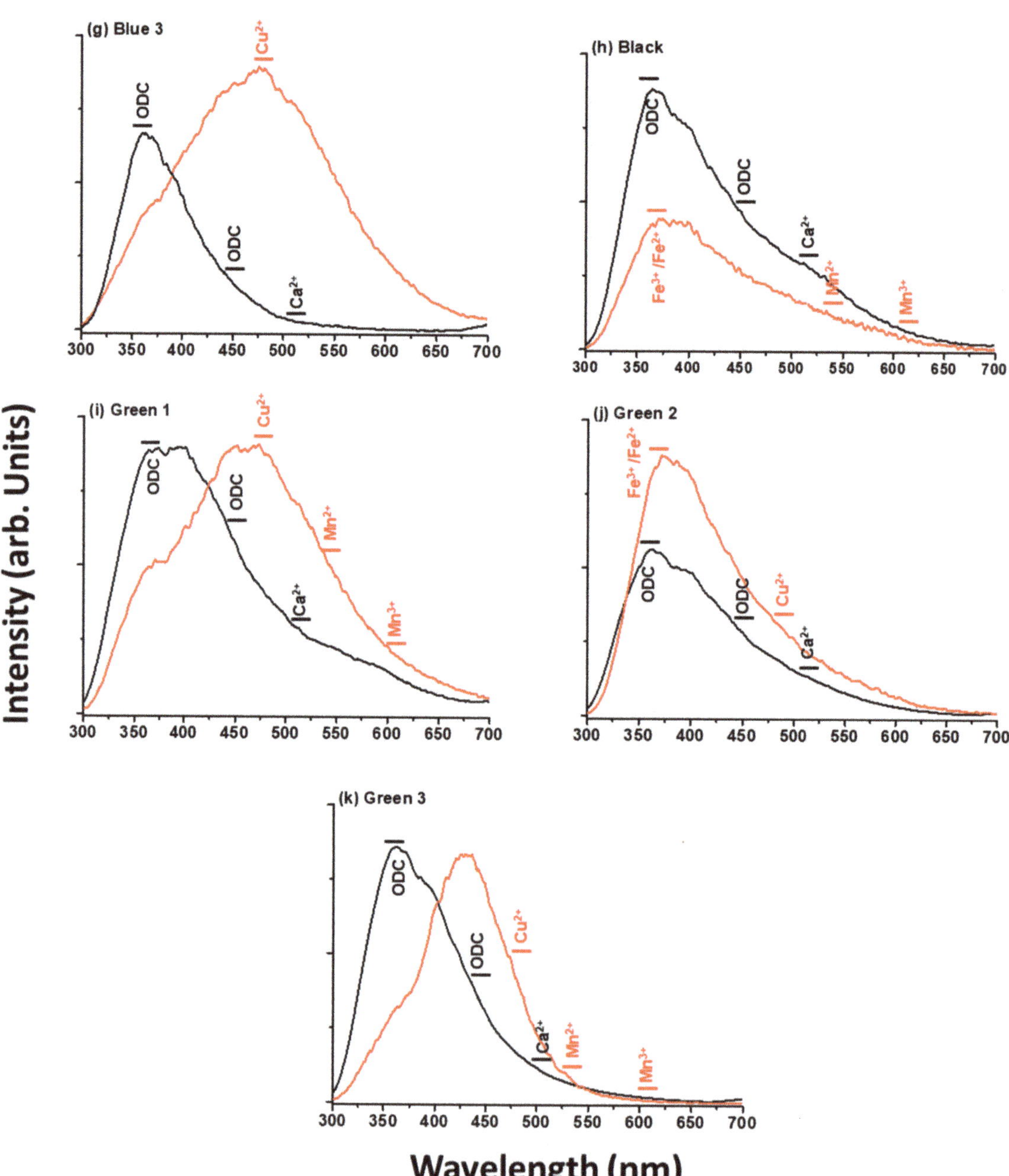

Figure 6. LIF spectra of base glass (lower black spectra) and colored layers (upper red spectra) from flashed glasses. ODC and (*) refer to bands due to oxygen deficiency centers and copper nanoparticles, respectively. Possible chromophores are tentatively assigned and marked by horizontal bars. (**a**–**k**) correspond to LIF spectra of samples Pink1, Pink2, Brown1, Brown2, Blue1, Blue2, Blue3, Black, Green1, Green2 and Green3, respectively.

The LIF spectra of colored layers (red lines in Figure 6) display some specific features in addition to those observed in the base glass. As already indicated and revealed by FESEM-EDS, UV-Vis-IR spectroscopy, and LIBS, chromophores based on Fe, Cu, Mn, Co, Ni, and Cr account for the different colorations. The contribution of these chromophores to fluorescence emissions of the colored layers is not much evident in the LIF spectra of the samples due to the expected reabsorption of the fluorescence from these compounds with absorption bands in the near-UV and visible spectral ranges. Despite this, the weak shoulders over-imposed on the main emission band and observed at 360, 475, 500, 540, and 610 nm have been tentatively attributed to the presence of Fe^{3+}/Fe^{2+}, Cu^{2+}, Fe^{3+}, Mn^{2+}, and Mn^{3+}, respectively [40,45]. On the other hand, the prominent bands observed around 580 nm, specifically for the dark-colored layers from the samples Pink2 and Brown2, may be assigned to emissions from copper nanoparticles [19].

The light coloration from samples Pink1 and Brown1 is due to the presence of Mn- and Fe-based chromophores, while the additional emissions attributed to copper nanoparticles account for the dark hue of Pink2 and Brown2 samples, in agreement with LIBS and FESEM-EDS results. Emissions attributed to different chromophores contribute to the black coloration in the Black sample, as also revealed by LIBS and FESEM-EDS. Regarding the green and blue samples, Cu, Fe, and Mn with different amounts and contributions are considered chromophores in these samples [19].

The combination of LIF and UV-Vis-IR spectroscopy techniques has been revealed to be very effective for the determination of the chromophores responsible for the coloration of glasses, although the latter has shown high sensitivity with respect to the former, revealing the presence of a very high number of chromophores.

4. Conclusions

Flashed glasses, constituted by two layers, one of a colorless glass with a thin colored layer applied on top, have been structurally and compositionally analyzed by optical microscopy, field emission scanning electron microscopy, and linear energy dispersive spectroscopy to determine the thicknesses of the colored layers. The elemental composition of the base glasses and their corresponding-colored layers was determined by laser-induced breakdown spectroscopy and field emission scanning electron microscopy-energy dispersive spectroscopy. Laser-induced fluorescence and UV-Vis-IR spectroscopy provided further information about the chemical nature of the base glasses and colored layers, signposting the presence of the main chromophores responsible for their colors. Laser-induced breakdown spectroscopy results allowed, in a micro-invasive way, the detection of light, major, minor, and trace elements, both in the base glasses and colored layers, with higher sensitivity for major and minor components as compared with the destructive cross-sectional technique of field emission scanning electron microscopy-energy dispersive spectroscopy, which in turn can yield valuable information on chronology, provenance, and manufacturing processes and adds extra value to studies in these type of glasses. Laser-induced breakdown spectroscopy has been revealed as a powerful technique to determine the elemental composition and to look for chemical differences or similarities in the composition of non-destructible samples with historical and heritage value. The speed of analysis achieved with this technique, added to the development and production of portable devices, facilitates the examination of a large number of samples in the field, at excavation sites, in museums, cathedrals, and other civil buildings. Finally, the combination of laser-induced fluorescence and UV-Vis-IR spectroscopy has been revealed to be very effective for the determining of the nature of chromophores responsible for glass coloration.

Author Contributions: Conceptualization, T.P. and M.O.; Methodology, M.O. and T.P.; Formal Analysis, T.P., M.M.-W., M.A., L.M.-G., M.C. and M.O.; Samples' preparation, T.P.; Writing—Original Draft Preparation, T.P. and M.O.; Writing—Review and Editing, T.P., M.M.-W., M.A., L.M.-G., M.C. and M.O.; Project Administration, M.C.; Funding Acquisition, M.C. All authors have read and agreed to the published version of the manuscript.

Funding: This research has been funded by the Spanish State Research Agency (AEI) through project PID2019-104124RB-I00/AEI/10.13039/501100011033, the Fundación General CSIC (ComFuturo Programme), by project TOP Heritage-CM (S2018/NMT-4372) from Community of Madrid, and by the H2020 European project IPERION HS (Integrated Platform for the European Research Infrastructure ON Heritage Science, GA 871034).

Institutional Review Board Statement: Not applicable.

Informed Consent Statement: Not applicable.

Data Availability Statement: This study did not report any supplementary data.

Acknowledgments: All authors acknowledge support from CSIC Interdisciplinary Platform "Open Heritage: Research and Society" (PTI-PAIS) and Fernando Cortés for supplying the glass samples.

Conflicts of Interest: The authors declare no conflict of interest.

References

1. Ceglia, A.; Meulebroeck, W.; Wouters, H.; Baert, K.; Nys, K.; Terryn, H.; Thienpont, H. *Using Optical Spectroscopy to Characterize the Material of a 16th c. Stained Glass Window*; Thienpont, H., Meulebroeck, W., Nys, K., Vanclooster, D., Eds.; Routledge: Kent, UK, 2012; p. 84220A.
2. Rich, C.; Mitchell, M.; Ward, R. *Stained Glass Basics: Techniques, Tools & Projects*; Sterling Publishing Company, Inc.: New York, NY, USA, 1997.
3. Weyl, W.A. *Coloured Glasses*; Society of Glass Technology: Sheffield, UK, 1951; ISBN 9780900682797.
4. Drünert, F.; Blanz, M.; Pollok, K.; Pan, Z.; Wondraczek, L.; Möncke, D. Copper-Based Opaque Red Glasses—Understanding the Colouring Mechanism of Copper Nanoparticles in Archaeological Glass Samples. *Opt. Mater.* **2018**, *76*, 375–381. [CrossRef]
5. Meulebroeck, W.; Wouters, H.; Nys, K.; Thienpont, H. Authenticity Screening of Stained Glass Windows Using Optical Spectroscopy. *Sci. Rep.* **2016**, *6*, 37726. [CrossRef] [PubMed]
6. Adlington, L.W.; Freestone, I.C. Using Handheld PXRF to Study Medieval Stained Glass: A Methodology Using Trace Elements. *MRS Adv.* **2017**, *2*, 1785–1800. [CrossRef]
7. Marchesi, V.; Negri, E.; Messiga, B.; Riccardi, M.P. Medieval Stained Glass Windows from Pavia Carthusian Monastery (Northern Italy). *Geol. Soc. Lond. Spec. Publ.* **2006**, *257*, 217–227. [CrossRef]
8. Alonso Abad, M.P.; Capel, F.; Valle Fuentes, F.J.; de Pablos Pérez, Á.; Ortega-Feliú, I.; Gómez-Tubio, B.M.; Respaldiza Galisteo, M.Á. Caracterización de Un Vidrio Rojo Medieval Procedente de Las Vidrieras Del Monasterio de Las Huelgas de Burgos. *Bol. Soc. Esp. Ceram. Vidr.* **2009**, *48*, 179–186.
9. Farges, F.; Etcheverry, M.P.; Scheidegger, A.; Grolimund, D. Speciation and Weathering of Copper in "Copper Red Ruby" Medieval Flashed Glasses from the Tours Cathedral (XIII Century). *Appl. Geochem.* **2006**, *21*, 1715–1731. [CrossRef]
10. Kunicki-Goldfinger, J.J.; Freestone, I.C.; McDonald, I.; Hobot, J.A.; Gilderdale-Scott, H.; Ayers, T. Technology, Production and Chronology of Red Window Glass in the Medieval Period—Rediscovery of a Lost Technology. *J. Archaeol. Sci.* **2014**, *41*, 89–105. [CrossRef]
11. Palomar, T. Chemical Composition and Alteration Processes of Glasses from the Cathedral of León (Spain). *Bol. Soc. Esp. Ceram. Vidr.* **2018**, *57*, 101–111. [CrossRef]
12. Machado, A.; Wolf, S.; Alves, L.C.; Katona-Serneels, I.; Serneels, V.; Trümpler, S.; Vilarigues, M. Swiss Stained-Glass Panels: An Analytical Study. *Microsc. Microanal.* **2017**, *23*, 878–890. [CrossRef]
13. Atrei, A.; Scala, A.; Giamello, M.; Uva, M.; Pulselli, R.M.; Marchettini, N. Chemical Composition and Micro Morphology of Golden Laminae in the Wall Painting "La Maestà" by Simone Martini: A Study by Optical Microscopy, XRD, FESEM-EDS and ToF-SIMS. *Appl. Sci.* **2019**, *9*, 3452. [CrossRef]
14. Legrand, S.; Van der Snickt, G.; Cagno, S.; Caen, J.; Janssens, K. MA-XRF Imaging as a Tool to Characterize the 16th Century Heraldic Stained-Glass Panels in Ghent Saint Bavo Cathedral. *J. Cult. Herit.* **2019**, *40*, 163–168. [CrossRef]
15. Bernady, E.; Goryl, M.; Walczak, M. XRF Imaging (MA-XRF) as a Valuable Method in the Analysis of Nonhomogeneous Structures of Grisaille Paint Layers. *Heritage* **2021**, *4*, 3193–3207. [CrossRef]
16. Van der Snickt, G.; Legrand, S.; Caen, J.; Vanmeert, F.; Alfeld, M.; Janssens, K. Chemical Imaging of Stained-Glass Windows by Means of Macro X-Ray Fluorescence (MA-XRF) Scanning. *Microchem. J.* **2016**, *124*, 615–622. [CrossRef]
17. Capobianco, N.; Hunault, M.O.J.Y.; Balcon-Berry, S.; Galoisy, L.; Sandron, D.; Calas, G. The Grande Rose of the Reims Cathedral: An Eight-Century Perspective on the Colour Management of Medieval Stained Glass. *Sci. Rep.* **2019**, *9*, 3287. [CrossRef] [PubMed]
18. Hunault, M.O.J.Y.; Bauchau, F.; Boulanger, K.; Hérold, M.; Calas, G.; Lemasson, Q.; Pichon, L.; Pacheco, C.; Loisel, C. Thirteenth-Century Stained Glass Windows of the Sainte-Chapelle in Paris: An Insight into Medieval Glazing Work Practices. *J. Archaeol. Sci. Rep.* **2021**, *35*, 102753. [CrossRef]
19. Palomar, T.; Grazia, C.; Pombo Cardoso, I.; Vilarigues, M.; Miliani, C.; Romani, A. Analysis of Chromophores in Stained-Glass Windows Using Visible Hyperspectral Imaging in-Situ. *Spectrochim. Acta Part A Mol. Biomol. Spectrosc.* **2019**, *223*, 117378. [CrossRef]

20. Rahrig, M.; Torge, M. 3D Inspection of the Restoration and Conservation of Stained Glass Windows Using High Resolution Structured Light Scanning. In *International Archives of the Photogrammetry, Remote Sensing and Spatial Information Sciences—ISPRS Archives;* International Society of Photogrammetry and Remote Sensing (ISPRS): Hanover, Germany, 2019; Volume 42, pp. 965–972.
21. Targowski, P.; Góra, M.; Wojtkowski, M. Optical Coherence Tomography for Artwork Diagnostics. *Laser Chem.* **2007**, *2006*, 35373. [CrossRef]
22. Liang, H.; Cid, M.G.; Cucu, R.G.; Dobre, G.M.; Podoleanu, A.G.; Pedro, J.; Saunders, D. En-Face Optical Coherence Tomography—A Novel Application of Non-Invasive Imaging to Art Conservation. *Opt. Express* **2005**, *13*, 6133. [CrossRef] [PubMed]
23. Filippidis, G.; Massaouti, M.; Selimis, A.; Gualda, E.J.; Manceau, J.M.; Tzortzakis, S. Nonlinear Imagig and THz Diagnostic Tools in the Service of Cultural Heritage. *Appl. Phys. A Mater. Sci. Process.* **2012**, *106*, 257–263. [CrossRef]
24. Anglos, D. Laser-Induced Breakdown Spectroscopy in Art and Archaeology. *Appl. Spectrosc.* **2001**, *55*, 186A–205A. [CrossRef]
25. Anglos, D.; Detalle, V. Cultural Heritage Applications of LIBS. In *Springer Series in Optical Sciences;* Springer: Berlin/Heidelberg, Germany, 2014; Volume 182, pp. 531–554. ISBN 9783642450846.
26. Nevin, A.; Spoto, G.; Anglos, D. Laser Spectroscopies for Elemental and Molecular Analysis in Art and Archaeology. *Appl. Phys. A Mater. Sci. Process.* **2012**, *106*, 339–361. [CrossRef]
27. Tognoni, E.; Palleschi, V.; Corsi, M.; Cristoforetti, G. Quantitative Micro-Analysis by Laser-Induced Breakdown Spectroscopy: A Review of the Experimental Approaches. *Spectrochim. Acta-Part B At. Spectrosc.* **2002**, *57*, 1115–1130. [CrossRef]
28. Melessanaki, K.; Mateo, M.; Ferrence, S.C.; Betancourt, P.P.; Anglos, D. The Application of LIBS for the Analysis of Archaeological Ceramic and Metal Artifacts. *Appl. Surf. Sci.* **2002**, *197–198*, 156–163. [CrossRef]
29. Giakoumaki, A.; Melessanaki, K.; Anglos, D. Laser-Induced Breakdown Spectroscopy (LIBS) in Archaeological Science-Applications and Prospects. *Anal. Bioanal. Chem.* **2007**, *387*, 749–760. [CrossRef]
30. Oujja, M.; Sanz, M.; Agua, F.; Conde, J.F.; García-Heras, M.; Dávila, A.; Oñate, P.; Sanguino, J.; Vázquez De Aldana, J.R.; Moreno, P.; et al. Multianalytical Characterization of Late Roman Glasses Including Nanosecond and Femtosecond Laser Induced Breakdown Spectroscopy. *J. Anal. At. Spectrom.* **2015**, *30*, 1590–1599. [CrossRef]
31. Martínez-Hernández, A.; Oujja, M.; Sanz, M.; Carrasco, E.; Detalle, V.; Castillejo, M. Analysis of Heritage Stones and Model Wall Paintings by Pulsed Laser Excitation of Raman, Laser-Induced Fluorescence and Laser-Induced Breakdown Spectroscopy Signals with a Hybrid System. *J. Cult. Herit.* **2018**, *32*, 1–8. [CrossRef]
32. Bai, X.; Syvilay, D.; Wilkie-Chancellier, N.; Texier, A.; Martinez, L.; Serfaty, S.; Martos-Levif, D.; Detalle, V. Influence of Ns-Laser Wavelength in Laser-Induced Breakdown Spectroscopy for Discrimination of Painting Techniques. *Spectrochim. Acta—Part B At. Spectrosc.* **2017**, *134*, 81–90. [CrossRef]
33. Muller, K.; Stege, H. Evaluation of the Analytical Potential of Laser-Induced Breakdown Spectrometry (LIBS) for the Analysis of Historical Glasses. *Archaeometry* **2003**, *45*, 421–433. [CrossRef]
34. Carmona, N.; Oujja, M.; Gaspard, S.; García-Heras, M.; Villegas, M.A.; Castillejo, M. Lead Determination in Glasses by Laser-Induced Breakdown Spectroscopy. *Spectrochim. Acta—Part B At. Spectrosc.* **2007**, *62*, 94–100. [CrossRef]
35. Carmona, N.; Oujja, M.; Rebollar, E.; Römich, H.; Castillejo, M. Analysis of Corroded Glasses by Laser Induced Breakdown Spectroscopy. *Spectrochim. Acta—Part B At. Spectrosc.* **2005**, *60*, 1155–1162. [CrossRef]
36. Palomar, T.; Oujja, M.; García-Heras, M.; Villegas, M.A.; Castillejo, M. Laser Induced Breakdown Spectroscopy for Analysis and Characterization of Degradation Pathologies of Roman Glasses. *Spectrochim. Acta—Part B At. Spectrosc.* **2013**, *87*, 114–120. [CrossRef]
37. Klein, S.; Stratoudaki, T.; Zafiropulos, V.; Hildenhagen, J.; Dickmann, K.; Lehmkuhl, T. Laser-Induced Breakdown Spectroscopy for on-Line Control of Laser Cleaning of Sandstone and Stained Glass. *Appl. Phys. A Mater. Sci. Process.* **1999**, *69*, 441–444. [CrossRef]
38. Gerhard, C.; Hermann, J.; Mercadier, L.; Loewenthal, L.; Axente, E.; Luculescu, C.R.; Sarnet, T.; Sentis, M.; Viöl, W. Quantitative Analyses of Glass via Laser-Induced Breakdown Spectroscopy in Argon. *Spectrochim. Acta—Part B At. Spectrosc.* **2014**, *101*, 32–45. [CrossRef]
39. Oujja, M.; Palomar, T.; Martínez-Weinbaum, M.; Martínez-Ramírez, S.; Castillejo, M. Characterization of Medieval-like Glass Alteration Layers by Laser Spectroscopy and Nonlinear Optical Microscopy. *Eur. Phys. J. Plus* **2021**, *136*, 859. [CrossRef]
40. Oujja, M.; Agua, F.; Sanz, M.; Morales-Martin, D.; García-Heras, M.; Villegas, M.A.; Castillejo, M. Multiphoton Excitation Fluorescence Microscopy and Spectroscopic Multianalytical Approach for Characterization of Historical Glass Grisailles. *Talanta* **2021**, *230*, 122314. [CrossRef]
41. Romani, A.; Clementi, C.; Miliani, C.; Favaro, G. Fluorescence Spectroscopy: A Powerful Technique for the Noninvasive Characterization of Artwork. *Acc. Chem. Res.* **2010**, *43*, 837–846. [CrossRef] [PubMed]
42. Fournier, J.; Néauport, J.; Grua, P.; Jubera, V.; Fargin, E.; Talaga, D.; Jouannigot, S. Luminescence Study of Defects in Silica Glasses under Near-UV Excitation. *Phys. Procedia* **2010**, *8*, 39–43. [CrossRef]
43. Reisfeld, R. Inorganic Ions in Glasses and Polycrystalline Pellets as Fluorescence Standard Reference Materials. *J. Res. Notionol Bureou Stondords A Phys. Chem.* **1972**, *76*, 613–635. [CrossRef]
44. Skuja, L. Optically Active Oxygen-Deficiency-Related Centers in Amorphous Silicon Dioxide. *J. Non. Cryst. Solids* **1998**, *239*, 16–48. [CrossRef]

45. Fernández Navarro, J.M. *El Vidrio*, 3rd ed.; Consejo Superior de Investigaciones Científicas; Madrid, Spain; Sociedad Española de Cerámica y Vidrio: Madrid, Spain, 2003; ISBN 8400081587.
46. Palomar, T.; Enríquez, E. Evaluation of the Interaction of Solar Radiation with Colored Glasses and Its Thermal Behavior. *J. Non. Cryst. Solids* **2022**, *579*, 121376. [CrossRef]
47. NIST Atomic Spectra Database. Available online: http://physics.nist.gov/asd (accessed on 15 January 2022).
48. Stevens-Kalceff, M.A. Cathodoluminescence Microcharacterization of Point Defects in α-Quartz. *Mineral. Mag.* **2009**, *73*, 585–605. [CrossRef]
49. Chen, H.; Peng, J.; Yu, L.; Chen, H.; Sun, M.; Sun, Z.; Ni, R.; Alamry, K.A.; Marwani, H.M.; Wang, S. Calcium Ions Turn on the Fluorescence of Oxytetracycline for Sensitive and Selective Detection. *J. Fluoresc.* **2020**, *30*, 463–470. [CrossRef] [PubMed]

MDPI

Article

The Influence of Raw Materials on the Stability of Grisaille Paint Layers

Carla Machado [1,2,*], Márcia Vilarigues [1,2], Joana Vaz Pinto [3] and Teresa Palomar [2,4,*]

1 Department of Conservation and Restoration, NOVA School of Science and Technology, 2829-516 Caparica, Portugal
2 Research Unit VICARTE–Glass and Ceramics for the Arts, NOVA School of Science and Technology, 2829-516 Caparica, Portugal
3 CENIMAT | i3N, Department of Materials Science and CEMOP/UNINOVA, NOVA School of Science and Technology, 2829-516 Caparica, Portugal
4 Institute of Ceramic and Glass, Spanish National Research Council (ICV-CSIC), 28049 Madrid, Spain
* Correspondence: cf.machado@campus.fct.unl.pt (C.M.); t.palomar@csic.es (T.P.)

Abstract: Grisaille is a glass-based paint made by mixing metal oxides (iron or copper) with ground lead-silica glass. The different materials used in the grisailles production (coloring agents, base glasses, or vehicles) can significantly impact their long-term stability along with the firing conditions. The main objective of this study was to achieve a better understanding of how raw materials influence the production and stability of these paints. To achieve this goal, 27 grisailles were produced, changing the raw materials, proportions, and firing conditions. The produced grisailles were characterized by X-ray fluorescence and diffraction, colorimetry, roughness measurement, and contact angle analysis. Adhesion and cleaning tests were also made. The use of different coloring agents has a significant impact on the final appearance and on the chemical and mechanical stability of the grisailles, but the latest is more affected by both firing temperature and the proportion between pigments and base glasses.

Keywords: grisaille; recipe; raw material; conservation; degradation

Citation: Machado, C.; Vilarigues, M.; Pinto, J.V.; Palomar, T. The Influence of Raw Materials on the Stability of Grisaille Paint Layers. *Appl. Sci.* **2022**, *12*, 10515. https://doi.org/10.3390/app122010515

Academic Editor: Claudia Mondelli

Received: 16 August 2022
Accepted: 14 October 2022
Published: 18 October 2022

1. Introduction

Grisaille was the first glass-based paint to be used in stained-glass windows. Its use spread throughout Europe from the 12th century onwards and is still used today for drawing the contours and outlines of images, called *grisaille à contourner*, and for the creation of shadows and textures, *grisaille à modéler* [1–4]. The grisailles are usually dark in color, mainly in different shades of brown and black, and are painted on colored or colorless glass supports, together with enamels, sanguine paint, and yellow silver stain [1].

The production of grisailles has changed slightly throughout history, being the first known grisaille recipe found in the Eraclius manuscript *"De coloribus et artibus Romanorum"* (10th–13th century). He describes the paint as a mixture of iron pieces that fall from the blacksmith anvil (oxidized iron) with Jewish glass (ground lead-rich silicate glass) [2,5]. This mixture of base glass and coloring agents is painted on a glass support and fired at temperatures between 650 °C and 700 °C. After the firing process, a thin and uneven layer of colorless glass matrix with the iron oxides dispersed is formed [4]. Recipes for grisaille paint can be found in historical written sources from the 10th to the 19th century [2]. The most common raw materials identified were base glasses and coloring agents [2]. The first ones permit the grisaille adhesion to the glass support and the second ones are responsible for the grisaille coloration.

The use of a high lead base glass continued to be described throughout the different sources, slightly varying the proportions between the silica and lead oxide or even changing the lead source in their production [2]. Initially, it is mainly described the use of burned lead, for example, in the Eraclius recipe for the Jewish glass [5], and later, as in the 19th-century treatise *Guide du verrier* by Georges Bontemps, is described the use of the mineral minium as lead source [6]. The lead is going to reduce the melting temperature of the glass [7], which explains the continuous use of high lead-based glasses in the grisailles formulation, as it is needed that grisailles have a lower melting temperature than the support glass in the way of not deforming it during the firing treatment of the paint layer.

Iron and/or copper oxides, obtained by firing these two metals, were the main coloring agents described throughout the centuries [2,4]. However, firstly proposed by Vasari in the 16th century [8] and later described by Bontemps [6], iron-based pigments, such as hematite, substituted these burned metals. Manganese is also punctually mentioned as a grisaille colorant in the 17th century, *Ars Vitraria Experimentalis* by Johannes Kunckel.

The recipes also describe vehicles and temporary binding agents as the materials that give the necessary plasticity to the mixture before being fired, allowing it to be applied on the glass support. The vehicles usually described are the common ones used in glass painting, such as gum arabic and water, wine, vinegar, egg white, and oils such as lavender oil [2].

Other materials, including various compounds, are also mentioned. The role of these compounds on the grisailles is uncertain, as it is not described in the historical sources. However, they can be understood as additives that can work as opacifiers and/or fluxes. For example, the use of the pigment lead white, described by Kunckel, can also help lower the melting temperature [2].

From the analytical results on historical grisaille compositions described in the literature, it was possible to confirm the use of the same raw materials described in the historical recipes. The grisailles generally present a high quantity of lead in their composition [9–11], which confirms the use of high lead-based glass in their production. Veritá et al. [12] analyzed the stained-glass windows of the Sainte Chapelle in Paris (France) and detected that the vitrified matrix had high lead contents, low silica, and small traces of alumina, lime, and potash [12]. Regarding the coloring agents, it is also possible to confirm the recurrent use of iron and copper oxide, being the hematite (α-Fe_2O_3) and tenorite (CuO), the main compounds identified in historical grisailles used individually or in combination, agreeing with the historical recipes [9,11,13]. When used in combination, mixed compounds can be formed as the cuprospinel ($CuFe_2O_4$), identified in the grisailles from the cathedral of St. Michael and Gudule in Brussels (Belgium) [14] and in the grisailles from the cathedrals of Avila and Segovia (Spain) [11].

All the changes in the raw materials can impact the grisaille paint layer's stability. As stated by Bettembourg [8] and further developed by others [1,9], the grisaille alteration can depend on its chemical composition and production methodology. Additionally, the firing conditions, the ratios between the different components, and the compatibility between the grisaille and the substrate glass can also influence the grisaille's durability [1,15,16]. For example, Schalm et al. [14] proposed that the pulverization of the granular grisailles from the windows of the cathedral of St. Michael and St. Gudule in Brussels (Belgium) is probably due to an unbalanced proportion between the coloring agents and base glass, which can lead to a poor attachment of the pigment grains to the glass support that can easily result in paint loss with small mechanical stress [1,14].

Hence, this work aims to understand better how raw materials and different production methodologies can influence the long-term conservation and stability of grisaille paint layers.

2. Materials and Methods

2.1. Samples Preparation

To better understand the influence of the raw materials on the stability, different grisailles were produced and compared, changing one variable each time, showed in Table 1. The variables considered were the coloring agents, base glass, vehicles and other materials, the firing temperature, the proportions between the components, and the glass support (Table 1).

Table 1. Produced grisailles formulation and image of the final grisaille for each formulation. (* grinding time, PA: practical grade).

Variables	Coloring Agents (CA)	Base Glass (BG)	(CA:BG) wt%	Vehicle	Other Materials	Substrate Glass	Temp.	Image of Final Grisaille
Model grisaille	Iron and copper (1:1 wt%)	Rocaille (SiO_2 + PbO (1:3 wt%))	(1:1)	Gum arabic + water	-	Glass slide	650 °C	
Coloring agents	Burned iron (10 min *)	Rocaille	(1:1)	Gum arabic + water	-	Glass slide	650 °C	
	Burned iron (5 min *)	Rocaille	(1:1)	Gum arabic + water	-	Glass slide	650 °C	
	Burned copper (10 min *)	Rocaille	(1:1)	Gum arabic + water	-	Glass slide	650 °C	
	Burned copper (5 min *)	Rocaille	(1:1)	Gum arabic + water	-	Glass slide	650 °C	
	Manganese (PA MnO_2)	Rocaille	(1:1)	Gum arabic + water	-	Glass slide	650 °C	
	Hematite (PA Fe_2O_3)	Rocaille	(1:1)	Gum arabic + water	-	Glass slide	650 °C	

Table 1. *Cont.*

Variables	Coloring Agents (CA)	Base Glass (BG)	(CA:BG) wt%	Vehicle	Other Materials	Substrate Glass	Temp.	Image of Final Grisaille
	Hematite (mineral)	Rocaille	(1:1)	Gum arabic + water	-	Glass slide	650 °C	
	Burned umber (pigment)	Rocaille	(1:1)	Gum arabic + water	-	Glass slide	650 °C	
Base glass	Iron and copper (1:1 wt%)	SiO_2 + PbO (1:2 wt%)	(1:1)	Gum arabic + water	-	Glass slide	650 °C	
	Iron and copper (1:1 wt%)	SiO_2 + PbO (1:4 wt%)	(1:1)	Gum arabic + water	-	Glass slide	650 °C	
Vehicles	Iron and copper (1:1 wt%)	Rocaille	(1:1)	Water	-	Glass slide	650 °C	
	Iron and copper (1:1 wt%)	Rocaille	(1:1)	Urine	-	Glass slide	650 °C	
	Iron and copper (1:1 wt%)	Rocaille	(1:1)	Wine	-	Glass slide	650 °C	

Table 1. *Cont.*

Variables	Coloring Agents (CA)	Base Glass (BG)	(CA:BG) wt%	Vehicle	Other Materials	Substrate Glass	Temp.	Image of Final Grisaille
	Iron and copper (1:1 wt%)	Rocaille	(1:1)	Vinegar	-	Glass slide	650 °C	
Other materials	Iron and copper (1:1 wt%)	Rocaille	(1:1)	Gum arabic + water	Alumina (PA Al_2O_3)	Glass slide	650 °C	
	Iron and copper (1:1 wt%)	Rocaille	(1:1)	Gum arabic + water	Antimony (PA Sb_2O_3)	Glass slide	650 °C	
	Iron and copper (1:1 wt%)	Rocaille	(1:1)	Gum arabic + water	Burned lead and tin (Pb_2SnO_4)	Glass slide	650 °C	
	Iron and copper (1:1 wt%)	Rocaille	(1:1)	Gum arabic + water	Burned lead (PA PbO)	Glass slide	650 °C	
	Iron and copper (1:1 wt%)	Rocaille	(1:1)	Gum arabic + water	Lead white ($2PbCO_3{\cdot}Pb(OH)_2$)	Glass slide	650 °C	
Temperature	Iron and copper (1:1 wt%)	Rocaille	(1:1)	Gum arabic + water	-	Glass slide	600 °C	

Table 1. *Cont.*

Variables	Coloring Agents (CA)	Base Glass (BG)	(CA:BG) wt%	Vehicle	Other Materials	Substrate Glass	Temp.	Image of Final Grisaille
CA:BG	Iron and copper (1:1 wt%)	Rocaille	(1:1)	Gum arabic + water	-	Glass slide	700 °C	
	Iron and copper (1:1 wt %)	Rocaille	(2:1)	Gum arabic + water	-	Glass slide	650 °C	
	Iron and copper (1:1 wt%)	Rocaille	(1:2)	Gum arabic + water	-	Glass slide	650 °C	
Substrate glasses	Iron and copper (1:1 wt%)	Rocaille	(1:1)	Gum arabic + water	-	Mixed-alkali glass (K-Na-Ca-Si)	650 °C	
	Iron and copper (1:1 wt%)	Rocaille	(1:1)	Gum arabic + water	-	Soda-lime glass (Na-Ca-Si)	650 °C	
	Iron and copper (1:1 wt%)	Rocaille	(1:1)	Gum arabic + water	-	Potash-lime glass (K-Ca-Si)	650 °C	

The initial grisaille (Model grisaille) formulation agrees with the most common raw materials identified in the historical treatises [2]. It was made by mixing burned iron and burned copper (1:1 wt%) as coloring agents, rocaille (SiO_2 + PbO (1:3 wt%)) as base glass, gum arabic (0.1 wt%) and water as temporary binder and vehicle. The base glass and the coloring agents were used in a 1:1 wt% ratio, painted on the non-tin side of commercial glass slides (Deltalab®), following the paint application methodology from Vilarigues et al. [3], and fired at 650 °C. A total of 26 grisailles were produced, changing one of the variables each time (see Table 1).

Practical grade (PA) reagents were used for some raw materials. All the brands and manufacturers of the commercial PA reagents are described in Appendix A.

Low-carbon steel and standard copper plates were used to obtain the burned iron and copper. Both metals were cut into small pieces and placed in crucibles in an electric furnace, during two heating cycles, to a maximum temperature of 850 °C for 1 h, promoting metal oxidation. Afterward, both metals were grounded into a fine powder for 5 and 10 min.

The SiO_2 and PbO for the base glasses were mixed in different proportions (Table 1). After 10 h of melting in an electric furnace (BARRACHA-model E6) at 1100 °C in ceramic crucibles, the mixtures were poured into water, dried, and grounded in an electric agate mortar.

Different substrate glasses were produced (Table 1) according to the compositions of the most common glasses from historical stained-glass windows [17]. They were melted in an electric furnace (TERMOLAB–BL) at 1400 °C for 3.5 h in ceramic crucibles and blown into the form of a roundel using the same procedure as the historical production of crown-window glass.

Most of the other materials added were practical grade reagents, except for the burned tin and lead and the lead white. The burned tin and lead were produced according to the descriptions from historical recipes; lead oxide (PbO) and tin oxide (SnO_2) were mixed in a 1:1 wt% ratio and burned at 900 °C for 2 h. This burned lead and tin was also known as a lead-tin-yellow pigment, widely used in oil painting [18].

The lead white pigment used was produced by exposing metallic lead to acetic acid (vinegar) [18].

The other materials (alumina, antimony, burned lead and tin, burned lead, and lead white) were added to the grisailles in proportions of 1:1 wt% with the coloring agents. Moreover, the sum of the coloring agents and the other materials was 1:1 wt% with the base glass. Leaving grisailles with 1:1:2 wt% proportions between the coloring agents, other materials, and base glasses. These ratios were chosen according to the quantities described in the historical recipes with these materials in their composition [2].

The different grisaille mixtures were painted and fired at different temperatures (600 °C, 650 °C, 700 °C) in a side-heated electric furnace (BARRACHA-model E1) with a temperature ramp of 3 °C/min up to the maximum temperature, followed by a dwell of 30 min and slow cooling. All the produced grisailles are shown in Table 1.

2.2. Analytical Techniques

The raw materials were characterized by X-ray fluorescence and X-ray diffraction. The produced grisailles were also characterized by X-ray diffraction, and the roughness, contact angle, and color were measured. Adhesion tests were conducted to assess the grisaille adhesion to the glass substrate. The tests were performed following the *European ISO Standard, Paints, and varnishes Cross cut test* (ISO 2409:1992). Following Wolbers' methods [11], a cleaning study was also performed to test the chemical solubility of these grisailles.

To identify the chemical composition of the different base and substrate glasses used in the sample preparation, analyses by X-ray fluorescence (XRF) were made. A PANalytical MagicX (PW-2424) wavelength-dispersed X-ray spectrometer equipped with a rhodium tube (SUPER SHARP) of 2.4 KW was used. Analytical determinations were carried out through the analysis of the IQ^+ curve, with a powder sample prepared in a fused pearl. The pearls were made in a Philips Perl'X3 equipment, melted at 1050 °C, in a platinum-gold crucible, from a homogeneous mixture of 0.3 g of the powder sample (<75 μm) and 5.5 g of $Li_2B_4O_7$ anhydrous and LiBr. The spectrometer can detect light elements starting at 18.998 atomic mass (Fluor) with an LOD of 0.1%.

The X-ray diffraction (XRD) of the coloring agents and additives identified the different crystallographic phases of the compounds. A Benchtop X-Ray Diffractometer RIGAKU model MiniFlex II, with a monochromatic X-ray source (Cu Ka line) operated at 30 kV of acceleration voltage and 15 mA current, was used. The spectra were acquired between 10 and 90° at 2°/min. X-ray diffraction was also performed on selected grisaille samples to identify the different crystallographic phases formed after the firing. A PANalytical X'Pert PRO MPD diffractometer equipped with an X'Celerator 1D detector and CuKα radiation was used. The XRD data were acquired in the 14°–90° 2θ range with a step size of 0.02°. The identification was made using the X'Pert HighScore Plus software and database and by comparison with the RRUFF database.

Roughness analyses were performed to compare and measure the surface morphology of the samples. The measurements were made with an optic rugosimeter TRACEiT from Innowep GmbH. Three-dimensional topographical maps (5 × 5 mm) with a resolution of 2.5 μm (Z-axis) and 2.5 μm (in X/Y axes). To compare the samples, the roughness maps were flattened, and the arithmetic average roughness (Ra) was measured with the software Gwyddion version 2.6 [19] and calculated following the method of Ariyathilaka et al. [20]

Contact angle measurements were made to attest to the hydrophilic and hydrophobic properties of the samples. Under laboratory conditions, the tests were performed using distilled water with the Easy Drop Standard "Drop Shape Analysis System" Kruss DSA 100 measurement apparatus. The diameter of the needle was 0.5 mm, and the drop volume was 2 μL. The contact angle of each sample was measured automatically by the equipment in the fitting mode three times, and the average and standard deviation were calculated.

A colorimetric study was performed to measure the color variations between the samples. A CM-700d Konica Minolta portable sphere spectrophotometer with vertical alignment with an 8 mm diameter mask was used. The data were recorded in SCI mode with a D65 Illuminant and processed by Color Data Software CM-1 in a CIELab color space. The analyses were performed in three areas on each sample, calculating the averages and standard deviation. Afterward, the results were converted to xy coordinates, and the results were expressed in a chromaticity diagram CIE 1931.

Finally, it was assessed the adhesion of the grisaille to the glass substrate by adhesion tests. Following the *European ISO Standard, Paints, and varnishes–Cross-cut test* (ISO 2409:1992) [21]. An Elcometer® cutter with six blades (equally spaced by 1 mm) was used, and the ISO Standard Adhesive Tape was applied. The test was performed three times in each sample to verify the uniformity samples and the reproducibility tests. Based on the Classification Table for the ISO Standard [21], the results were classified from 0 to 5, corresponding to:

0—Only the superficial layer was cut without detachment;

1—Only the superficial layer was cut, with residual detachment visualized in the adhesive tape;

2—Residual detachment was visualized in the adhesive tape, and between 5% and 15% of the cross-cut area was affected;

3—Residual detachment was visualized in the adhesive tape at 15%, and 35% of the cross-cut area was affected;

4—Residual detachment was visualized in the adhesive tape at 35%, and 65% of the cross-cut area was affected;

5—Total detachment of the painted layer.

A cleaning study was made on the studied samples to test their solubility and solvent resistance. The methodology for this cleaning study was adapted from Wolbers' methods for painted surfaces of temperas and oils [22]. The solubility was evaluated with four solutions: distilled water (H_2O), distilled water (H_2O) + ethanol (C_2H_6O) (1:1), ethanol (C_2H_6O), and acetone (C_3H_6O). These solutions are the most common ones used for cleaning stained-glass windows [23]. The samples' surfaces were cleaned with each solution (Figure 1) under the observation of the Dino-Lite Edge digital portable microscope, model AM7915MZTL.

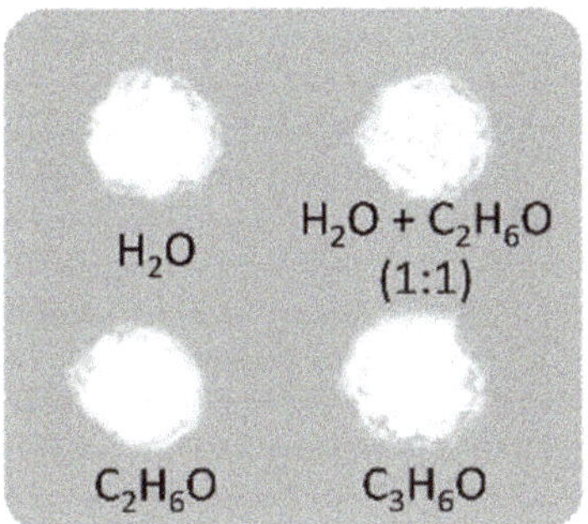

Figure 1. Representation of the order and position in which the solvents were tested in the sample cleaning study.

3. Results and Discussion

3.1. Raw Materials Characterization

The chemical composition of base and substrate glasses was analyzed by X-ray fluorescence, as shown in Tables 2 and 3.

Table 2. Base glasses composition (wt.%) obtained by XRF.

	Na_2O	Al_2O_3	SiO_2	K_2O	Fe_2O_3	PbO
SiO_2:PbO (1:2)	-	0.57	33.9	-	-	65.5
SiO_2:PbO (1:3)	-	1.97	26.7	-	-	71.3
SiO_2:PbO (1:4)	0.56	2.26	28.6	0.15	0.22	68.0

Table 3. Substrate glasses composition (wt.%) obtained by XRF.

	Na_2O	MgO	Al_2O_3	SiO_2	P_2O_5	K_2O	CaO	MnO	Fe_2O_3
Glass slide	11.8	4.43	1.58	72.9	<0.05	0.72	8.12	<0.05	<0.10
Soda-lime glass (Na-Ca-Si)	12.3	1.78	3.78	67.8	<0.05	3.80	9.20	<0.05	<0.10
Potash-lime glass (K-Ca-Si)	<0.10	1.79	2.97	52.5	1.75	20.7	19.9	<0.05	<0.10
Mixed-alkali glass (K-Na-Ca-Si)	9.49	3.89	3.29	62.1	0.14	6.88	13.4	0.37	0.40

Despite the increasing amounts of lead added, it was impossible to see a significant change in the chemical composition of the different base glasses. The glass with a 1:4 proportion between SiO_2 and PbO experienced a high lead loss due to lead volatilization, shifting the proportion of SiO_2:PbO to (1:2.5). In the binary SiO_2-PbO systems, a sharper deflection of PbO volatilization occurs with the increase in lead content, mainly above 80 wt% PbO [7,24]. It is also possible to see small amounts of aluminum contamination through the different glasses and sodium, potassium, and iron in the SiO_2:PbO (1:4) glass. All these elements can appear due to contamination from the crucible during the melting process. The increasing amount of these elements in the SiO_2:PbO (1:4) glass can be related to the higher amount of lead in the initial formulation, which leads to a higher contact time of the molten glass with the crucible walls as the glass reaches the melting point faster and at a lower temperature. As these contaminations represent less than 3 wt% of the base glasses composition, they would not influence the future results of this study.

The composition of the different substrate glasses is shown in Table 3. These glasses represent typical compositions from historical stained-glass windows: soda-lime silicate glass, potash-lime silicate glass, and mixed-alkali glass [17]. The glass slide is also classified as soda-lime silicate glass.

The X-ray diffraction results showing the crystallographic phases of the coloring agents and other materials tested are represented in Table 4. All the diffractograms can be consulted in Appendix B, Figures A1–A11. For the burned iron, it was only possible to identify it in one state of oxidation after the burning, iron (III) oxide (hematite (Fe_2O_3)). On the other hand, the burned copper is composed of a mixture of copper oxides in two different oxidation states, copper (I) oxide (cuprite (Cu_2O)) and copper (II) oxide (tenorite (CuO)). For the mixture of burned tin and lead, independent oxides of lead (massicot (PbO) and minium (Pb_3O_4)) and tin (cassiterite (SnO_2)) were formed instead of a mixed compound as a lead stannate (Pb_2SnO_4). It was also possible to confirm the crystallography of the chosen practical grade (PA) compounds: burned lead (PbO), manganese (MnO_2), hematite (Fe_3O_4), alumina (Al_2O_3), and antimony (Sb_2O_3)). Iron (III) oxide (hematite (Fe_2O_3)) was also identified for the pigments/earths of natural hematite and burned umber. This was unexpected, as usually, these natural pigments have a significant number of impurities, such as aluminum oxide in the hematite and manganese in the burned umber [18]. Hydrocerussite (($Pb_3(CO_3)_2(OH)_2$), which is a hydrated form of cerussite ($PbCO_3$), was identified in the lead white pigment [18].

Table 4. Crystallographic phases of the coloring agents and other materials obtained by XRD.

	Crystallographic Phases
Burned iron	Hematite (Fe_2O_3)
Burned copper	Cuprite (Cu_2O), Tenorite (CuO)
Manganese PA	Pyrolusite (MnO_2)
Hematite PA	Hematite (Fe_2O_3)
Natural hematite	Hematite (Fe_2O_3), Quartz (SiO_2)
Burned umber	Hematite (Fe_2O_3)
Alumina PA	Corundum (Al_2O_3)
Antimony PA	Senarmontite (Sb_2O_3)
Burned SnPb	Cassiterite (SnO_2), Massicot (PbO), Minium (Pb_3O_4)
Burned lead (PbO PA)	Massicot (PbO)
Lead white	Hydrocerussite ($Pb_3(CO_3)_2(OH)_2$)

3.2. Grisaille Properties

3.2.1. Crystallographic Characterization

Table 5 shows the crystalline phases of the model grisaille, which, as described in Table 1, was produced with a mixture of burned iron and copper. The results showed the presence of hematite (Fe_2O_3), cuprite (Cu_2O), and tenorite (CuO). These three compounds were previously identified in the raw material characterization (Table 4). However, the tenorite (CuO) has a much higher intensity in the model grisaille (Figure A12) than in the burned copper diffraction result (Figure A2). This indicates that some cuprite oxidizes into tenorite during the grisaille's firing. At these firing conditions, tenorite is also a more stable compound [25].

The results for the grisailles (Figures A13–A17) where different coloring agents were used and the other materials added are also shown in Table 5. The identified components were the same as the ones identified in the raw materials (Table 4), except for the burned lead and lead white that formed the compound called iron barysilite ($Pb_8Fe(Si_2O_7)_3$) after the firing process of the grisailles.

The crystalline phases identified correspond to previously identified components in historical grisaille samples characterized and found in the literature [11,12,26–28].

Table 5. Crystallographic phases of selected grisaille samples obtained by XRD.

Grisailles	Crystallographic Phases
Model	Hematite (Fe_2O_3); Cuprite (Cu_2O); Tenorite (CuO)
Hematite PA	Hematite (Fe_2O_3)
Natural hematite	Hematite (Fe_2O_3); Quartz (SiO_2)
Burned umber	Hematite (Fe_2O_3)
Burned lead	Hematite (Fe_2O_3); Cuprite (Cu_2O); Tenorite (CuO); Iron barysilite ($Pb_8Fe(Si_2O_7)_3$)
Lead white	Hematite (Fe_2O_3); Cuprite (Cu_2O); Tenorite (CuO); Iron barysilite ($Pb_8Fe(Si_2O_7)_3$)

3.2.2. Colorimetry

The results from the colorimetric study are represented in Figure 2. It is possible to observe that the results for almost all the samples agree with the model grisaille, where similar color was observed. The use of hematite (PA or natural) and burned umber changes the color toward a warmer hue (red). These color alterations are directly related to the original colors of raw materials, as the hematite and burned umber are red-brownish pigments [18,29]. In addition, after the firing process, it was not possible to see any changes in the crystalline compounds used for these grisailles, as visible in Table 5 results, where hematite (Fe_2O_3) was identified as the main compound for these three grisailles.

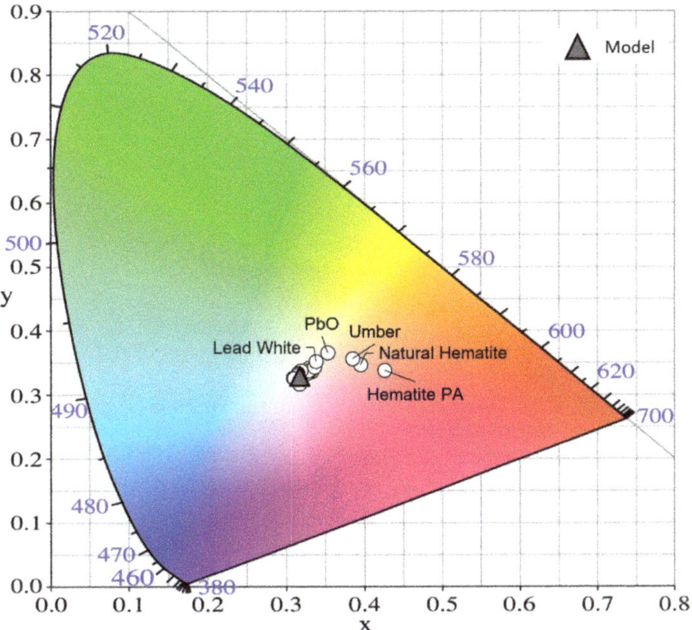

Figure 2. Colorimetric xy diagram for the produced samples.

Additives can also influence the final color of the grisaille. The lead white and the PbO turn the grisailles toward a yellowish color, as shown in Figure 2. This change can occur because when these two lead compounds are added to the grisaille mixture, a new compound called iron barysilite ($Pb_8Fe(Si_2O_7)_3$) is formed (Table 5). This compound is a lead-rich iron silicate that can be a low-temperature precursor of the melanotekite ($Pb_2Fe_2(Si_2O_7)O_2$) with a yellowish/greenish hue, which can influence the final color of the grisaille [11,30,31].

3.2.3. Roughness

The results from the roughness analyses are represented in Figure 3.

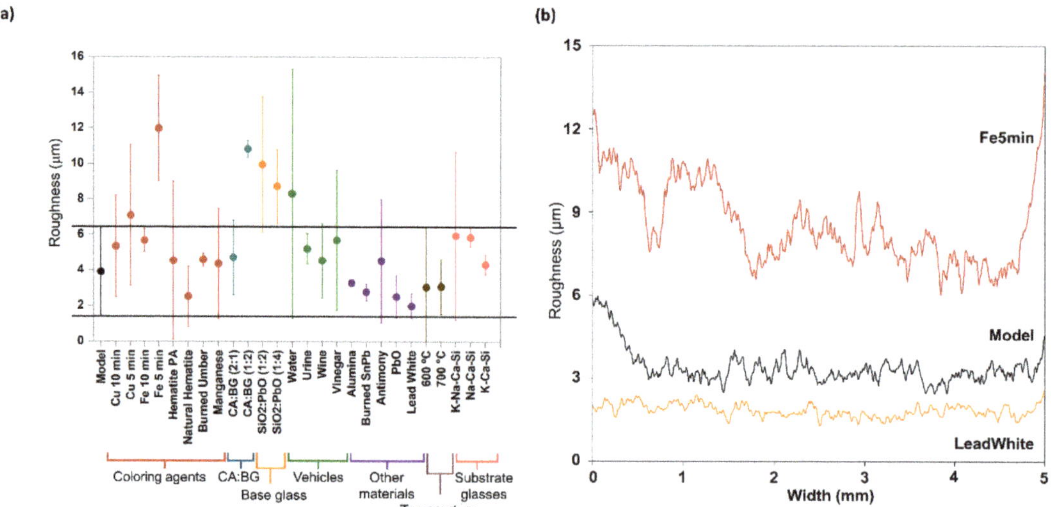

Figure 3. Results from the roughness analysis, (**a**) graphical representation from the average roughness (Ra*) and their standard deviations of the produced grisaille samples, (**b**) graphical representation of the roughness variation throughout the sample length.

The interpretation of these results must consider that despite the painting technique for applying the paint layers into the substrate glass was the same, and it was made by hand, so the final grisaille layers are susceptible to handcraft-related variations.

Considering what was described before, it is possible to see that the average roughness of the different grisaille samples does not present a significant variation from the model grisaille ($1.99 \pm 0.28 < Ra^* < 11.96 \pm 1.20$ µm), being the average of almost all the samples inside the range of the model grisaille standard deviation, as shown in Figure 3a. Nevertheless, the samples where the copper and iron were only grounded for 5 min presented a higher rugosity (7.08 ± 1.59 µm and 11.96 ± 1.20 µm, respectively) because the size of the metal's oxides was larger (Figure 3a). Nonetheless, iron presents a higher rugosity than copper because iron has higher hardness [20] in comparison and, therefore, needs more grinding time to obtain a thinner powder.

In the grisaille, with more base glass (CA:BG (1:2)), the roughness almost doubles the value (10.81 ± 0.20 µm) compared with the model sample and with the one with less base glass (CA:BG (2:1)) (4.71 ± 0.85 µm). This can indicate that the time or temperature used during the firing was insufficient to soften this higher quantity of base glass and even out and flatten the painted layer.

The different base glasses used (SiO_2:PbO (1:2) and SiO_2:PbO (1:4)) also give higher rugosity (9.95 ± 1.54 µm and 8.71 ± 0.84 µm, respectively) to the grisaille when compared with the one used for the model grisaille (SiO_2:PbO (1:3)) (3.90 ± 0.99 µm). These glasses are the ones that have less lead in their composition, as shown in Table 2. Therefore, they have a slightly higher melting temperature being less flattened.

An increase in rugosity (8.30 ± 2.82 µm) and its range of standard deviation was also observed when water without any binding agent was used as a vehicle (Figure 3a). This is a consequence of the difficult manipulation and application of the paint only using water as a vehicle. Less plastic paint is obtained during the application, which leads to a struggle to achieve a final smooth and even layer.

The standard deviation of the results also gives information about the samples because smaller ranges, such as the samples with burned umber, alumina, and burned tin and lead, can indicate a homogeneous surface throughout the sample.

Higher rugosities can increase the grisailles porosity, which, combined with bubbles and fissures, can increase the susceptibility of water penetration through the paint layers, accelerating the degradation process [11–13,16,32–35]. This is visible in the samples from the Czech Republic, studied by Cílová et al., where it was possible to visualize vertical and horizontal cracks throughout the paint layer [35]. These cracks probably facilitated the water penetration and were responsible for the high corrosion observed in the substrate glasses under the grisaille layers, ultimately leading to the paint layers' pulverization.

Figure 3b shows the roughness variation throughout the length of the samples, representing the samples with the higher (Fe 5 min sample) and lowest (lead white sample) rugosity measured, as well as the model sample result. It is possible to observe that the grisaille with a higher rugosity (Fe 5 min) also presents higher variation throughout the length of the sample with its uneven line, and the grisaille with the lowest rugosity (lead white) presents a more even line throughout the length of the sample.

3.2.4. Contact Angle

Figure 4 represents the results of the contact angle measurement. The higher the contact angle, the more hydrophobic the sample is. In Figure 4a, it is possible to visually compare the water drops' behavior when in contact with the model sample and with a less hydrophobic one (burned umber).

The results showed a general decrease in the contact angle value when comparing the samples with the model grisaille (Figure 4b), which is related to a decrease in the hydrophobic characteristics of the samples, demonstrating the impact that the different variables can have in the grisailles water affinity.

The samples that showed a higher affinity with water were the ones where hematite (PA or Natural) and burned umber were used, as well as the grisaille fired at 700 °C and the one painted on a mixed-alkali glass (Figure 4b).

These results do not match the roughness results (Figure 3). It is expected that the samples with higher rugosity were the more hydrophilic ones. However, the water fixation and penetration in the grisaille paint layers depend on several factors. For example, unsuitable firing temperatures and incompatibilities between the grisaille and glass support can create bubbles and tensions, which can lead to the formation of fissures helping the water penetration [16,35,36]. This is visible in grisailles from the windows of the church of S. Giovani and Paolo in Venice (Italy) [16], which showed fissures in the grisaille layers parallel to the glass support, which were caused by a wrong firing temperature. This can justify the lower contact angles presented by the grisaille fired at 700 °C and the one painted on a mixed-alkali glass. The intrinsic properties of some raw materials can also influence the hydrophilic characteristic of the painted layers. This can be observed in the grisailles where hematite (PA and Natural) and burned umber were used, which also presented lower angles. Shrimali et al. previously studied hematite ores used to have contact angles between 60 and 10 degrees [37] depending on the surface hydroxylation, which agrees with the results presented in Figure 4b. The burned iron did not present similar results despite also being hematite. This could happen because it was produced by burning steel pieces, which creates a more aggregated and less thin powder with a less exposed contact surface.

The contact angles from the different support glasses without grisaille paint were also measured, and the results are shown in Figure 4b. Comparing the results from the support glasses with the ones obtained for the grisailles painted on them, it is possible to understand that most support glasses present lower contact angles, being more hydrophilic than the grisaille layers. This indicates that the grisaille layer, in its majority, can become a protective layer, diminishing the water affinity of the surface. However, this also depends on the raw materials' intrinsic characteristics, the firing temperature, the compatibility between the grisaille and the substrate glasses, and the presence of fissures or bubbles.

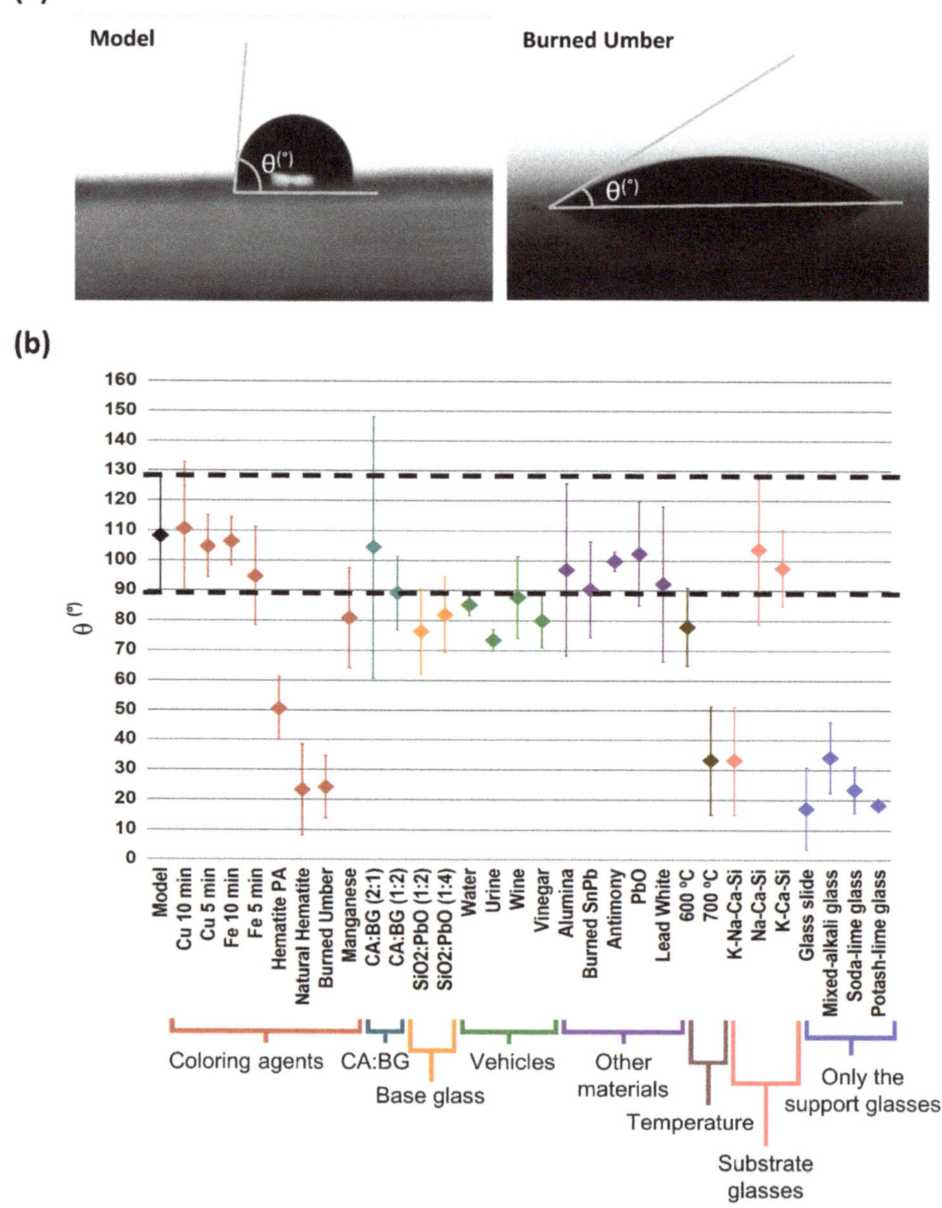

Figure 4. Contact angle results (**a**) examples of pictures taken during the test and (**b**) graphical representation of contact angle for all the samples (model sample in black).

3.3. Adherence Tests

The adhesion tests were made to analyze the compatibility between the grisailles and substrate glasses, shown in Table 6. The model sample was classified as 1, where only the superficial layer was cut, as it is practically impossible to observe the marks in the painted sample, but with residual detachment visualized in the adhesive tape (Table 6). Eighteen of the samples (Fe 10 min, Fe 5 min, Cu 10 min, Cu 5 min, BG (SiO_2 + PbO 1:2), BG (SiO_2 + PbO 1:4), water, urine, wine, vinegar, alumina, antimony, burned PbSn, 700 °C, CA:BG (1:2), mixed-alkali glass, soda-lime glass, and potash-lime glass) were also classified as 1 showing similar results as the one for the model grisaille. Indicating that these variables will not significantly impact the adhesion of the grisaille paint layers. Some of these variables' results are shown in Table 6.

Table 6. Results from the adhesion test.

Variable	Painted Samples	Adhesion Test		
	(Before Adhesion Tests)	(Reflected Light)	Adhesive Tape	Classification
Model grisaille				1
Fe 5 min				1
Cu 5 min	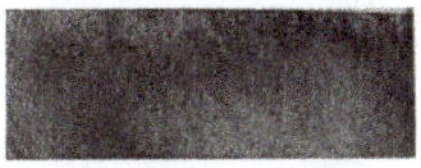		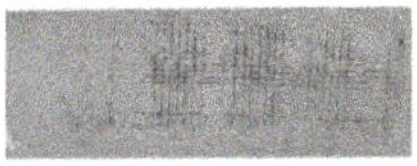	1
Manganese			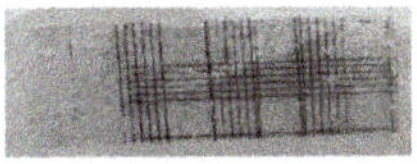	3
Hematite PA				4
Natural hematite			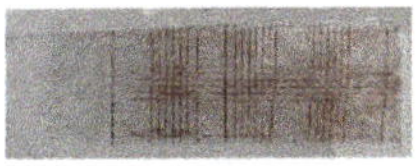	3
Burned umber				4
Burned SnPb		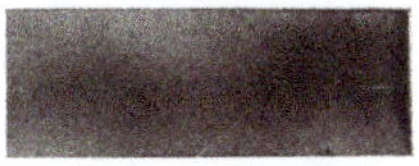	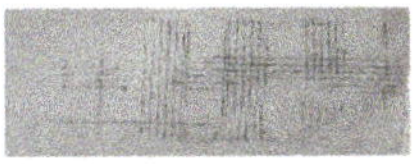	1
PbO				2
Lead white		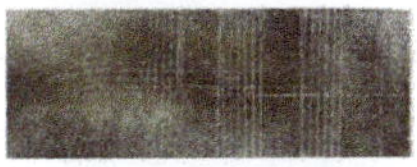	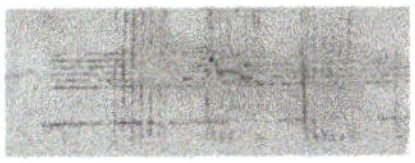	2
600 °C		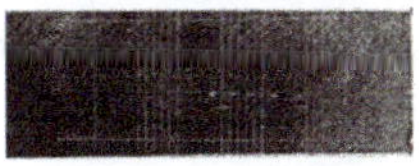	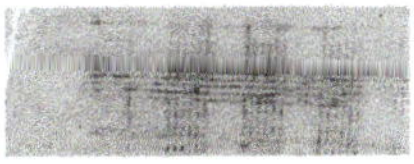	2
700 °C			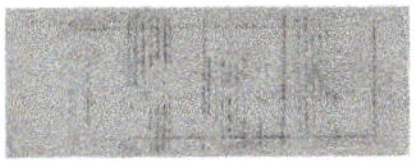	1

Table 6. *Cont.*

Variable	Painted Samples (Before Adhesion Tests)	Adhesion Test (Reflected Light)	Adhesive Tape	Classification
CA:BG (2:1)				2
CA:BG (1:2)				1

The samples where pigments were used as coloring agents were the ones that showed less adhesion of the grisaille (Table 6). The samples with manganese and natural hematite had an intense detachment, classified as 3, and the hematite PA and burned umber grisailles were classified as 4 (Table 6). In general, the pigments have smaller particle sizes than the burned metals leading to an increase in the contact surface between the coloring agents and the base glass, needing more base glass to obtain a suitable fixation of the coloring agents. The proportion between the base glass and coloring agents must be enough to cover the metal oxide grains to guarantee suitable adhesion to the substrate, as described by Bettembourg [32]. These results also agree with the contact angle results (Figure 4) where hematite (PA and natural) and burned umber were added and also showed a higher affinity with water, reinforcing that the addition of these raw materials seems to create less resistant grisailles.

The grisailles where burned lead and lead white were used also showed poor adhesion and were classified as 3 (Table 6). These compounds can also create unbalanced grisailles, altering the ratios between coloring agents and base glasses.

A lower temperature (600 °C) can also be insufficient to create a suitable adhesion, being probable that the softening temperature was not reached, as well as a higher proportion of coloring agent to base glass.

Table 7 shows some examples of the cleaning test. The results showed that the model grisaille is not susceptible to solvents, with only some fibers of the cotton swab being visible on the surface, and the cotton swabs did not present any grisaille trace (Table 7). Similar results were observed in 18 of the samples (Fe 10 min, Fe 5 min, Cu 10 min, Cu 5 min, BG (SiO_2 + PbO 1:2), BG (SiO_2 + PbO 1:4), water, urine, wine, vinegar, alumina, antimony, burned PbSn, 700 °C, CA:BG (1:2), mixed-alkali glass, soda-lime glass, and potash-lime glass) the same stable samples as in the adhesion tests.

The grisailles that showed a higher susceptibility to the solvents are the ones where burned umber, burned lead, and lead white were used, with the total removal of the painted layers. The grisailles where manganese and hematite were used showed traces of the grisaille on the cotton swabs but without removal marks on the painted layers (Table 7). These results show that a superficial layer was susceptible to the solvents, but the underneath layer was more resistant to cleaning. The solvents that have a higher impact are ethanol and acetone.

The cleaning test also proved that a temperature of 600 °C is insufficient to create a cohesive and well-adhered grisaille. The solvents can remove the painted layer, and the ones based on ethanol seem more aggressive, leaving intense marks on the surface (Table 7). The sample with a higher proportion of coloring agent can also interfere with the grisaille adhesion, showing some traces of paint on the ethanol cotton swab (Table 7).

The cleaning and adhesion tests (Table 6) agree with each other, as the less adhered grisailles are more susceptible to the solvents tested.

Table 7. Results from the cleaning test.

Variable	Before	After	Swabs
Model	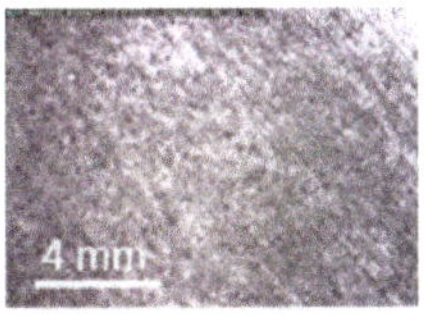	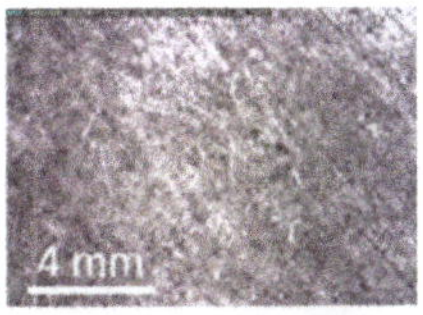	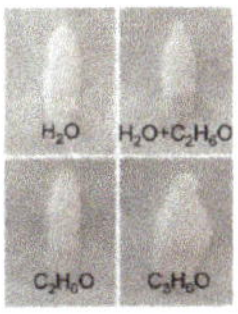
Fe 5 min			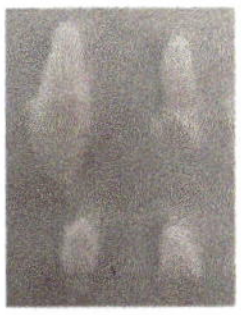
Cu 5 min		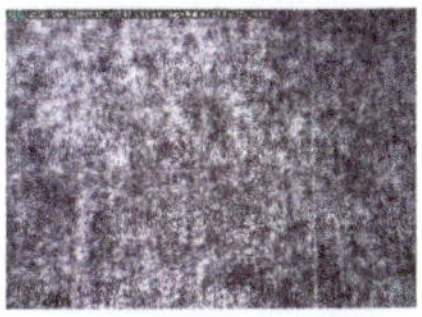	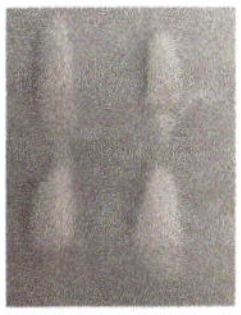
Manganese			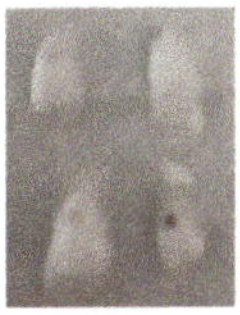
Hematite PA			
Natural hematite			
Burned Umber		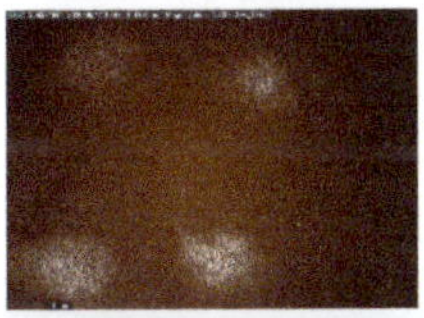	
Burned SnPb	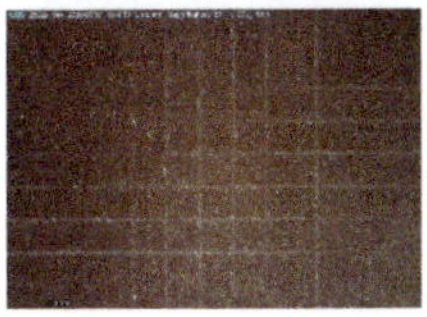	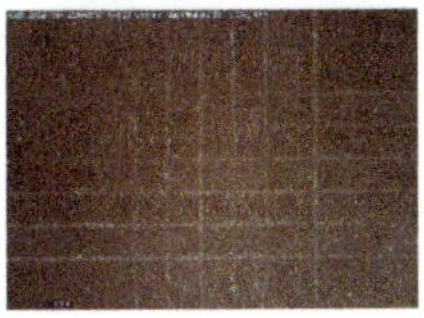	

Table 7. *Cont.*

Variable	Before	After	Swabs
PbO			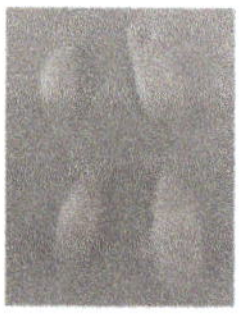
Lead White			
600 °C	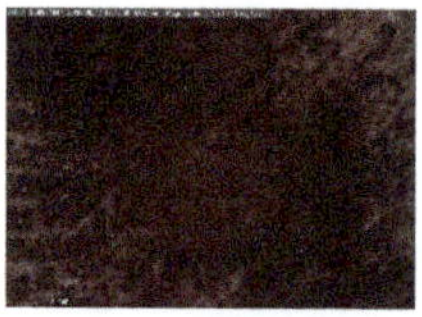	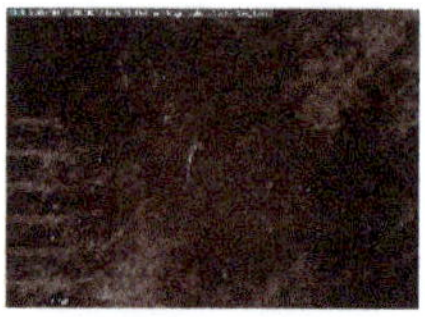	
700 °C			
CA:BG (2:1)	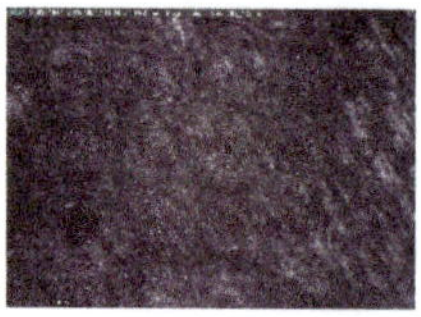	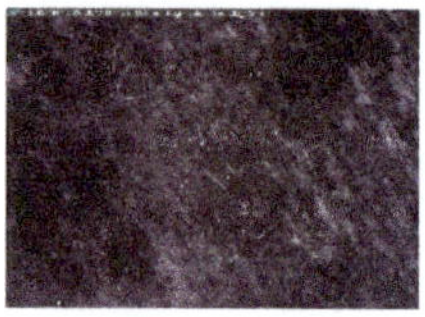	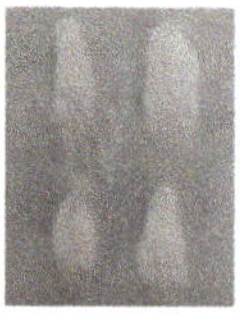
CA:BG (1:2)		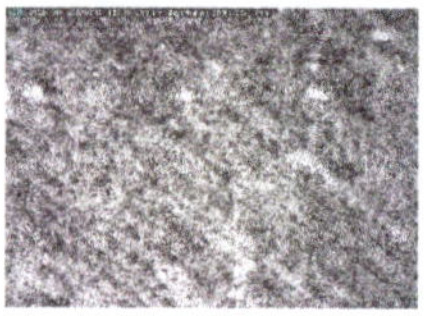	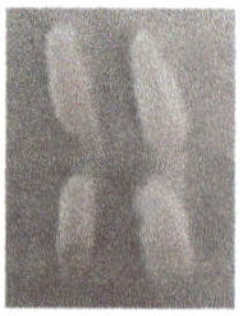

4. Conclusions

This work has demonstrated that raw materials directly affect the properties and appearance of grisailles.

Iron and copper produce similar hues; however, pigments, such as hematite (PA or natural) and burned umber, and other materials, such as lead oxide or lead white, can give specific characteristics or appearances to the final paints.

The raw materials also affect the affinity to water. Less hydrophobic materials, such as hematite and burned umber, showed an increase in the hydrophilic properties of the samples. A significant impact was also visible at temperatures of 700 °C and when a less compatible substrate glass (mixed-alkali glass) was used. These reinforce the idea that the water absorbance and penetration depend not only on the rugosity, as these results do not agree with the roughness analysis.

Considering the analysis, the average roughness (Ra*) did not show significant changes in the majority of cases. Concluding that the most important factor is the size of the coloring agents, and it is essential to adapt the grinding time to the hardness of each raw material in the grisailles production.

Another factor that can significantly impact the long-term stability of the grisailles is the unbalanced volume between base glass and coloring agent particles. The adhesion and cleaning test results proved this. As in both cases, the grisailles with hematite (PA or natural), burned umber, and manganese showed less resistance to the cutting blades and solvents, justified by the small particle size of these pigments, which created unbalanced grisailles in volume. Additionally, the grisailles with hematite (PA and natural) appeared to have a superficial layer susceptible to the solvents and another one that was more cohesive and resistant underneath.

With the characterization and tests carried out in this study, it was possible to understand that the grisaille stability can be affected by different reasons. The variables chosen to be tested allowed for establishing the main factors that must be considered while choosing the raw materials to produce grisaille. The coloring agents used and their treatments before being added to the grisailles can greatly affect the final paint layer, not only in its color but also in its rugosity, affinity with water, and adhesion and chemical resistance. The different base glasses mainly affected the grisailles rugosity, as their lead content will influence the firing conditions needed. The different vehicles tested did not greatly impact the final grisailles. It was only possible to observe the rugosity affected by the use of water due to the difficult manipulation of the grisaille and its application before firing. Paired with the coloring agents, the other materials can also significantly affect the grisailles, mainly the burned lead and lead white. Not only was it verified that they would impact the color and the grisailles adhesion and chemical resistance, but also that during the firing process, they will decompose and link themselves with components from the coloring agents and base glass, forming new mixed compounds. The temperature is another variable to consider as unsuitable firing conditions affected the grisailles, demonstrated by the contact angle measurements and the results of the adhesion and cleaning tests. As described in the literature and verified in this study, the proportions between the different grisaille components can also be a key factor in the grisailles stability. The tests verified that unbalanced proportions could affect the grisailles rugosity, mainly their physical and chemical adhesion to the glass support. The different substrate glasses tested showed suitable compatibility with the grisailles. Nevertheless, it is always a factor to take into consideration.

Furthermore, aging tests on the produced samples will significantly contribute to understanding the long-term stability and possible corrosion mechanisms of these grisailles, as well as to verify the real impact that each of the variables tested in this study will have on the grisaille layers degradation.

Author Contributions: Conceptualization, M.V. and T.P.; methodology, C.M.; validation, T.P.; investigation, C.M. and J.V.P.; writing—original draft preparation, C.M.; writing—review and editing, C.M., M.V., J.V.P. and T.P.; supervision, M.V. and T.P. All authors have read and agreed to the published version of the manuscript.

Funding: This research was funded by Fundação para a Ciência e Tecnologia de Portugal (projects UIDB/00729/2020, UIDP/00729/2020, UIDB/50025/2020-2023, researcher grant CEECIND/02249/2021 and doctoral grant ref. PD/BD/136673/2018).

Institutional Review Board Statement: Not applicable.

Informed Consent Statement: Not applicable.

Data Availability Statement: Not applicable.

Acknowledgments: The authors wish to thank María José Velasco (ICV-CSIC, Spain) for the XRF analysis, Daniela Hernandez (ICV-CSIC, Spain) for her help during the contact angle analysis, and Mónica Álvarez de Burgo (IGEO-CSIC-UCM, Spain) for the access to the colorimetric and roughness analysis equipment.

Conflicts of Interest: The authors declare no conflict of interest.

Appendix A

Table A1. Brands/manufacturers of the commercial materials used in the samples production.

Compound	Brand/Manufacturer
SiO_2	Sigma-Aldrich Chemistry
PbO	Sigma-Aldrich Chemistry0
SnO_2	Alfa Aesar Chemicals
Fe_2O_3	Riedel-de Haën–Honeywell Research Chemicals
MnO_2	Panreac Química SLU–ITW Reagents
Al_2O_3	Fluka–Honeywell Research Chemicals
Sb_2O_3	Alfa Aesar Chemicals
Burned Umber	Winsor and Newton
Gum Arabic	Debitus–Peintures pour verre

Appendix B

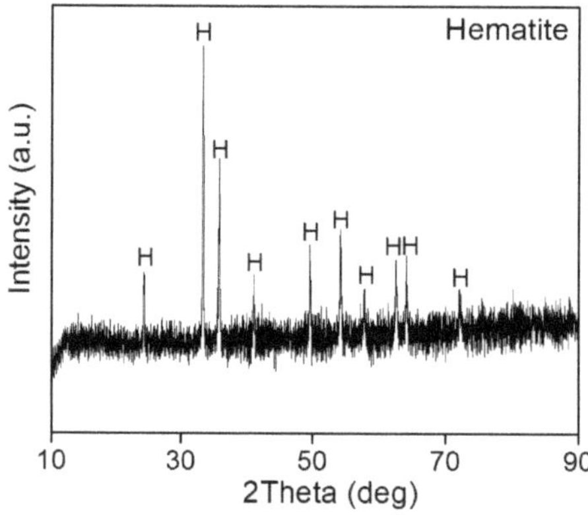

Figure A1. XRD result for burned iron. Hematite (H) was identified.

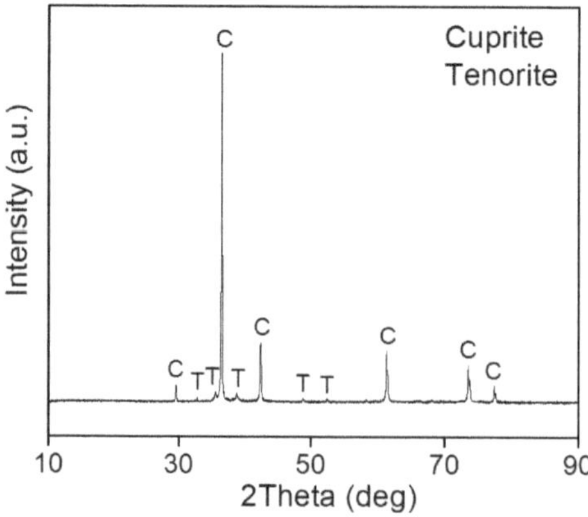

Figure A2. XRD result for burned copper. Cuprite (C) and tenorite (T) were identified.

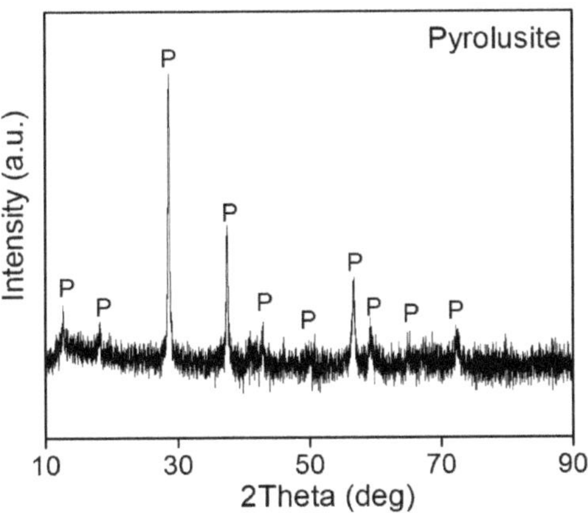

Figure A3. XRD result for manganese PA. Pyrolusite (P) was identified.

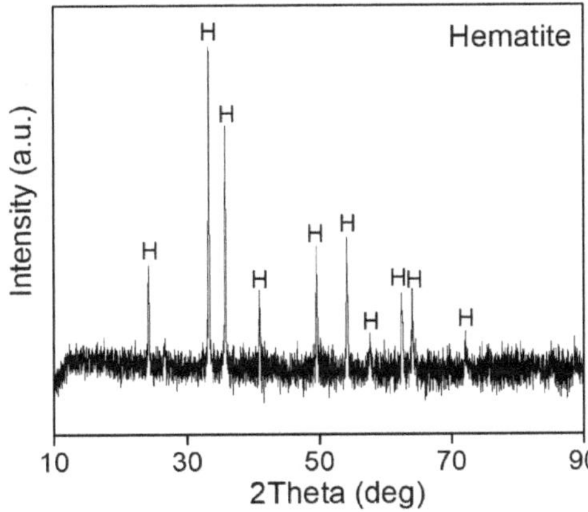

Figure A4. XRD result for hematite PA. Hematite (H) was identified.

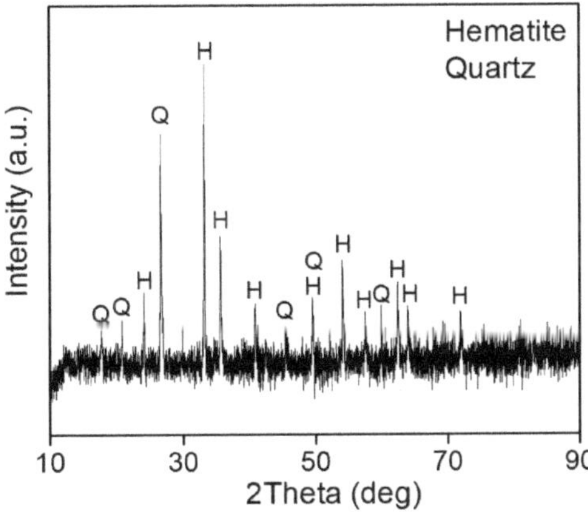

Figure A5. XRD result for natural hematite. Hematite (H) and quartz (Q) were identified.

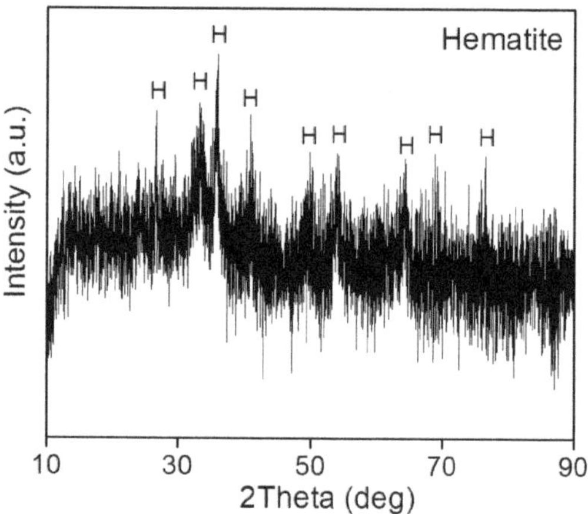

Figure A6. XRD result for burned umber. Hematite (H) was identified.

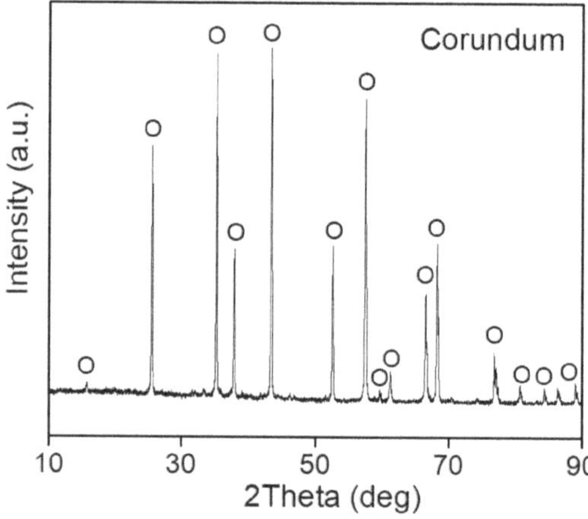

Figure A7. XRD result for alumina PA. Corundum (O) was identified.

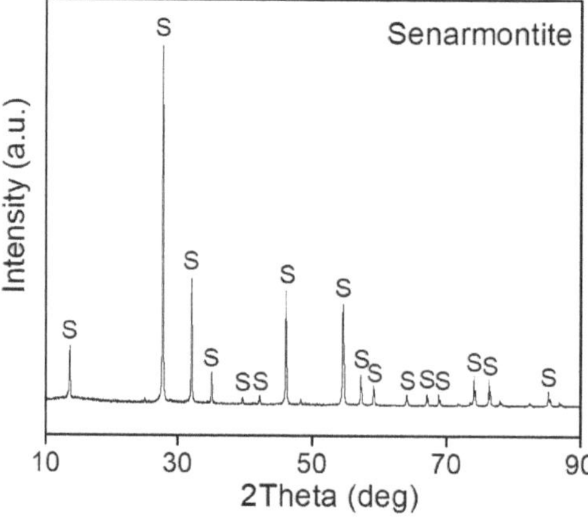

Figure A8. XRD result for antimony PA. Senarmontite (S) was identified.

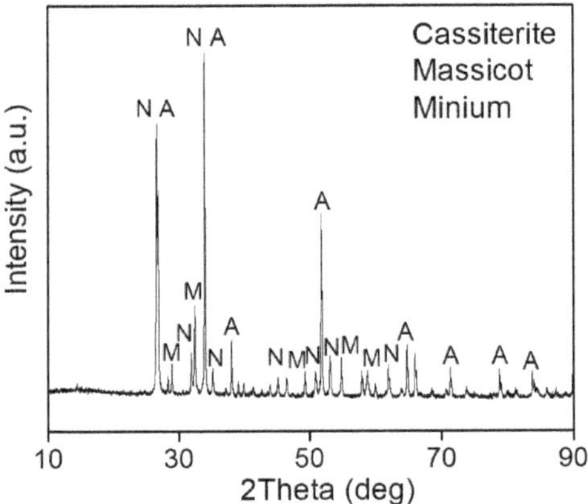

Figure A9. XRD result for burned tin and lead. Cassiterite (A), massicot (M), and minium (N) were identified.

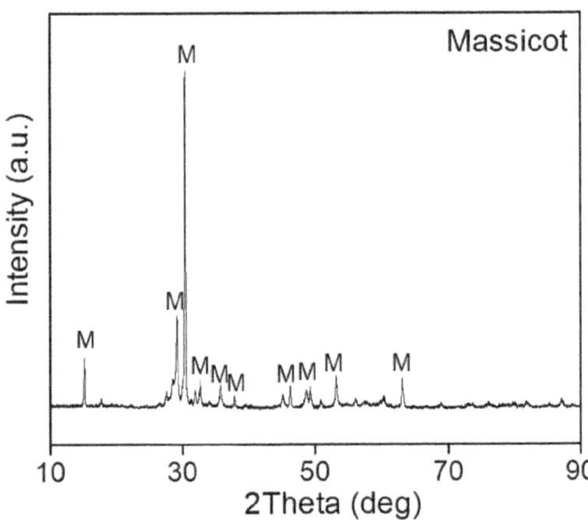

Figure A10. XRD result for burned lead. Massicot (M) was identified.

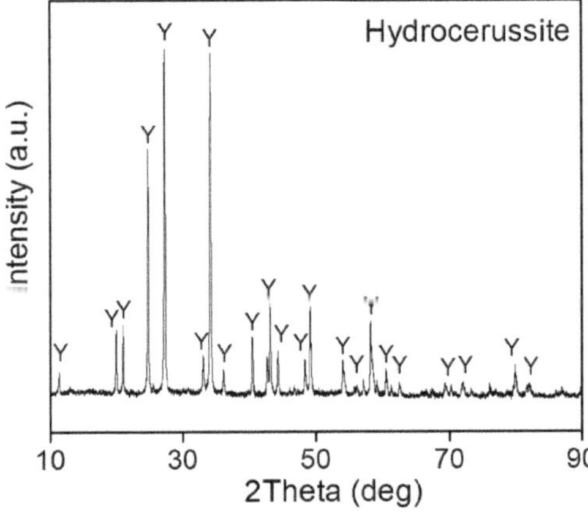

Figure A11. XRD result for lead white. Hydrocerussite (Y) was identified.

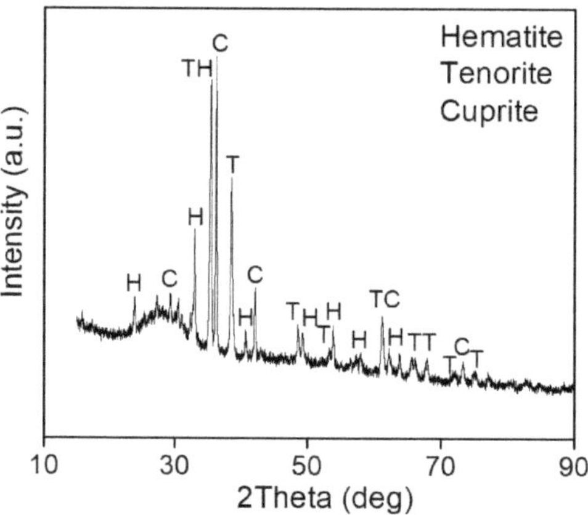

Figure A12. XRD result from the model grisaille. Hematite (H), tenorite (T), and cuprite (C) were identified.

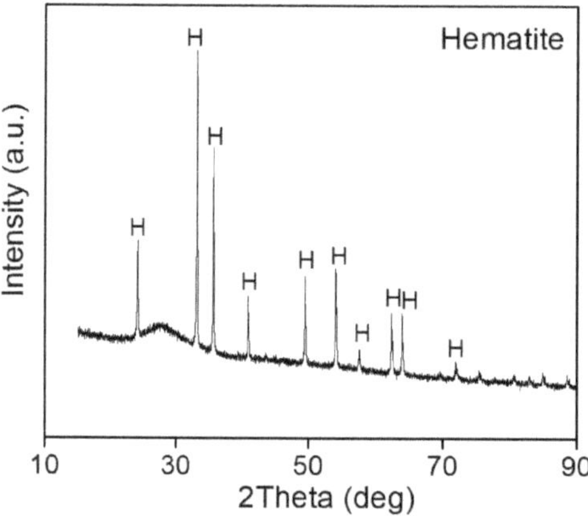

Figure A13. XRD result from the grisaille where hematite PA was used. Hematite (H) was identified.

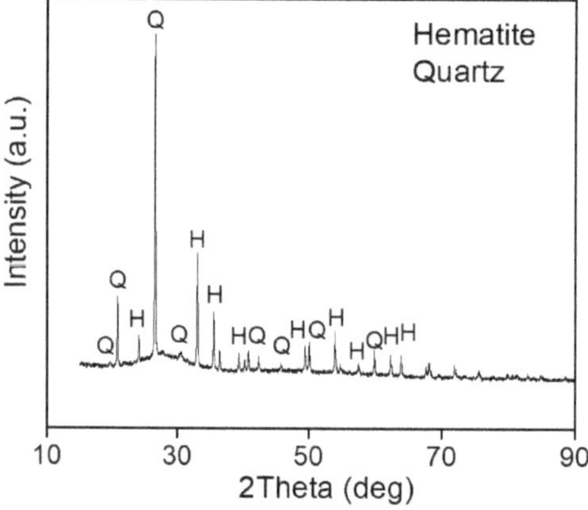

Figure A14. XRD result from the grisaille where natural hematite was used. Hematite (H) and quartz (Q) were identified.

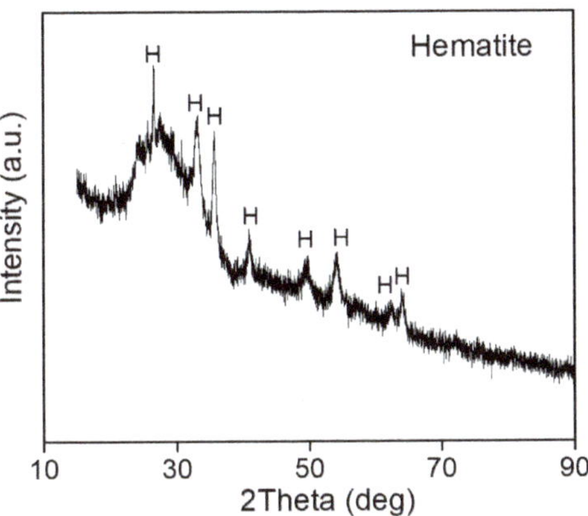

Figure A15. XRD result from the grisaille where burned umber was used. Hematite (H) was identified.

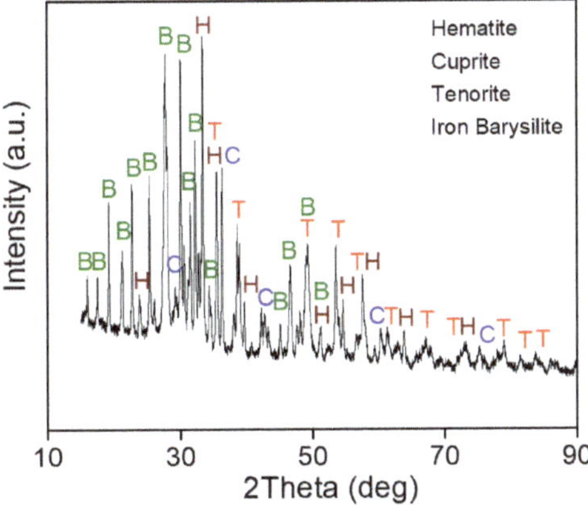

Figure A16. XRD result from the grisaille where burned lead was used. Hematite (H), cuprite (C), tenorite (T), and iron barysilite (B) were identified.

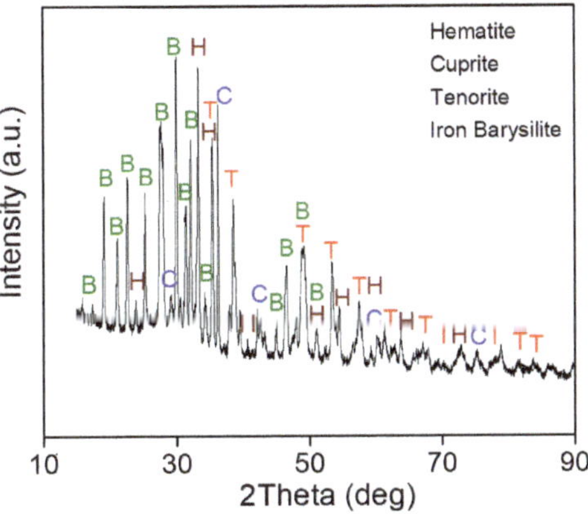

Figure A17. XRD result from the grisaille where lead white was used. Hematite (H), cuprite (C), tenorite (T), and iron barysilite (B) were identified.

References

1. Schalm, O. *Characterization of Paint Layers in Stained-Glass Windows: Main Causes of the Degradation of Nineteenth Century Grisaille Paint Layers*; Universiteit Antwerpen: Antwerpen, Belgium, 2000.
2. Machado, C.; Machado, A.; Palomar, T.; Vilarigues, M. Grisaille in Historical Written Sources. *J. Glass Stud.* **2019**, *61*, 71–86.
3. Vilarigues, M.; Machado, C.; Machado, A.; Costa, M.; Alves, L.C.; Cardoso, I.P.; Ruivo, A. Grisailles: Reconstruction and Characterization of Historical Recipes. *Int. J. Appl. Glas. Sci.* **2020**, *11*, 756–773. [CrossRef]
4. Machado, C.; Vilarigues, M.; Palomar, T. Historical Grisailles Characterisation: A Literature Review. *J. Cult. Herit.* **2021**, *49*, 239–249. [CrossRef]
5. Merrifield, M. *Original Treatises on the Arts of Paintings*; Dover Publications: Mineola, NY, USA, 1849; Volume 1.
6. Bontemps, G. *Guide du Verrier: Traité Historique et Pratique de la Fabrication des Verres, Cristaux, Vitraux*; Librairie du Dictionnaire des Arts et Manufactures: Paris, France, 1868.
7. Volf, M.B. *Chemical Approach to Glass*, 2nd ed.; Elsevier: Amsterdam, The Netherlands, 1984.
8. Foster, J. *Lives of the Most Eminent Painters, Sculptors, and Architects: Translated from the Italian of Giorgio Vasari*; Bohn, H.G., Ed.; Modern Library: New York, NY, USA, 1855.
9. Verità, M. Paintwork in Medieval Stained-Glass Windows: Composition, Weathering, and Conservation. In *Proceedings of the Forum for the Conservation and Restoration of Stained Glass, the Art of Collaboration: Stained-Glass Conservation in the Twenty-First Century, New York City, NY, USA, 1–3 June 2009*; Shepard, M.B., Pilosi, L., Stroble, S., Eds.; Harvey Miller: London, UK, 2010; pp. 210–216.
10. Rodrigues, A.; Coutinho, M.L.; Machado, C.; Alves, L.C.; Machado, A.; Vilarigues, M. An Overview of Germanic Grisailles through the Stained-Glass Collection at Pena Palace. *Heritage* **2022**, *5*, 1003–1023. [CrossRef]
11. Pradell, T.; Molina, G.; Murcia, S.; Ibáñez, R.; Liu, C.; Molera, J.; Shortland, A. Materials, Techniques and Conservation of Historic Stained Glass "Grisailles". *Int. J. Appl. Glass Sci.* **2016**, *7*, 41–58. [CrossRef]
12. Verità, M.; Nicola, C.; Sommariva, G. The Stained Glass Windows of the Sainte Chapelle in Paris: Investigations on the Origin of the Loss of the Painted Work. *AIHV Ann. 16 Congrès* **2003**, 347–351.
13. Vilarigues, M.; Da Silva, R.C. Ion Beam and Infrared Analysis of Medieval Stained Glass. *Appl. Phys. A Mater. Sci. Process.* **2004**, *79*, 373–378. [CrossRef]
14. Schalm, O.; Janssens, K.; Caen, J. Characterization of the Main Causes of Deterioration of Grisaille Paint Layers in 19th Century Stained-Glass Windows by J.-B. Capronnier. *Spectrochim. Acta Part B At. Spectrosc.* **2003**, *58*, 589–607. [CrossRef]
15. Bettembourg, J.-M. Altération et Problèmes de Conservation Des Grisailles. *Corpus Vitr. News Lett.* **1984**, *37/38*, 5–7.
16. Verità, M. Composition, Structure et Mécanisme de Détérioration Des Grisailles. *Doss. Comm. R. Monum. Sites Fouill.* **1996**, *3*, 61–68.
17. Palomar, T.; Redol, P.; Almeida, I.C.; Pereira, E.; Vilarigues, M. The Influence of Environment in the Alteration of the Stained-Glass Windows in Portuguese Monuments. *Heritage* **2018**, *1*, 365–376. [CrossRef]
18. Eastaugh, N.; Walsh, V.; Chaplin, T.; Siddall, R. *Pigment Compendium: A Dictionary of Historical Pigments*; Elsevier Butterworth-Heinemann: Oxford, UK, 2004.
19. *DIN EN ISO 4287*; Surface Roughness—Terminology: Part 1, Surface and Its Parameters. ISO: Geneva, Switzerland, 1998.
20. Ariyathilaka, P.; Haddock, D.; Spindloe, C.; Tolley, M.K. Surface Roughness of NaCl Coating Used as Release Layers in Thin Film Production. *J. Phys. Conf. Ser.* **2018**, *1079*, 012017. [CrossRef]
21. *ISO 2409*; European ISO Standard, Paints, and Varnishes—Cross Cut Test. ISO: Geneva, Switzerland, 1992.
22. Wolbers, R. *Cleaning Painted Surfaces*; Archetype Publications Ltd.: London, UK, 2000.
23. Redol, P. *O Mosteiro Da Batalha e o Vitral Em Portugal Nos Séculos XV e XVI*; da Batalha, C.M., Ed.; Câmara Municipal da Batalha: Batalha, Portugal, 2003.
24. Eberhard, Z. *Chemische Technologie des Glases*; Cable, M., Translator; Society of Glass Technology: Sheffield, UK, 2013.
25. Schmidt-Whitley, R.D.; Martinez-Clemente, M.; Revcolevschi, A. Growth and Microstructural Control of Single Crystal Cuprous Oxide Cu_2O. *J. Cryst. Growth* **1974**, *23*, 113–120. [CrossRef]
26. García-Heras, M.; Carmona, N.; Gil, C.; Villegas, M.A. Neorenaissance/Neobaroque Stained Glass Windows from Madrid: A Characterisation Study on Some Panels Signed by the Maumejean Fréres Company. *J. Cult. Herit.* **2005**, *6*, 91–98. [CrossRef]
27. Beltrán, M.; Schibille, N.; Brock, F.; Gratuze, B.; Vallcorba, O.; Pradell, T. Modernist Enamels: Composition, Microstructure and Stability. *J. Eur. Ceram. Soc.* **2020**, *40*, 1753–1766. [CrossRef]
28. Beltran, M.; Schibille, N.; Gratuze, B.; Vallcorba, O.; Bonet, J.; Pradell, T. Composition, Microstructure and Corrosion Mechanisms of Catalan Modernist Enamelled Glass. *J. Eur. Ceram. Soc.* **2021**, *41*, 1707–1719. [CrossRef]
29. Weber, F.W. *Artists' Pigments: Their Chemical and Physical Properties*; D. Van Nostrand Company: New York, NY, USA, 1923.
30. Hudson Institute of Mineralogy. Melanotekite (mindat.org). Available online: https://www.mindat.org/min-2632.html (accessed on 30 June 2022).
31. Ito, J.; Frondel, C. Syntheses of Lead Silicates: Larsenite, Barysilite and Related Phases. *Am. Mineral.* **1967**, *52*, 1077–1084.
32. Bettembourg, J.-M. Composition Er Durabilité Des Grisailles. *Sci. Technol. Conserv. Restaur. Oeuvres d'Art Patrim.* **1991**, *2*, 47–55.
33. Silvestri, A.; Molin, G.; Pomero, V. The Stained Glass Window of the Southern Transept of St. Anthony's Basilica (Padova, Italy): Study of Glasses and Grisaille Paint Layers. *Spectrochim. Acta Part B At. Spectrosc.* **2011**, *66*, 81–87.

34. Machado, C.; Vilarigues, M.; Palomar, T. Characterization of the Alteration of Debitus Grisailles. *Stud. Conserv.* **2021**, *67*, 413–422. [CrossRef]
35. Cílová, Z.Z.; Kučerová, I.; Knížová, M.; Trojek, T. Corrosion Damage and Chemical Composition o Czech Stained Glass from 13th to 15th Century. *Glass Technol. Eur. J. Glass Sci. Technol. Part A* **2015**, *56*, 153–162.
36. Palomar, T. Chemical Composition and Alteration Processes of Glasses from the Cathedral of León (Spain). *Bol. la Soc. Esp. Ceram. Vidr.* **2018**, *57*, 101–111. [CrossRef]
37. Shrimali, K.; Jin, J.; Hassas, B.V.; Wang, X.; Miller, J.D. The Surface State of Hematite and Its Wetting Characteristics. *J. Colloid Interface Sci.* **2016**, *477*, 16–24. [CrossRef]

Article

Environmental pH Evaluation in Exhibition Halls of Museo Nacional de Ciencias Naturales (CSIC, Madrid)

Daniel Morales-Martín [1], Fernando Agua [1], Josefina Barreiro [2], Angel Luis Garvía [2], Manuel García-Heras [1] and Maria Angeles Villegas [1,*]

[1] Instituto de Historia, CSIC, Calle Albasanz, 26-28, 28037 Madrid, Spain; daniel.morales@cchs.csic.es (D.M.-M.); fernando.agua@cchs.csic.es (F.A.); manuel.gheras@cchs.csic.es (M.G.-H.)

[2] Museo Nacional de Ciencias Naturales, CSIC, Calle José Gutiérrez Abascal, 2, 28006 Madrid, Spain; jbarreiro@mncn.csic.es (J.B.); garvia@mncn.csic.es (A.L.G.)

* Correspondence: mariangeles.villegas@cchs.csic.es; Tel.: +34-916022672

Abstract: Optical sol-gel environmental pH sensors have been applied for air evaluation in exhibition halls of the Museo Nacional de Ciencias Naturales (Madrid, Spain). Sensor synthesis and calibration was undertaken following a previous patent by some of the present authors. Monitoring was carried out for one full year to check the influence of meteorological seasons as well as quality of surrounding outdoor air to a big avenue located in downtown Madrid and close to the museum. Particular sites selected for sensor positions were inside showcases, under the free environment of exhibition halls, and outdoor façades of the building, for comparison purposes. pH recordings showed that exhibition halls near the outdoor air entrance had slightly low pH values, which can be attributed to outdoor pollution. However, halls located far from air entrance had neutral conditions. Concerning showcases tested, some of them showed slightly acidic pH while others were moderately acidic due to natural goods exhibited and/or to materials with which showcases were made. pH values recorded allowed the museum to make some decisions on its preventive conservation strategy.

Keywords: environmental pH; exhibition halls; sensors; museum; preventive conservation

Citation: Morales-Martín, D.; Agua, F.; Barreiro, J.; Garvía, A.L.; García-Heras, M.; Villegas, M.A. Environmental pH Evaluation in Exhibition Halls of Museo Nacional de Ciencias Naturales (CSIC, Madrid). *Appl. Sci.* **2021**, *11*, 10091. https://doi.org/10.3390/app112110091

Academic Editors: Cesareo Saiz-Jimenez and Min-Suk Bae

Received: 28 September 2021
Accepted: 24 October 2021
Published: 28 October 2021

1. Introduction

Preventive conservation in exhibition halls is a complex challenge frequently conditioned by items exhibited, visitors' regimes, and the near environmental characteristics. Reference literature has usually dealt with issues concerning environmental protection of collections [1] as well as conservation concepts for curators and natural history collections [2,3]. The effects of pollutants in the museum environment (storerooms and exhibition halls) have been studied [4,5], and guidelines on preventive conservation have also been published [6]. Very often, environmental conditions needed by goods do not fit well with those more comfortable for visitors, even though items exhibited are located inside showcases or in some kind of protected places. Moreover, outdoor environment, air quality, and potential pollution risk from urban locations could also affect long-term conservation [7,8]. Indoor air pollution and its control has been analyzed to design adequate conservation strategies [9], and microclimates generated by pollutants in museums have been monitored to propose preventive conservation measures [10]. In addition, the use of museum spaces as places for socialization and interaction between the general public and the museum culture makes the museum environment a fragile system that needs to be stabilized for proper preventive conservation of exhibited items [11,12]. Accordingly, the evaluation of the museum air quality and the limit of pollutants has been also investigated [13].

In fact, these concepts and starting assumptions are the same or very close to those considered for storerooms of museums where many items, which are even more abundant than those exhibited, must be properly conserved [14]. Environmental monitoring by using sensors has proved to be a useful strategy for protecting cultural and natural items, especially those sensitive to pollutants and susceptible to deterioration or corrosion [15]. Within this subject, previous research was conducted in the Museo Nacional de Ciencias Naturales (MNCN) using environmental pH sensors. Resulting data pointed out the meaning and relevance of the deviation from neutral conditions and how it should be considered a direct methodology to evaluate air quality and real preventive conservation conditions.

Environmental pH sensors are based on sol-gel technology; they consist of a thin coating applied upon a common soda lime glass slide. The coating matrix is composed of a polysiloxane network obtained from a silicon alkoxide in which a sensitive organic dye, against pH, is immobilized into the porous coating structure. Partial thermal densification of the whole coating allows one to obtain a material able to exhibit sensitivity to air chemical acid/base species (due to its residual porosity) and is stable against water, chemicals and handling [16].

Some former investigations carried out in different museums demonstrated that these sol-gel sensors are useful to indicate the current status of environmental pH in exhibition halls, showcases, storerooms, etc. They are also informative about pH variations due to natural or controlled ventilation, the impact of an unusual number of visitors, or when a special event took place in the museum spaces or in its surroundings [17,18]. Therefore, environmental pH monitoring by using these sensors is a useful tool to check preventive conservation conditions quickly, since no more than 24 h is needed for sensitization of sensors with accuracy high enough to detect small variations on air quality (sensor accuracy is estimated to be ± 0.1 units of the pH scale) [19].

Other properties and characteristics of pH sensors are reversibility, reuse, low cost, and miniaturization possibility, or adaptation for every real need in museum spaces. In addition, sensors can be used to check outdoor pH, which allows an air quality comparison indoors and outdoors, as well as an assessment of the impact of potential urban pollution on museum collections. One of the most practical properties of sensors is reversibility. They are able to determine environmental pH in a dynamic way, i.e., they detect all the acid/basic species affecting the whole pH at a given moment, since they are continuously sensitized, and indicate the air pH value quantitatively. Therefore, they can be regenerated in the laboratory by using pH-buffered solutions, which makes the sensors reusable for many evaluations until their lifetime is ended (at about 9 months for indoor use and 4 months for outdoor use). Among other papers on preventive conservation based on air quality monitoring and evaluation of air acidity [20–23], a work by some of the authors has reviewed other pH sensors currently available and the outstanding advantages of the present pH optical sensors [19].

The aim of the present research was to evaluate environmental pH in several exhibition spaces of the MNCN. This main objective was essential to revise preventive conservation actions necessary to be taken into account for integral conservation of the MNCN collections and, in particular, of natural items exhibited, which are visited by many people, due to their high biological and cultural significance. This conservation interest is even higher because of the nature of these items which are, in part, dissected animals from varying geographical origins, dissection methods, sizes, and natural components (skin, hair, feather, squama, bone, etc.) [24].

From previous work with environmental pH sensors carried out in the MNCN, the methodology applied consisted of sensor installation in the most critical positions and close to the most important specimens to be monitored, comparison with pH values detected outdoors, study of the ventilation regime, and assessment of the influence of meteorological parameters in the one full year evaluation. In this way, the interpretation of the overall results can be of great help for making decisions on a preventive conservation strategy of the MNCN.

2. Materials and Methods

Sol-gel pH-sensitive coatings upon common soda line glass slides were prepared following a previous patent by some of the present authors [25]. The silica matrix was provided from tetrathoxysilane ($Si(OC_2H_5)_4$, TEOS), absolute ethanol (EtOH), and hydrochloric acid (37 wt. % HCl). All reagents used were of analysis grade purity. The pH-sensitive phase added to the prehydrolized silica matrix was chlorophenol red ($C_{19}H_{12}Cl_2O_5S$, 3′-3′-dichlorophenolsulfonaphthalein, CR), an organic dye that changes its optical absorption depending on pH: yellow below pH 4.8, violet above pH 6.7, pink around neutral pH. Details on sol synthesis are provided in [19]. Coatings upon glass slides were obtained by dipping at room temperature and at a drawing rate of 1.35 mm $\times$ s^{-1}. Thermal partial densification of coatings was achieved at 60 °C for 3 days in a forced air stove. After that, the coated area of the glass slide was cut from the uncoated one, and the glass edges were then polished. Figure 1 shows the aspect of one of the sensors obtained.

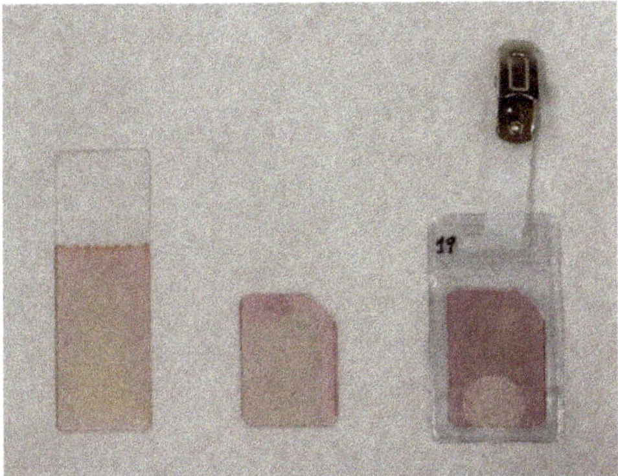

Figure 1. Aspect of a pH sensor obtained from the glass slide, coated, cut, polished and finally protected by a polyethylene bag and provided with a gripper.

Sensors were calibrated one by one using pH-buffered solutions (Buffer Hydrion Salt, Sigma Aldrich, St. Louis, MO, USA) in which they were immersed for 10 min. After that immersion time, a visible spectrum was recorded within the 380–750 nm wavelength range. An Ocean Optics model HR 4000 CG spectrophotometer was used. Sensor calibration was mainly carried out at pH 5, 6, 7 and 8. These pH values were selected because the most important pH range to be evaluated is around neutrality. Values of the absorption band intensity at 575 nm (characteristic band of CR) for each pH nominal calibration were used to set up the calibration curve. Figure 2 shows the calibration curve for one of the sensors synthesized and the error bars corresponding to standard deviation. On the basis of individual calibration curves for all sensors synthesized, estimated sensor accuracy is $\pm$0.1 units of the pH scale.

Once the sensors were calibrated, they were installed in different positions of exhibition spaces of the MNCN, according to criteria of specimens' importance and/or potential risk of urban pollution impact as well as of visitors' impact. Table 1 summarizes sensor positions during the evaluation carried out in February 2019.

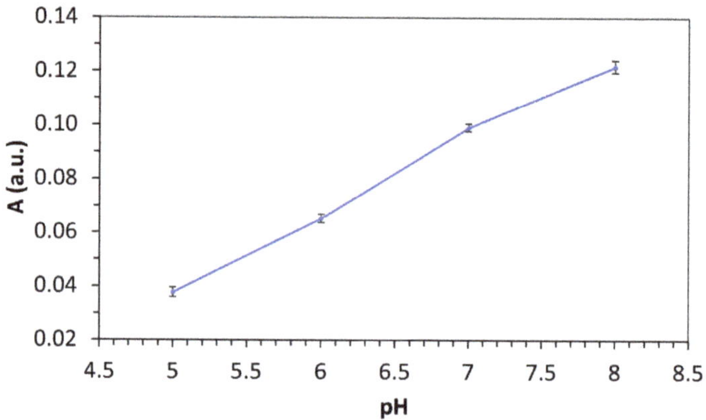

Figure 2. Calibration curve for sensor number 1 obtained on 22 November 2019.

Table 1. Sensor position during the evaluation carried out in February 2019.

Sensor nr.	Inside (I) Hall Free Environment (E) Outdoor (O)	Particular Position
1	I	Showcase, common moorhen, Biodiversity area
2	E	On the showcase ceiling, common moorhen, Biodiversity area
3	I	Showcase, rhinoceros (complete specimen), Biodiversity area
4	E	Entrance on the guard's space (on the ceiling of showcase V18), Biodiversity area
5	I	Showcase, kangaroo skeleton, Biodiversity area
6	E	On the whale skeleton, Biodiversity area
7	I	Showcase, birds, National Parks area
8	E	On the first joist, National Parks area
9	I	Showcase, little bustard, Terrestrial Mediterranean area
10	E	On the diorama ceiling, Pyrenean chamois group, Terrestrial Mediterranean area
11	O	Main façade gate, Biodiversity area
12	O	Rear façade gate, Biodiversity area

Sensors were sensitized for 24 h, and then their optical response was recorded in situ using the patented portable measure unit [25], which was connected to a laptop. The computer was provided with the specific software for handling the records [26]. Methodology for recordings consisted of taking the sensor from its sensitization position and, immediately, introducing it in the measure unit. Then, the pH values detected appear in the laptop screen. This routine was repeated for all sensors installed, and, after that, all sensors were regenerated one by one at the laboratory by using pH-buffered solutions and repeating the calibration procedure explained above. Following this procedure, the sensors could be reused many times in the same or in different positions, according to needs and criteria of the museum collection curators. In general, sensors were installed once again in the same position to check potential pH changes due to meteorological seasons and/or to visitors, but in other cases in which pH showed noticeable stability near neutrality, the corresponding sensors were installed in other positions to obtain additional data on environmental pH. Recordings were taken after the first 24 h of sensor exposition and then at both 48 and 72 h. These later recordings were taken to confirm the results of the first recordings since 24 h is enough for sensors to be fully sensitized. After two or three confirmation recordings, sensors were uninstalled, regenerated, and recalibrated in the laboratory up to the next evaluation that could be undertaken immediately or after some days or weeks.

3. Results

Evaluation with pH optical sensors has been accomplished for the four seasons of a full year from February 2019 to November 2019. Monitoring covered four different exhibition halls in which 12 positions were checked. A total of 236 recordings were made. In exhibition halls, the sensor positions were inside showcases as well as under the hall free environment. In addition, two sensors were installed outdoors for comparison purposes. One of them was installed in the main façade (West orientation) and another one in the rear façade (East orientation). The main façade is near Paseo de la Castellana, an important avenue in downtown Madrid with heavy road traffic. The rear façade is somehow protected from pollution and is very close to a backyard without road traffic. Figure 3 shows some sensor positions.

(a)

(b)

(c)

Figure 3. Some sensor positions. (**a**) Under the hall free environment. (**b**) Inside a showcase. (**c**) Outdoor in the main MNCN façade.

Winter evaluation (Table 1) was carried out from 4 February up to 18 February. Six recordings were done for each sensor from the day after installation up to the end of evaluation when sensor recordings were stabilized. Outdoor sensors were installed on 6 February and they were exposed up to 18 February. Figure 4 summarizes the average results obtained.

Although the initial pH values before sensor sensitization were plotted in Figure 4, these values are not taken into account for average calculations. Sensor number 4 was lost during the evaluation period; therefore, its recordings of 14 and 18 February are missing. As can be seen in Figure 4, all indoor sensors are stabilized after the first 24 h of exposition, which confirms, as expected, their high sensitivity and fast response time. However, sensors installed outdoors showed a decreasing habit that indicates an intensely acidic pH as well as insufficient exposition time. Probably they would reach stabilization after more days of exposition.

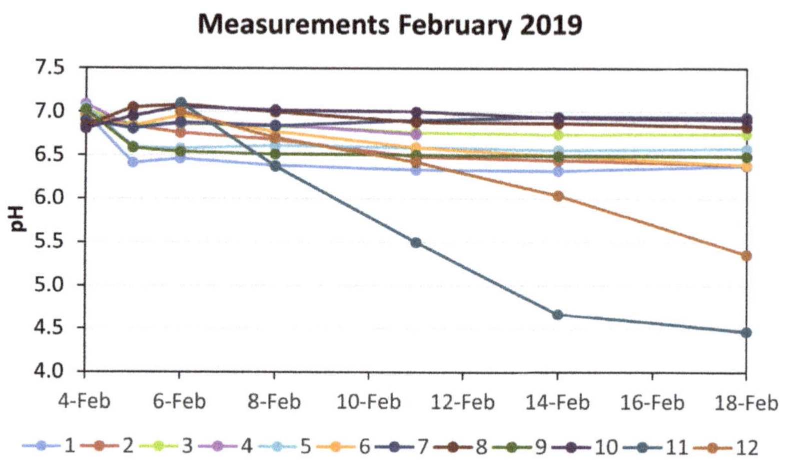

Measurements February 2019

Figure 4. pH average results obtained during the evaluation carried out on February 2019. Numbers at the bottom of the X axis are the sensors numbers.

The second evaluation was undertaken from 20 May up to 27 May. Four recordings were taken for each sensor from the day after installation up to the end of evaluation. Sensor positions were the same as for the former evaluation. pH average results obtained are shown in Figure 5. In this case, an initial stabilization of pH recordings from outdoor sensors can be observed.

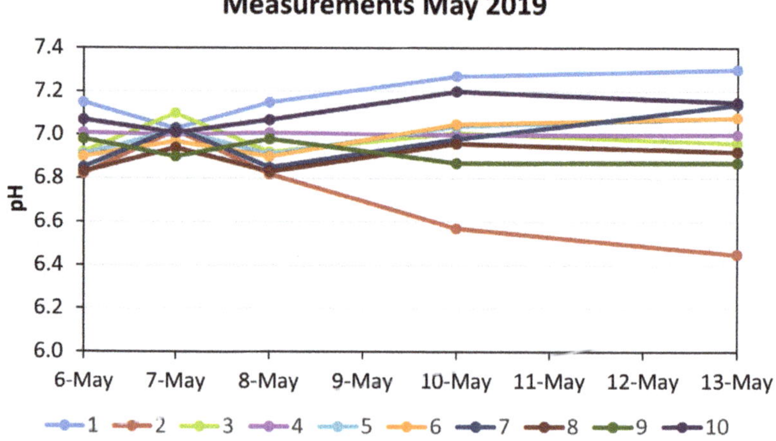

Figure 5. pH average results obtained during the evaluation carried out on May 2019. Numbers at the bottom of the X axis are the sensor numbers.

Summer evaluation was carried out from 23 July up to 31 July. Four measurements were recorded from the day after installation to the evaluation ending. All sensor positions were changed except those of showcase V18 near the MNCN entrance and the two sensors installed outdoors. Figure 6 shows the results obtained, and Table 2 summarizes the new sensor positions.

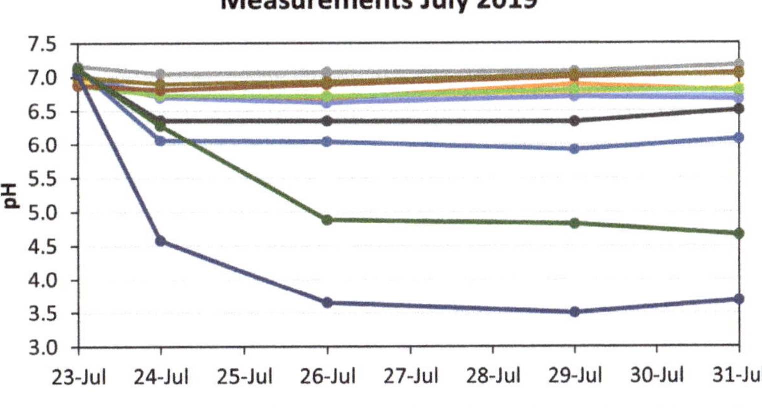

Figure 6. pH average results obtained during the evaluation carried out in July 2019. Numbers at the bottom of the X axis are the sensor numbers.

Table 2. Sensor positions during the evaluation carried out in July 2019.

Sensor nr.	Inside (I) Hall Free Environment (E) Outdoor (O)	Particular Position
1	I	Showcase, African fauna, Biodiversity area
2	E	On the ceiling showcase, geographical distribution, Biodiversity area
3	I	Showcase, rhinoceros skull, Biodiversity area
4	E	Entrance on the guard's space (on the ceiling of showcase V18), Biodiversity area
5	I	Showcase, okapi, Biodiversity area
6	E	On the totem close to thylacine, Biodiversity area
7	I	Diorama Imperial Eagle Group, National Parks area
8	E	On the second joist (on diorama Badgers Group), National Parks area
9	I	Diorama Shoveler Ducks Group, Terrestrial Mediterranean area
10	E	On the diorama ceiling Cranes Group, Terrestrial Mediterranean area
11	O	Main façade gate, Biodiversity area
12	O	Rear façade gate, Biodiversity area

Finally, the fourth evaluation corresponding to fall was carried out from 25 November up to 29 November. Sensor positions were the same as in the former summer evaluation (Table 2). Three recordings were taken for each sensor, and the corresponding results are summarized in Figure 7. It should be noted that in this evaluation the pH values from outdoor sensors reached stabilization during the evaluation period.

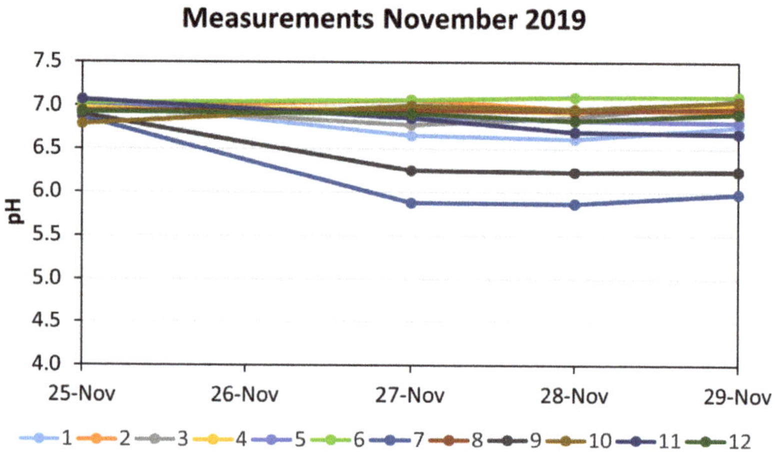

Figure 7. pH average results obtained during the evaluation carried out on November 2019. Numbers at the bottom of the X axis are the sensor numbers.

Comparison of all pH recordings from the different meteorological seasons are summarized in Table 3. For clarity, the results corresponding to different positions of the same sensor (e.g., 1a, 1b) are indicated in separated cells.

Table 3. pH average results obtained during the four evaluations.

Sensor nr.	Feb-2019	May-2019	Jul-2019	Nov-2019	Average pH
1a	6.4	6.5	-	-	6.5
1b	-	-	6.7	6.7	6.7
2a	6.6	6.9			6.8
2b	-	-	6.8	7.0	6.9
3a	6.8	6.8			6.8
3b	-	-	7.1	6.9	7.0
4	6.8	6.9	6.8	7.0	6.9
5a	6.6	6.8	-	-	6.7
5b	-	-	6.7	6.8	6.8
6a	6.7	6.8	-	-	6.8
6b	-	-	6.8	7.1	7.0
7a	6.9	6.7			6.8
7b	-	-	6.0	5.9	6.0
8a	7.0	6.9	-	-	7.0
8b	-	-	6.9	7.0	7.0
9a	6.5	6.7	-	-	6.6
9b	-	-	6.4	6.2	6.3
10a	7.0	7.0	-	-	7.0
10b	-	-	7.0	7.0	7.0
11	5.3	5.3	3.9	6.8	5.3
12	6.1	5.9	5.2	6.9	6.0

Orange = Inside showcase. Blue = Hall free environment. Red = Outdoor. - Not measured.

4. Discussion

Results obtained indicate that pH values recorded vary to a great extent depending on the particular location of exhibition halls and showcases as well as on meteorological seasons.

Related to the influence of sunlight, environmental humidity, and temperature in sensing results, sunlight, in this case, does not affect the sensor response since they were installed indoors and were always illuminated by LED systems. It was widely demonstrated in former research [19] that environmental humidity affects the sensor response (obviously, pH depends on the amount of H_3O^+ and OH^-), while temperature affects results as far as it influences relative humidity. Temperature and relative humidity are simultaneously measured and controlled during pH sensor sensitization. The patented portable device that measures the sensors' optical responses [25] includes both relative humidity and temperature sensors. Either the calibration curves or the specific software for the corresponding results management take into account humidity and temperature effects. Such effects were widely investigated in [27–29] and revised in [19].

Concerning sensors two, four and six installed in the so-called Biodiversity hall (main greatest hall and the first one after the museum entrance), the positions of sensors two and six were changed during the third and fourth evaluation. Two other sensors were installed in two smaller halls (number eight in the National Parks hall and number ten in the Mediterranean hall); these were somehow protected from the urban air coming from the outdoors. Sensors two, four and six of the Biodiversity hall recorded an average pH between 6.7 and 6.9 (slightly acidic). Since some sensors were not stabilized after the evaluation period, the exposition time was then extended. This could be due to the particular environmental conditions of the Biodiversity hall, which is directly exposed to the urban air that enters through the very big main door oriented to Paseo de la Castellana, a big avenue with heavy road traffic. No control of such ventilation was provided during the sensor sensitization. Moreover, intake of potential polluted air could be intensified by the high number of visitors accessing the museum by this main door, which is the regular entrance for museum personnel also. However, sensor numbers eight and ten installed in other exhibition halls far from the main entrance recorded pH average values near neutrality (6.9–7.0). Therefore, the connection between urban air, probably polluted, and the slightly acid recordings at the Biodiversity hall seems to be clear.

As far as the five sensors installed in showcases (numbers one, three, five, seven and nine) are concerned, all their positions were changed during the third and fourth evaluations. Except the sensor located in the showcase containing a rhinoceros skull, which recorded neutral pH, all the other sensors recorded a slightly acidic pH in the 6.5–7.0 range. Among them, three sensors (numbers 1a, 7b and 9b), which were installed inside ancient wooden showcases with small size and abundant organic matter (e.g., dried plants, feathers, paper, wood, etc.), recorded pH average values below 6.5. Combination of these kinds of materials and showcase tightness could promote generation of an acidic microenvironment, probably due to the emission of volatile organic compounds (e.g., formaldehyde, formic acid and acetic acid).

Concerning the sensors installed outdoors, recordings show a very acidic environment due to the heavy road traffic of Paseo de la Castellana as is mentioned above. The main acidic pollutants in urban areas are SO_2 and NO_x, which are combined with moisture generated sulfuric and nitric acid [30]. In addition, when relative humidity is low, for instance during warm seasons, a dry acid deposition phenomenon takes place [31–33]. Outdoor positions of sensors 11 and 12 are shown in Figure 8, and the corresponding pH average recordings are summarized in Table 3.

Figure 8. Location of the MNCN in downtown Madrid and position of outdoor sensors 11 and 12.

Although pH average values recorded in both main and rear façades of the MNCN are strongly acidic, pH values corresponding to the main façade are lower due to its direct exposure to road traffic, while the rear façade is oriented to the opposite direction towards a backyard. Influence of outdoor environmental conditions on indoor pH can be analyzed from the data of Table 3 (sensors number 2a, 4, 6a and 11) and Figure 9.

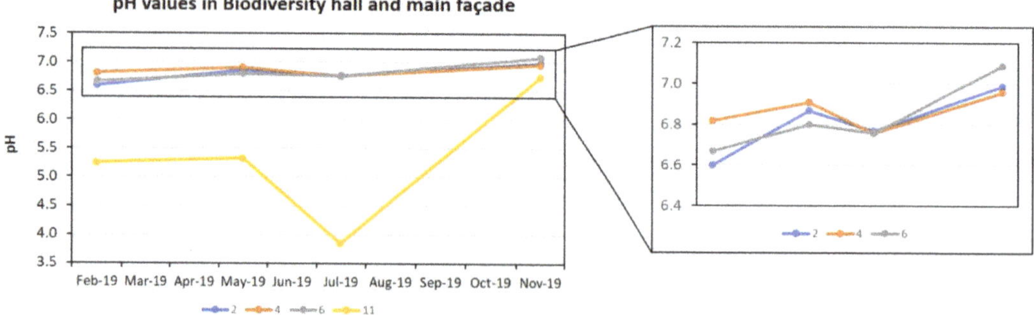

Figure 9. Evolution of pH average values recorded in the main Biodiversity hall and the main façade. Numbers at the bottom of the X axis are the sensor numbers.

The same behavior pattern can be observed both indoors and outdoors, which means that the slightly acidic pH recorded in the Biodiversity hall is due to the influence of the nearby polluted urban environment. The fact that indoor pH is noticeably less acidic than outdoor pH is explained by the effect of the building as an absorbing surface for gaseous pollutants (e.g., porous materials: masonry, bricks, painted walls, tiles, etc.).

According to the present environmental evaluation results, the main entrance of the MNCN, including the circulation path of visitors and staff, has been redesigned to mitigate, as far as possible, the urban pollution which are combined with moisture and then generate sulfuric and nitric acid effects on the Biodiversity hall. Likewise, the pandemic situation generated by COVID-19 required a new organization of the museum entrance and new security protocols. Main access to the MNCN is composed of three big doors of the same size in central, right and left positions. These doors are directly connected to the Zoology area in which biodiversity exhibitions are located, where environmental pH measures indicated some acidic values. To minimize the direct entrance of outdoor air in this area (probably polluted), the visitor circulation was changed in such a way that the entrance was restricted to one of the side doors and the exit was restricted to the other, while the central one remained closed. Nowadays, this central door is reserved and only opened in case of exceptional events and emergency situations. From 5 June 2020, the museum reopened after the general confinement due to the country alarm situation. Since that date, the remodeled access will contribute to the preservation of a clean and safe environment for both natural items and human beings. Likewise, some other preventive actions concerning natural items in showcases were adopted, which include substances that absorb pollutants [34,35]; even more preventive measures are now under discussion and are being validated by interdisciplinary points of view, taking into account visitors and staff health and optimal conservation of natural and cultural items of collections and exhibitions.

Further pH measurements in the MNCN, to check the influence of mitigation measures carried out, are the subject of an ongoing research, and results will be published elsewhere.

5. Conclusions

Field research undertaken with sol-gel pH sensors in exhibition spaces of the MNCN and the corresponding outdoor locations has pointed out a very acidic pH near the building's façades, especially in the main one oriented near a heavily trafficked avenue, which directly affects pH values recorded indoors in the adjacent hall. Climate and meteorological seasons influence pH values recorded. Acidity recorded during long dry seasons is stronger than that during rainy seasons, which are scarce in Madrid. This fact is related to the dry acid deposition phenomenon.

Comparison of pH average values of the main hall close to the MNCN entrance and pH values recorded in other smaller halls far from the main entrance indicates that uncontrolled ventilation promotes a more acidic environment. In showcases, pH average values vary from slightly acidic to moderately acidic, and are mainly due to natural goods exhibited and/or the kind of materials with which showcases were made.

Bearing in mind the former results, some recommendations could be stated. In exhibition halls located far from the main entrance, no special preventive conservation actions should be adopted. However, in the main Biodiversity hall, some preventive measures could be considered, such as those oriented to diminish the entrance of urban polluted air, especially during the long dry seasons, e.g., control of the main door opening, installation of some physical–chemical barrier in the entrance area by means of panels, etc. The slightly acidic pH detected inside showcases could be neutralized by periodic ventilation or tightness release, or by introducing absorbing substances to neutralize volatile organic compounds, e.g., active carbon.

Among other preventive measures, the main entrance to the MNCN through the façade big door has now been protected and changed to a lateral entrance that deadens urban pollution impact. Further research will include evaluation of air acidity in this MNCN area to re-check environmental conditions.

Author Contributions: Conceptualization, M.A.V.; Methodology, F.A.; Formal Analysis, F.A. and D.M.-M.; Investigation, M.A.V., M.G.-H., A.L.G. and J.B.; Resources, M.G.-H., A.L.G. and J.B.; Writing—Original Draft Preparation, M.A.V.; Writing—Review & Editing, all coauthors; Project Administration, M.G.-H.; Funding Acquisition, M.G.-H. and M.A.V. All authors have read and agreed to the published version of the manuscript.

Funding: This research was funded by: Project HERICARE "Sistematización de métodos y protocolos de conservación integral de materiales y colecciones del Patrimonio Cultural, su caracterización arqueométrica y valorización social", (ref. PID2019-104220RB-I00) from the Spanish Ministry of Science and Innovation; Program TOP Heritage "Technologies in Heritage Sciences" (ref. CM S2018/NMT-4372) from the Regional Government of Madrid and European Social Funds. D. Morales-Martín is contracted under the TOP Heritage program.

Institutional Review Board Statement: Not applicable.

Informed Consent Statement: Not applicable.

Acknowledgments: This research has been sponsored by the CSIC Interdisciplinary Platform "Open Heritage: Research and Society" (PTI-PAIS). The authors acknowledge technical assistance of O. Ramos-Lugo (MNCN) and professional support from the TechnoHeritage Network of Science and Technology for the Conservation of Cultural Heritage.

Conflicts of Interest: The authors declare no conflict of interest.

References

1. Appelbaum, B. *Guide to Environmental Protection of Collections*; Sound View Press: Madison, WI, USA, 1991.
2. Bachmann, K. (Ed.) *Conservation Concerns: A Guide for Collectors and Curators*; Smithsonian Institution: Washington, DC, USA, 1992.
3. Carter, D. *Care and Conservation of Natural History Collections*; Angrete, K., Ed.; Butterworth-Heinemann: Oxford, UK, 1999.
4. Hatchfield, P.B. *Pollutants in the Museum Environment: Practical Strategies for Problem Solving in Design, Exhibition and Storage*; Archetype: London, UK, 2002.
5. Thomson, G. *The Museum Environment*, 2nd ed.; Routledge: New York, NY, USA, 1994.
6. García, I. *La Conservación Preventiva de Bienes Culturales*; Alianza Forma: Madrid, Spain, 2013.
7. Tétreault, J. *Airborne Pollutants in Museums, Galleries and Archives: Risk Assessment, Control Strategies and Preservation Management*; Canadian Conservation Institute: Otawa, ON, USA, 2003.
8. Grzywacz, M. Monitoring for gaseous pollutants in museum environments. In *Tools for Conservation*; Getty Conservation Institute: Los Ángeles, CA, USA, 2006; pp. 86–90.
9. González-Martín, J.; Kraakman, N.; Pérez, C.; Lebrero, R.; Muñoz, R. A state-of-the-art review on indoor air pollution and stratagies for indoor air pollution control. *Chemosphere* **2020**, *262*, 128376. [CrossRef]
10. Barbosa, K.; Ferreira, T.; Moreira, P.; Vieira, E. Monitoring pollutant gases in museum microclimates: A relevant preventive conservation strategy. *Conserv. Patrim.* **2021**, in press. [CrossRef]
11. Brimblecombe, P. The composition of museum atmospheres. *Atmos. Environ.* **1990**, *24B*, 1–8. [CrossRef]
12. Valentín, N.; Muro, C.; Montero, J. *Métodos y Técnicas Para Evaluar la Calidad del Aire en Museo: Museo Nacional Centro de Arte Reina Sofía*; CARS—IIC Grupo Español: Madrid, Spain, 2010; pp. 63–81.
13. Tétreault, J. The evolution of specifications for limiting pollutants in museums and archives. *J. Can. Assoc. Conserv.* **2018**, *43*, 21–37.
14. Lee, L.R.; Thickett, D. *Selection of Materials for the Storage and Display of Museum Objects, in Preventive Conservation in Museums*; Caple, C., Ed.; Routledge: New York, NY, USA, 2011; pp. 239–257.
15. Prosek, T.; Kouril, M.; Dubus, M.; Taube, M.; Hubert, V.; Scheffel, B.; Degres, Y.; Jouannic, M.; Thierry, D. Real-time monitoring of indoor air corrosivity in cultural heritage institutions with metallic electrical resistance sensors. *Stud. Conserv.* **2013**, *58*, 117–128. [CrossRef]
16. García-Heras, M.; Gil, C.; Carmona, N.; Faber, J.; Kromka, K.; Villegas, M.A. Optical behaviour of pH detectors based on sol-gel technology. *Anal. Chim. Acta* **2005**, *1*, 147–152. [CrossRef]
17. Peña-Poza, J.; García-Heras, M.; Palomar, T.; Laudy, A.; Modzelewska, E.; Villegas, M.A. Environmental evaluation with chemical sensors in the Palace Museum of Wilanów. *Bull. Pol. Acad. Sci. Tech. Sci.* **2011**, *3*, 247–252. [CrossRef]
18. Peña-Poza, J.; Gálvez Farfán, J.M.; González Rodrigo, M.; García Ramírez, S.; Villegas Broncano, M.A.; García Heras, M. Propuesta de protocolo de valoración de la acidez ambiental en salas y vitrinas de la exposición temporal "El último viaje de la fragata Mercedes. La razón frente al expolio" (Museo Naval, Madrid). *GE-Conservación* **2015**, *8*, 4–26. [CrossRef]
19. Villegas, M.A.; Peña-Poza, J.; García-Heras, M. Sol-Gel environmental sensors for preventive conservation of Cultural Heritage. In *Handbook of Sol-Gel Science and Technology*, 2nd ed.; Klein, L.C., Jitianu, A., Aparicio, M., Eds.; Springer International Publishing: Cham, Switzerland, 2018; pp. 1–32. [CrossRef]
20. Gibson, L.T.; Watt, C.M. Acetic and formic acids emitted from wood samples and their effect on selected materials in museum environments. *Corros. Sci.* **2010**, *52*, 172–178. [CrossRef]
21. Krupinska, B.; Van Grieken, R.; De Wael, K. Air quality monitoring in a museum for preventive conservation: Results of a three-year study in the Plantin-Moretus Museum in Antwerp, Belgium. *Microchem. J.* **2013**, *110*, 350–360. [CrossRef]
22. Agbota, H.; Mitchell, J.E.; Odlyha, M.; Strlic, M. Remote assessment of cultural heritage environments with wireless sensor array networks. *Sensors* **2014**, *14*, 8779–8793. [CrossRef]
23. Hackney, S. Colour measurement of acid-detector strips for the quantification of volatile organic acids in storage conditions. *Stud. Conserv.* **2016**, *61*, 55–69. [CrossRef]
24. Doadrio, I.; Araujo, R.; Sanchez-Álmazán, J. (Eds.) *Las Colecciones del Museo Nacional de Ciencias Naturales. Investigación y Patrimonio*; Consejo Superior de Investigaciones Científicas: Madrid, Spain, 2019.
25. Villegas Broncano, M.A.; García Heras, M.; Peña Poza, J.; de Arcas Castro, G.; Barrera López de Turiso, E.; López Navarro, J.M.; Llorente Alonso, A. *Spanish Patent: Sistema Para la Determinación de Acidez Ambiental y Método que Hace uso del Mismo*; ref. P201031071; OEPM: Madrid, Spain, 2010.

26. Llorente-Alonso, A.; Peña-Poza, J.; de Arcas, G.; García-Heras, M.; López, J.M.; Villegas, M.A. Interface electronic system for measuring air acidity with optical sensors. *Sens. Actuators A-Phys.* **2013**, *194*, 67–74. [CrossRef]
27. Villegas, M.A.; Fernández Navarro, J.M. Characterization and study of Na_2O-B_2O_3-SiO_2 glasses prepared by the sol-gel method. *J. Mater. Sci.* **1988**, *23*, 2142–2152. [CrossRef]
28. Villegas, M.A.; Fernández Navarro, J.M. Characterization of B_2O_3-SiO_2 glasses prepared via sol gel. *J. Mater. Sci.* **1988**, *23*, 2464–2478. [CrossRef]
29. Villegas, M.A.; Pascual, L. Sol-gel silica coatings doped with a pH-sensitive chromophore. *Thin Solid Films* **1999**, *351*, 103–108. [CrossRef]
30. Yang, L.; Stulen, I.; De Kok, L.J.; Zheng, Y. SO_2, NO_X and acid deposition problems in China—Impact on agriculture. *Phyton Spec. Issue Glob. Chang.* **2002**, *42*, 255–264.
31. Hicks, B.B. *Conservation of Historic Buildings and Monuments. Wet and Dry Surface Deposition of Air Pollutants and Their Modeling*; National Academy Press: Washington, DC, USA, 1982; pp. 183–196.
32. Baergen, A.M.; Donaldson, D.J. Photochemical renoxification of nitric acid on real urban grime. *Environ. Sci. Technol.* **2013**, *47*, 815–820. [CrossRef]
33. Peña-Poza, J.; Conde, J.F.; Agua, F.; García-Heras, M.; Villegas, M.A. Application of sol-gel based sensors to environmental monitoring of Mauméjean stained glass windows housed in two different buildings at downtown Madrid. *Bol. Soc. Esp. Ceram.* **2013**, *6*, 268–276. [CrossRef]
34. Cruz, A.J.; Pires, J.; Carvalho, A.P.; Brotas de Carvalho, M. Comparison of adsorbent materials for acetic acid removal in showcases. *J. Cult. Herit.* **2008**, *9*, 244–252. [CrossRef]
35. Schieweck, A. Adsorbent media for the sustainable removal of organic air pollutants from museum display cases. *Herit. Sci.* **2020**, *8*, 1–18. [CrossRef]

Article

Use and Management in the Heritage Conservation of the Historic Water Supply of Canal de Isabel II, Madrid

Jorge Bernabéu-Larena [1,*], Beatriz Cabau-Anchuelo [1], Pedro Plasencia-Lozano [2] and Patricia Hernández-Lamas [1]

[1] Miguel Aguiló Foundation (FMA), School of Civil Engineering, Technical University of Madrid (UPM), 28040 Madrid, Spain; beatriz.cabau@upm.es (B.C.-A.); patricia.hlamas@upm.es (P.H.-L.)

[2] Department of Construction and Manufacturing Engineering, University of Oviedo, 33600 Mieres, Spain; plasenciapedro@uniovi.es

* Correspondence: jorge.bernabeu@upm.es

Abstract: The historic water supply to large cities constitutes a constructed heritage characterised by comprising a range of public structures—dams, canals, tanks, siphons and aqueducts—over a large geographical area. Within this international context, this paper looks at the case of Canal de Isabel II (CYII) and its historic infrastructure, built in the second half of the 19th century and the early 20th century. The purpose of this study is to analyse how these water supply public works, which maintain their original use, have also taken on new functions through the conversion of some of their parts and added new values to the existing ones. In order to do this, an inventory was drawn up with the location and cultural value of each structure based on its historic, technological, landscape and symbolic features, as well as its use. The results establish the significance of the overall system, not only in functional terms but also as a cultural resource. It is essential to understand the historic water supply infrastructure as a whole, not just as individual components but rather as pieces of a network. This is also essential for the management and preservation of the system, both where the structures are still in use as part of the water supply and where they have been converted for other uses.

Keywords: public works; dams; aqueducts; siphon; reservoir; overall value; heritage system; network; new uses

Citation: Bernabéu-Larena, J.; Cabau-Anchuelo, B.; Plasencia-Lozano, P.; Hernández-Lamas, P. Use and Management in the Heritage Conservation of the Historic Water Supply of Canal de Isabel II, Madrid. *Appl. Sci.* **2022**, *12*, 6731. https://doi.org/10.3390/app12136731

Academic Editor: Cesareo Saiz-Jimenez

Received: 30 May 2022
Accepted: 27 June 2022
Published: 2 July 2022

1. Introduction

The heritage of public infrastructure must take into consideration its relationship with its surroundings. A connection with location lies in the original meaning and function of a structure and in turn forms part of the building up of the cultural landscape of the environment in which it is situated [1]. Historic water supply systems for large towns and cities necessitated the construction of a number of highly significant public works, particularly in terms of both size and cultural value.

The arrival of railways in the mid-19th century represented a major transformation of the function of the landscape, the growth of urban areas due to the phenomenon of rural exodus and greater industrialisation; this led to increased demand for water for the population. During that period, most large European and American cities (London, Paris, New York, Boston, etc.) had to rethink their water supply systems, as they were insufficient in terms of both quantity and quality (Table 1); in Spain, Madrid approved modern engineering projects, such as those in Jerez [2], El Puerto de Santa María [3], Santander [4] or Valladolid [5], which were completed over subsequent decades.

It was not only a case of increasing the available flow due to the growth in the urban population; the purpose was also to increase the amount of water available per person per day. With some exceptions, European cities did not receive more than 100 L per person per day, yet at the end of the 19th century, there was some consensus on the need to provide more than that: Parker recommended 157 L/person/day, Prouts recommended 200 L/person/day, and Pettenkofer even suggested as much as 300 L/person/day (Table 1).

As a result, projects in smaller provincial capitals such as Cáceres or Lugo were designing networks based on 200 L/person/day by the end of the century [6].

The projects submitted and completed during the 19th century also supported ongoing research into the flow of water through pipes and canals that dated back a century: in 1728, Pitot proved that resistance to flow varies inversely to the diameter. In 1732, Couplet was credited with the first known gauging of flow-through pipes, and in 1777, Bossut developed a practical formula for calculating water head loss due to curves. Later, notable contributions made by Darcy or Bazin in the 19th century helped to develop better and more ambitious water supply projects [7] (p. 108).

Table 1. Some supply and systems data from different towns and cities in Europe and America. Authors' own creation from [8–14].

Town/City	System	Supply (Litres per Person per Day)	Inhabitants	Year per Data	Source
Albany	Dam and transport via man-made canal to storage tank	757	60,000	1855	[8]
New York	Dam and transport via man-made canal to storage tank	272	500,000	1855	[8]
New York	Dam and transport via man-made canal to storage tank	405	1,000,000	1876	[9]
Boston	Mixed system with ground and elevated tanks and pumping to storage tanks.	219	140,000	1855	[8]
Philadelphia	-	200	400,000	1885	[10]
Lisbon	Mixed: water drawn from a spring, indoor sources, etc.	4, 46 in summer	350,000	1856	[11]
Berlin	River water pumped to different distribution tanks.	104	437,000	1857	[12]
Chicago	-	180	Calculated for when the city has 1 million inhabs.	1874	[13]
Paris	-	67		1863	[14]
Rome	Spring water and water supplied by aqueducts	1105		1863	[14]
London	-	112		1863	[14]

Madrid was no exception, as its population had reached 200,000 (Figure 1), and the "water-trips" from the Arab Era making up the means of water supply until that time were gradually closed down due to a lack of acceptable hygiene conditions [15].

This study looks at the case of Canal de Isabel II in the historic supply to the city of Madrid, undertaken in the mid-19th century. The purpose of the paper is to analyse how these water supply public works, which maintain their original purpose, have also taken on new uses through the conversion of some of their parts, added new value to the existing ones and positioned the infrastructure as an identifying mark on the landscape. The system's cultural value—understood to be an integral vision encompassing constructions, technology, activities and landscapes—helps us to understand this system in terms of its historic, cultural and social significance.

The existing literature and studies on supply works generally focus on their historical development as shapers of the modern city at the beginning of the 20th century [16–18] or their technological development [19,20], mainly centred on nodal elements such as dams [21–23], reservoirs [24–26], aqueducts [27] or pumping stations [28,29], without considering structures such as wells and cisterns [30,31].

Recently, Douet, in a study commissioned by the International Committee for the Conservation of the Industrial Heritage (TICCIH) on the "water industry", listed the different values of public water works [32] (p. 61) and, through different case studies, reflected on how "the heritage of the modern water industry is almost totally absent" in the UNESCO World Heritage List, "despite its indisputable relevance for human development" [32] (p. 9).

Hunter also argued for the need to enhance the value of this heritage of the water industry, where the management of historic waterworks from a heritage point of view is a challenge, both because they are designed to be "silent and unseen" and because they are active systems as part of everyday life that are "frequently ignored by the public unless it stops working" [33] (p. 21).

Both Douet, who devoted a section to examples of cities with "distant gravity supply", and Hunter included examples of the reuse and revaluation of water industry elements with the aim of establishing evaluation criteria. Although they recognised their relevance within cultural, industrial and maritime landscapes, both urban and rural, they did not generally cover the territorial extent of the system, only looking at different elements in isolation (dams, tanks, pumping stations, water treatment plants, etc.) rather than the whole system.

The best-known example of heritage recognition of a water supply system as a whole is the Old Croton Aqueduct in New York (1837–1842). It was designated a National Historic Landmark in 1992 from the Croton dam until the touchdown of the High Bridge in Manhattan (41.8 km). It also has several individual structures listed individually on the New York State Register of Historic Places [32] (pp. 89–93).

However, its recognition as a landmark of civil engineering came once it was no longer used due to insufficient supply (1955) and was replaced by the New Croton Aqueduct, which tripled its dimensions. In 1968, it became the Old Croton Aqueduct State Historic Park, thus adding a new recreational and cultural use to its heritage and symbolic dimension.

Another similar case of heritage recognition is the Lisbon water supply system (1731–1799). It was designated as a National Monument in 1910 and included in the UNESCO "Tentative List". It remained in service until 1968. Its most iconic elements, namely, the Aguas Livres aqueduct bridge (1748), the Mãe d'Água das Amoreiras cistern (1746), the Patriarchal Reservoir (1864) and the Barbadinhos pumping station (1880), are managed by the Water Museum, an institution responsible for research and dissemination of water resources [34].

Fortunately, in both cases, functional obsolescence has not led to their abandonment, but rather, these public works have acquired new values and meaning not only as engineering landmarks but also as identity elements of the landscape that they form. The management of an operational supply system as a historical resource, as in the case of Canal de Isabel II, is more complicated, as utility usually takes preference over heritage, historical, cultural and landscape values. The study forms part of the research project "Analysis and definition of territorial scale strategies to characterise, restore and heighten esteem for the public works heritage" [35].

A number of sources were used for this research: academic research material, UNESCO documents, media reports and field work on several sections of Canal de Isabel II.

Under the Section 2, there is an analysis of the historic construction of Canal de Isabel II and its subsequent development. We identified the "visible" parts of the system, its structures and its constructions that demonstrate the values described by Douet [32]. We address a management system that allows the supply network to be run alongside the use or transformation of its cultural elements. Under the Section 3, we emphasise the overall value of the linear public works, their interest as networks connecting different places and actions, and their value as cultural heritage beyond their practical use. In Canal de Isabel II, we highlight newly acquired values and the new uses of some of its components despite the obsolescence of their original function. The final Discussion resumes the matter of the territorial nature of public works, the importance of discipline over the landscape in order to understand and value it, opportunities and new uses, concluding with the importance of considering their value as a whole.

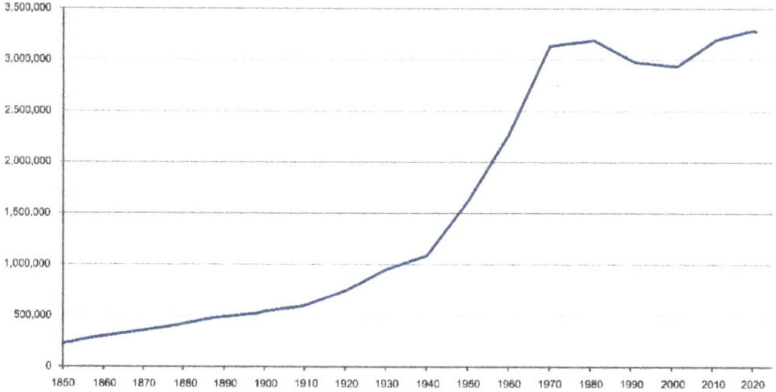

Figure 1. Evolution of the population of Madrid: year–inhabitants. Source: Created by the authors based on data from the Spanish National Institute of Statistics, INE 2022.

2. Materials and Methods

2.1. Historic Construction

In around 1850, Madrid was still supplied with drinking water through the same system that had been in use since the 16th century. There were 77 public water fountains within the urban perimeter, plus around 1000 water carriers. It is estimated that the amount of water supplied would have ranged between 2000–2500 m^3 per day for a population of about 235,000 [36] (p. 27). In other words, there were between 50 and 100 litres available per person per day. The projected urban growth was also notable, as can be seen in the approval of the project for the urban development of Madrid before the end of the decade in 1857.

Under the reign of Isabel II and the government of Bravo Murillo, a water supply project was undertaken for Madrid. It was first decided that water would be brought down from the Guadalix and Lozoya rivers due to the purity of their waters. After several different proposals, in 1848, the project put forward by Rafo and Ribera was accepted. After the appointment of a team of engineers (José Garcia Otero, Lucio del Valle, Juan de Ribera and Eugenio Barron), in 1851, work started on a new dam over the bridge called Pontón de la Oliva (Figure 2) and the dam built by Cabarrús (1775) for an irrigation canal linking the Lozoya and Jarama river basins.

Figure 2. Construction of the Pontón de la Oliva dam, 1855. Photo: Charles Clifford. (Public domain. Biblioteca virtual del Patrimonio Bibliográfico, https://bvpb.mcu.es/es/consulta/registro.cmd?id=403102) (accessed on 15 May 2022).

At the same time, construction work began on the canal and storage tanks in the city itself (Figure 3). When mapping out the routes, it was decided to build large infrastructures in order to shorten the length of the system. In other words, instead of following the curvature of the landscape, it was decided to transport the water as directly as possible (Figure 3), meaning the construction of tunnels, aqueducts, siphons, etc., to bridge valleys and ravines. Irrigation canals, siphons, regulating tanks and related constructions were built up into this network, at the same time becoming milestones across the Sierra Madrileña.

Work was complicated and full of ups and downs, not just because of the socio-economic context (lack of income, scarcity of resources, non-specialist manpower, illness, etc.) but also because it was a considerable engineering challenge at the time, without the precedent of similar projects. Cities all around the world (Washington, New York, Manchester, Paris, Lisbon, etc.) were all facing the same problem: how to create a water supply that could allow for population growth.

In addition, there were issues with seepage through the limestone on which the El Pontón de la Oliva dam was located, which meant that the water was not rising high enough to reach the entrance to the canal [37] (p. 113). A provisional connection from the Guadalix river was put in place with a new aqueduct, which eventually became permanent. At the same time, the Azud de Navarejos diversionary dam was built upstream of the Pontón de la Oliva to act as the new connection to the head of the canal.

With a total length of 70 km, made up of 29 bridges/aqueducts, 4 siphons and 41 shafts, on 24 June 1858, the Canal Bajo was officially opened by the Queen. The new water supply started to transform the city with decorative fountains in the streets, plazas and gardens, improving not only hygiene but also people's way of life.

Due to the increased demand for water, low pressure in some areas and the poor functioning of the Pontón dam due to water seepage, Canal de Isabel II ordered the construction of a new dam further upstream in a narrow gorge around 50 metres below an old bridge called El Villar.

Although water flow increased with the new dam, the issue of murky water for several days a year continued, and so work began on the transverse canal or Canal del Villar (22 km), which entered into service in 1911. This 22-km canal ran from the Villar dam and joined the Canal Bajo at the Aldehuela aqueduct, taking out of service the older stretch of the canal running from the La Parra dam to the aqueduct. Taking advantage of the difference in altitude (150 m) between the start and the end of this new stretch, the Torrelaguna hydroelectric plant was built.

Figure 3. Ground route of the first Canal de Isabel II, 1868. Source: Created by the authors.

To completely resolve the murky water issue, the Puentes Viejas dam was built (1914–1929) upstream of El Villar, which used sedimentation to clean the water before it was transferred to the first dam.

In 1921, an extension was proposed for the transversal canal with what was known as the Canal Nuevo (1929–1945). This was built almost entirely underground and ends at the storage tanks in Chamartín (Figure 4).

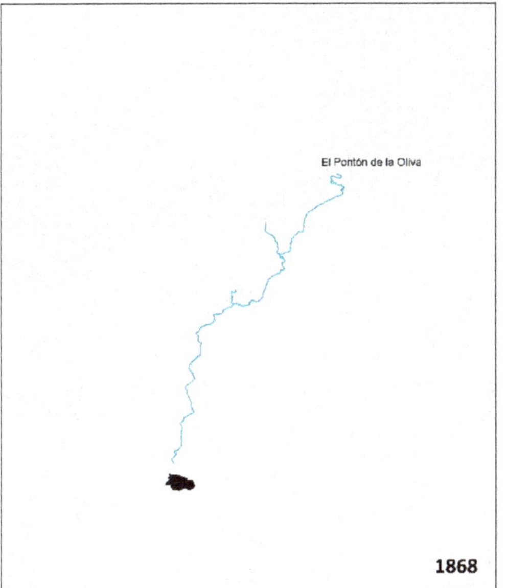

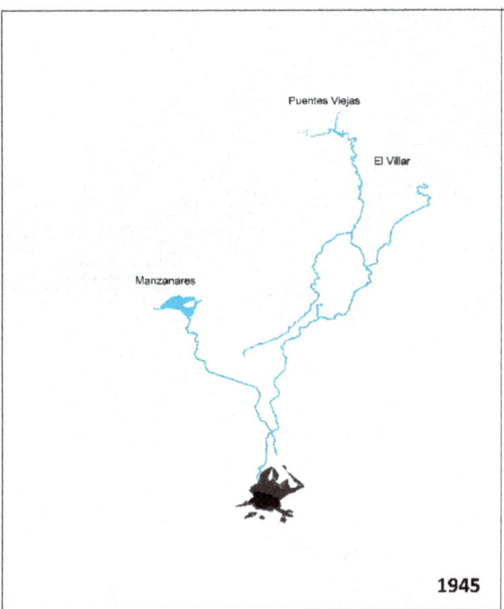

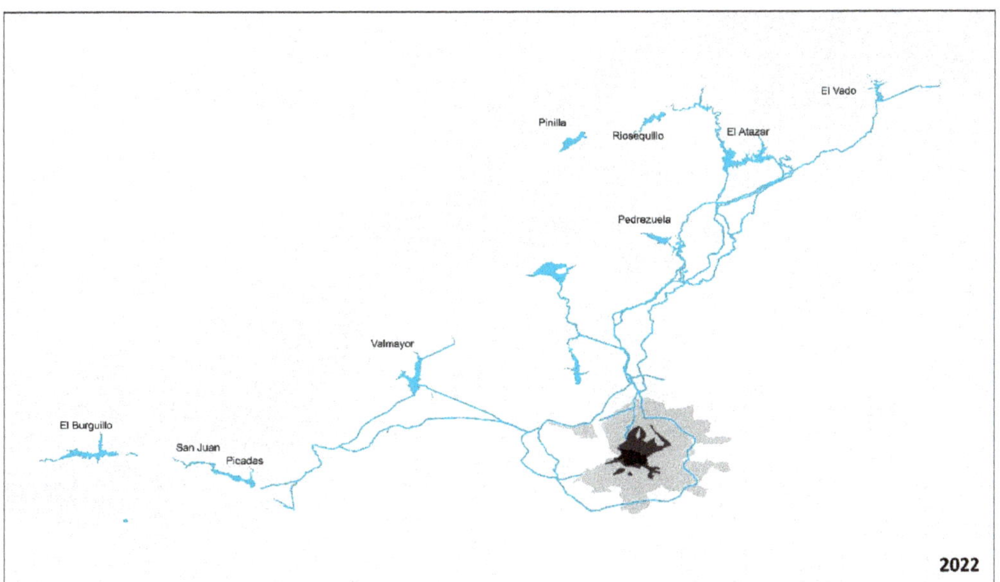

Figure 4. Evolution of the water supply network and the urban nucleus of Madrid: 1868, 1945 and 2022. Source: Created by the authors.

2.2. System Components

It is important to understand the canal as a whole: its material and functional components (layout, structures, materials, etc.) are not isolated and unrelated elements, but rather, they make up a network that is not always visible, which can prevent it from being perceived in this way.

The combination of all of these components is what makes the water supply work and, at the same time, is a reflection of the history of Madrid, which would not be the same without its water. These components include dams (Figure 5), bridges/aqueducts (Figure 6), storage tanks (Figure 7) and siphons (Figure 8).

Figure 5. Pontón de La Oliva (1858–1860): straight solid masonry gravity-based dam and narrow-lipped spillway. Villar Dam (1869–1882) Engineers Elzeario Boix and José Morer: first curved gravity dam built worldwide with a lateral spillway system. Puentes Viejas Dam (1914–1929): curved plan gravity dam. Source: FMA.

Figure 6. Canal Bajo Aqueducts: La Sima Aqueduct (1855), La Retuerta Aqueduct (1855) and El Espartal Aqueduct (1852). Source: FMA.

Figure 7. Tanks: Primer Depósito-Campo de Guardias (1858), Santa Engracia water tower (1912), Tercer Depósito (1893) and Plaza de Castilla water tower (1939; relocated in 1952). Source: Charles Clifford and FMA.

Twenty-nine narrow aqueducts were built, combining structural elegance with savings in construction material [37] (p. 67). In the higher sections of the system, they are masonry-built using local stone, while in lower sections, they are constructed in brickwork, only using masonry for angles and edges.

The canal project is important because it overlaps with the long tradition of structures of this kind built in Spain: Roman aqueducts such as those in Tarragona (Las Ferreras), Segovia or the whole of Mérida [38] had continuity over the centuries in constructions such as the Morella aqueduct (13th century) [39], the Arcos de San Antón in Plasencia (16th century) [40], the Seventeen Arcs in Lorca (18th century) [41] or the Arroyo Quintana aqueduct, part of the San Telmo water pipeline in Malaga (18th century) [42]; this list continued into the 20th century, and Eduardo Torroja designed some notable constructions, such as Tempul in Jerez de la Frontera (1925) or Alloz in Navarra (1939), which were ground-breaking in their typological concept thanks to the use of new materials [43].

Figure 8. Siphons: Guadalix Siphon; Morenillo Siphon. Source: Clifford, 1858; FMA.

The aqueduct also links us to countries of the Mediterranean environment, such as Italy, which to its collection of Roman aqueducts would add others in the Modern Age, such as those of Eleuterio in Palermo (1443) or Bigini in Castelvetrano (16th century) [44], or Portugal, with the 16th-century aqueducts Água da Prata in Évora and Amoreira in Elvas [45]. Similarly, the bridge aqueduct technology would also spread to America and, in particular, to present-day Mexico, where major works were undertaken in the 16th century, such as the Padre Tembleque Aqueduct crossing the Tepeyahualco ravine, a magnificent hybrid of European and pre-Hispanic technology [46], or the aqueduct in Valladolid, today renamed Morelia [47]. In other words, bridge aqueducts in the canal's system are links in a triple chain connecting the land where they are located with the constructive history of Spain, with the Mediterranean basin and with colonial America. In short, we note that all of the aqueducts mentioned are still standing today, and they are core parts of the landscape; some have even been declared World Heritage Sites, such as Padre Tembleque, Segovia, Elvas, Tarragona or Mérida.

The Roman-inspired underground storage tanks are formed by archwork with barrel vaults and brick pillars built on masonry plinths. The first tank (Primer Depósito) also presented seepage problems and was therefore closed in 1894, when the second and third tanks were opened. A project for three water towers was drawn up for the supply of areas on higher ground; only the one in Calle Santa Engracia was built [37] (p. 208). The structure had a central elevator mechanism designed by Ramon de Aquinaga, which raised the water to a tank 36 metres above the ground (Figure 7).

Siphons are also an interesting part of the project (Figure 8). Although they have been used since ancient times, the ceramic material that limited diameters and advanced knowledge of the hydraulic principles required made it more common to opt for the construction of aqueducts. For the Canal de Isabel II project, cast iron was chosen as the material for the siphons. At the time, they were considered some of the most notable examples: Bodonal, Malacuero or Guadalix [1] (Table 2).

Table 2. Main siphons in Europe. Note the maximum load of those designed in Madrid, exceeded only by the ones in Liverpool [2].

Population	Siphon	Length (m)	Maximum Load (m)	Interior Diameter (m)	Thickness (m)
Madrid	Guadalix	323.50	53.60	0.92	0.018
	Malacuero	843.00	45.00	0.92	0.018
	Bodonal	1410.00	21.00	0.92	0.018
New York	Manhattan	1254.00	31.50	0.92	0.018
Glasgow	Glasgow	3500.00	25	1.12	0.19
Liverpool	Aspull a Montrey	13,000	86.50	1.12	0.025
Jerez	Albaladejo	10,200	80.00	0.61	0.019 and 0.025
	Guadalete	18,000	90.00	0.61	0.019 and 0.025

2.3. *Water Management for Water Supply*

Water management policies are indissolubly associated with the construction of water supply public works. They are at the base of their promotion, construction, regulation and exploitation. The management model determines the growth of the city and the sociocultural signification of water. This section identifies the turning points in water management that have influenced urban development.

At the beginning, Canal de Isabel II was a public company run by a Board made up of State representatives, the City Council and a small group of stakeholders. However, in 1868, it became a public company, reporting to the Ministry of Public Works and Transport. It was not until hydropower arrived in the region and the private company Hidráulica Santillana was constituted that this management scheme changed. Hidráulica Santillana opened the Colmenar hydraulic plant in 1902 and obtained a royal concession to supply water to the northern areas of the city, supported by the City Council, in 1906 [48] (p. 20). These two events created a conflictive situation due to the competitiveness that forced an administrative reorganisation of the company in 1907. CYII became an independent organisation using an industrial business model, still reporting to the Ministry of Public Works [49] (p. 214). It was at this point that CYII initiated the modernisation of its infrastructures and its organisation [50] (p. 483). In 1976, it became a public limited company with its own assets and independent management, reporting to the Ministry of Public Works and Transport until 1984, at which point it started reporting to the regional government, the Comunidad Autónoma de Madrid.

The crucial turning point came in 2002 when its objectives and the composition of its management board were amended by law. Since then, CYII has had permission to provide services outside the Comunidad Autónoma de Madrid and may carry out any kind of commercial or industrial activity directly related to its functions, including taking a minority or majority shareholding in companies with the permission of the regional government [51] (p. 144). In this way, CYII and the city of Madrid can form part of the virtual network system that runs new international-scale economic processes. This new concept of CYII as a multinational company widens its geographical influence, making it difficult to describe [52].

A Regional Government Act in 2008 went one step further with this idea of the liberalisation and extension of the influence of CYII: it permitted the creation of a public limited company for the provision of services relating to water and allowed the privatisation of 49% of shares. In June 2012, the new company "Canal de Isabel II Gestión, S.A." was created, completely separate from the original CYII, which remains a regulation body. Canal de Isabel II Gestión, S.A took over the operation of water supply, sanitation and hydraulic works for the next 50 years. Even though today, over 80% of its shares are held by CYII and the rest are held by local authorities [53], there is still tension between the people and the regional government. A hydraulic technique has replaced water culture in Madrid, as happened in the 19th century when Canal de Isabel II was created [49] (p. 208).

3. Results

3.1. Value of the Overall System

As a work of engineering, Canal de Isabel II was a fundamental part of the development and modernisation of Madrid. However, it is not only functional; as heritage, it also "contains knowledge of the past with specific present uses" [54]. Its management should be analysed from a functional point of view not only as a water resource but also as a cultural resource.

The difficulty in valuing the heritage of this kind of linear hydraulic work is that it is not always visible. Through individual structures (aqueducts, siphons, irrigation canals, water tanks, etc.) (Figure 9), its value as a whole or as a system must be understood insofar as these components are part of the greater network. They cannot be understood in isolation but only as part of a system, a network with great historic, technical and cultural value. Although some elements of the canal no longer form part of the system in functional terms, they must still be perceived as an integral part of it. It is important to stress the transformative power of this entire network on the landscape.

In addition to its value as a system, the canal and its different components also have other heritage values:

- Historic: It is proof of a significant activity from history, the bringing of water to supply the people of Madrid.
- Technological: The canal has played a role in the evolution of engineering and in significant elements such as the design of dams, aqueduct design, pump technology, etc. These are technical milestones and technological challenges overcome from the double perspective of typology and the construction process, bearing in mind their innovative nature. They have the capacity to adapt to different uses over time while maintaining their original character.
- Social or identity: The utilitarian purpose (value of use) tends to prevail over cultural value, as this is a large infrastructure for the service of the population (allowing the elimination of diseases transmitted in water, improved comfort, etc.) and has made a positive economic contribution (allowing urban and geographical growth, industrial development, etc.). However, it also has a strong presence in our society and culture (literature, songs, films, stamps, folklore, etc.). The canal has become a cultural reference in collective memory. It can evoke and stage and establish links to the viewer. It has taken on a symbolic character.
- Singularity: From a technical point of view, these are integral and authentic structures (not imitations). Some of the elements are singular typologies or are first examples of other more common structures. The canal is also unique from the perspective of the historic moment, as it set a precedent for the construction of later systems.
- Aesthetic, landscape and environmental: Its linear nature creates networks that transform the landscape on both urban and geographical scales, giving it unity through its structures, materials, etc.

  Canal de Isabel II, as a public company reporting to the Ministry of Public Works and Transport since 1977 [55] (p. 138), not only is the authority in terms of water/water resources but, through its Foundation, seeks to provide know-how in innovation, bring the environment and culture closer to citizens and also promote taking care of water resources [56].

  Canal de Isabel II, as a piece of cultural heritage, is included under the National Industrial Heritage Plan, which defines heritage as "an integral whole including the landscape in which the different components that make it up" "are included and related" (covering its declaration as a Bien de Interés Cultural).

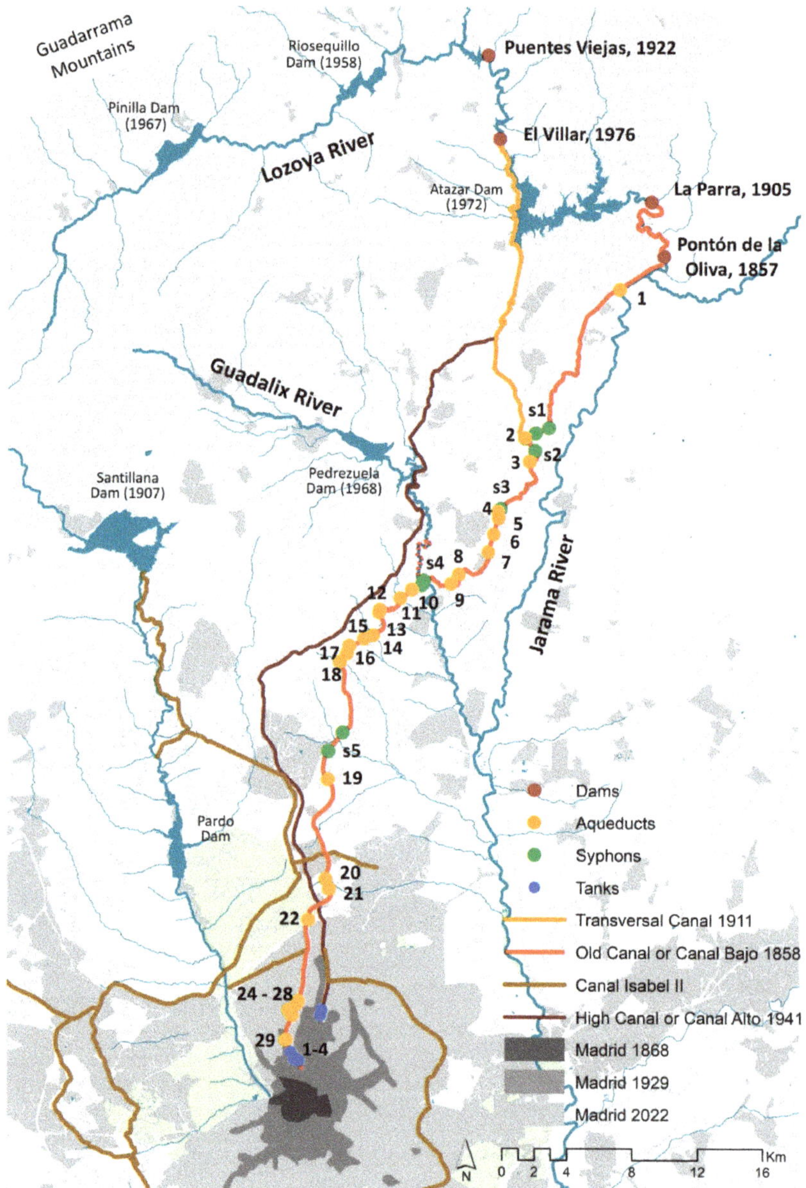

Figure 9. Canal de Isabel II as a whole system. Source: Created by the authors.

3.2. New Values and New Uses

Water supply systems often maintain the utilitarian function for which they were created, subject to ongoing transformations and extensions in order to cover the growing demand for water by the urban population. However, as Sáenz de Ruidrejo claimed, changes for multiple uses over time are a typical feature of large public works that often far exceed the expectations of their creators [57] (p. 27). In this way, some of the components of Canal de Isabel II are no longer used as originally intended and yet, at the same time, have acquired new uses and values, remaining part of the overall system (Table 3).

Table 3. Heritage elements of the canal, types and original and current uses. Note the correspondence with Figure 9.

Type	Number	Name	Year	In Service	New Use
Aqueducts	1	Cuevas	1852	Yes	
	2	La Aldehuela	1852	Yes	
	3	Espartal	1852	Yes	
	4	Bajada Morenillo	1852	Yes	
	5	Regachuelo	1855	Yes	
	6	de la Cerca de Gavino	1855	Yes	
	7	Fuente del Palo	1855	Yes	
	8	Valmayor	1855	Yes	
	9	Barbotoso	1855	Yes	
	10	Navalperal	1855	Yes	
	11	Retuerta	1855	Yes	
	12	Sima	1855	Yes	
	13	Valcaliente	1855	Yes	
	14	Colmenarejo	1855	Yes	
	15	Cabeza-Cana	1855	Yes	
	16	Mojapán	1855	Yes	
	17	El Cerrillo	1855	Yes	
	18	La Parrilla	1855	Yes	
	19	Valdealeas	1858	Yes	
	20	Valle de la Fuente	1858	Yes	
	21	Vallegrande o Valdelatas	1858	Yes	
	22	El Sotillo	1858	Yes	
	23	Valdeperales	1858	Disappeared	
	24	Los Pinos	1858	Yes	
	25	La Traviesa	1858	Yes	
	26	Valdeacederas	1858	Yes	
	27	Los Barrancos	1858	Yes	
	28	La Huerta del Obispo	1858	Yes	
	29	Amaniel	1858	Yes	
Syphons	s1	La Malacuera	1858	Yes	
	s2	Los Yesos	1858	Yes	
	s3	El Morenillo	1858	Yes	
	s4	Guadalix	1858	Yes	
	s5	El Bodonal, Viñuelas	1858	Yes	
Tanks	1	First buried tank	1858–1894	No	Archive, exhibitions
	2	Second buried tank	1879	Yes	Park
	3	Third buried tank	1915	Yes	Park
	4	First elevated tank	1912–1952	No	Exhibitions
Dams	1	Pontón de la Oliva	1858	No	Landmark
	2	Navarejos	1859	Yes	
	3	El Villar	1882	Yes	
	4	La Parra	1904	Yes	
	5	Puentes Viejas	1929	Yes	

- Pontón de la Oliva dam. Although it no longer performs the function for which it was built, the dam remains standing and is part of the historic heritage of the Ayllón mountain range. The mountain wall on the left bank of the dam is used by climbing enthusiasts. The dam is also part of a popular trekking route.
- Recreational area at Riosequillo. Located on the right bank of the Riosequillo reservoir is one of the largest swimming pools in the Madrid region, with a capacity of more than 2000 people. It is operated by the town hall of Buitrago de Lozoya.
- Cervera Marina. Located on the El Atazar reservoir, this recreational area is the only nautical base in the region of Madrid. It is home to an adapted sailing school, and people can practice windsurfing and rowing, as well as rent berths.
- First underground tank at Campo de Guardias (1858). It was converted into an archive (1990) and then provisionally into an exhibition hall (2001).
- Second tank, or the first water tower in Santa Engracia (1907) with a capacity of 1500 m^3. It was taken out of service in 1952 and reopened as an exhibition in 1985, maintaining its unique character (Figure 10).

Figure 10. First water tank (1907) converted into exhibition hall (1985). Source: FMA.

- Third tank (J.E. Ribera, 1915) converted into the Santander Park (2007). The tank was waterproofed, and the pillars and archwork were reinforced to build a large park with sport and leisure facilities on the roof (80,000 m^2) of the tank still in use (capacity 500,000 m^3).
- Fourth tank in Plaza de Castilla or second water tower (1935) with a tank capacity of 3800 m^3. One of the underground tanks was converted into an exhibition hall in 2000, maintaining its special and characteristic layout of 1447.5 m high brick arches (2500 m^2); a park has been created over the roof of the structure (45,000 m^2).

All of these new uses are managed differently, but always through the Canal de Isabel II Foundation.

All of these elements making up Canal de Isabel II, both those still in use as originally intended and those that have been adapted for other uses, are key parts of the collective memory and of the conservation and recovery of the urban landscape. They constitute an immense heritage to be rehabilitated and reused, with great potential for the socio-economic and cultural development of the areas where they are located. These new actions must always seek to maintain the character of the canal as key testimony to understanding and documenting this great work of engineering in the city of Madrid.

Nevertheless, as the infrastructure is still in use, some clashes arise with new uses, such as water quality problems due to recreational activities or overuse, the deterioration of vegetation and the environment, the surrounding area with the opening of new paths or routes, etc. In order to avoid this, a balance must be found between the different parts of the system and the new uses, with the new activities having virtually zero effects on the canal and its geography.

4. Discussion

The landscape is the sum of the many dialogues between people and the land, adding layer after layer to make up our current reality. These layers are the images of that territory, which humans have collected throughout their existence, and they blend together into the mental image—the landscape—that humans have of that land [58] (Lynch said that "the landscape serves as a vast mnemonic system for the retention of group history and ideals" [59], p. 126). Furthermore, a phrase from Waterman (humans have always left their mark on the landscape, such as the first cave paintings or the "great feats of engineering such as Stonehenge" [60], p. 15) shows that not all layers of the landscape have the same depth or weight. Without a doubt, certain public works are major protagonists of the character of the landscape and therefore make a distinctive mark on it [61,62]; Canal de Isabel II is the protagonist of the territory where it is located because its "layer" overshadows other man-made or natural elements present along its path; it can therefore be considered the cultural landscape of engineering.

With this idea in mind, three types of cultural landscapes of engineering can be distinguished: one or more elements in an urban location; a singular isolated work or one in a rural environment; and a set of interrelated elements with common characteristics, located across a large geographical area. Given that it is a work comprising a set of singular elements, Canal de Isabel II falls under the third category and should therefore be protected and managed on a territorial scale, integrating the whole environment [63].

Similarly, in recent times, there has been an increase in a new way of integrating into the landscape and perceiving it from within, as a part of it. One example of this is the spread of new viewpoints or observation sites appearing across the landscape, which demonstrates the need to relate to places [64]. A work of engineering can therefore be a special place for looking out over the landscape. There are notable examples of this reality in geographical routes established in recent years, associated with linear public works: within Spain, the Caminito del Rey [65], the Canal de Castilla [66] or the numerous green routes [67,68].

Indeed, tourism is an attractive option for the valuation of these heritage sites with such a historic background. We believe that aqueducts and visitable civil works should be considered places with the potential for receiving visitors interested in the history of engineering techniques or old structures. As pointed out by Medina Lasansky: "Everything, from historical monuments to exotic holiday destinations has been redesigned and packaged up for tourist consumption." "As a result we now have a new conceptualisation of the history of specific buildings, spaces and places" [69].

Regarding the tourist use—but not exclusively linked to it—it is worth referring to the assumption that the space crossed by the canal is a cultural landscape of engineering. Just as Aldo Rossi said that architecture is a place, an event and a symbol all at the same time [70] (p. 7), we can apply the idea to large engineering works and attribute to the entire system the idea of a single place, as well as that of the technical symbol of a specific moment in time and, therefore, an event.

To conclude, it is essential to ensure the cultural preservation of public works. It should be remembered that, despite its monumental features, in the mid-20th century—and after the first declarations of historical and artistic monuments in Spain, dating from 1844, after the protection of the Segovia Aqueduct in 1884 and the Tarragona Aqueduct in 1905, and after Riegl published The Modern Cult of Monuments in 1903—in around 1915, the 16th-century Los Pilares aqueduct in Oviedo, with 41 arches and 3190 m in length, was partially pulled down to widen the bay of tracks for the railway station. Only five lonely arches remain in an uneventful area of the town. This, and other cases—such as the Eiffel bridge over the Tagus, recognised at the time as one of its 10 best bridges [71], yet demolished in 1932—must alert us to the risk posed to public works, which is always difficult to protect from an administrative point of view once they fall into disuse.

The greatest enemy of public works is ignorance of their existence, in conjunction with open-air deterioration due to natural causes: sun, frost, wind, rain and biological activity [72,73]. Registration, conservation and rehabilitation are necessary for the preservation of these works, but their recognition as heritage must be based on the significance of public works as a cultural value. The conservation of heritage must be supported by a system of protection that allows for renovation. As we have seen, one of the greatest threats to public works is obsolescence. For this reason, the best approach is one that deals not only with the conservation and repair of works but also—fundamentally—with their adaptation and rehabilitation for new uses.

In the case of historic supply systems, it is essential to understand these works as a whole, not just as individual components but rather as pieces of a network. This is also essential for the management and preservation of the canal, both where the structures are still in use as part of the water supply and where they have been converted for other uses.

5. Conclusions

In this work, the case of the Canal de Isabel II water supply to Madrid was used to analyse how these public works, which maintain their original purpose, have also taken on new uses through the conversion of some of their parts, added new values to the existing ones and positioned the infrastructure as a landmark. The main features of the original project are described in the Section 2, including the administrative and technical contexts, or the main constructions (dams, tunnels, aqueducts, siphons, etc.), and the importance of considering the water supply as a whole unit and not as a group of isolated elements is highlighted. Moreover, the relevance of aqueducts in the history of water supply projects from Roman times to the 20th century in Spain and the tradition of aqueduct construction in western Mediterranean countries or in Spanish America is underlined to understand the cultural relevance linked to this kind of public work in its geographical context. Additionally, the evolution of the water management model from the origin of the CYII to today is described.

The results establish the significance of the overall system, not only in functional terms but also as a cultural resource. Thus, several heritage values are stated: historic, technological, social or identity reference, singularity, aesthetic, landscape and environmental. These characteristics make the canal a relevant item in terms of historic, cultural and social significance, especially given the almost total absence of the heritage of the modern water industry in the UNESCO World Heritage List.

Another key point of the research is the knowledge of the new uses given to some elements of the water supply scheme. Thus, some of the components currently out of service, such as the Pontón de la Oliva dam or some old tanks, hold new uses linked to the cultural sector; however, Canal de Isabel II still manages them through its Foundation. This administrative solution, which is uncommon, allows the canal to continue its history as a singular and vast individual element regarding its unit.

Overall, the main finding of the research is the relevance of the whole water supply system. This fact is decisive not only in the use and management of the infrastructure, in which each element is part of the system, but also in the enhancement of its heritage value and in its eventual transformation. Its consideration should not be limited to the individual element but extend to the overall network. It is essential to maintain the potential of the whole in its use, management, reconversion and heritage value.

As future lines of research, we would like to point out two lines of interest. On the one hand, the study methodology and valuation proposal can be applied to other public works that also have an important territorial extension and historical value and use, for example, roads, railways or ports. In addition to their linear development in the territory or along the coast, they also have similarities to water supply systems in their important territorial identity and to the historical evolution of the cities. Finally, the analysis of tourist opportunities as social and cultural activators is a line with great potential. Again, it will be essential to maintain the overall character of public works in their heritage appreciation.

Author Contributions: Conceptualisation, J.B.-L. and P.P.-L.; methodology, B.C.-A. and P.H.-L.; research, J.B.-L., B.C.-A. and P.H.-L.; data curation, B.C.-A. and P.H.-L.; writing—original draft preparation, P.P.-L. and P.H.-L.; writing—review and editing, J.B.-L. and B.C.-A.; visualisation, B.C.-A.; supervision and project administration, P.H.-L. All authors have read and agreed to the published version of the manuscript.

Funding: This research was funded by Ministerio de Ciencia, Innovación y Universidades, Gobierno de España. Proyectos I + D+i, 2020–2022, (PID2019-105877RA-I00). Análisis y definición de estrategias para la caracterización, recuperación y puesta en valor del patrimonio de las obras públicas. Una aproximación desde la escala territorial. Coordinator: Rita Ruiz, Universidad de Castilla La Mancha.

Institutional Review Board Statement: Not applicable.

Informed Consent Statement: Not applicable.

Conflicts of Interest: The authors declare no conflict of interest. The funders had no role in the design of the study; in the collection, analyses or interpretation of data; in the writing of the manuscript; or in the decision to publish the results.

References

1. Hernández Lamas, P.; Cabau Anchuelo, B.; de Castro Cuartero, O.; Bernabéu Larena, J. Mobile Applications, Geolocation and Information Technologies for the Study and Communication of the Heritage Value of Public Works. *Sustainability* **2021**, *13*, 2083. [CrossRef]
2. Editorial Office. Acueducto de Jerez. *Rev. Obras Públicas* **1865**, *13*, 166–171.
3. Pardo, M. Abastecimiento de aguas del Puerto de Santa María. *Rev. Obras Públicas* **1878**, *26*, 111–114.
4. Barron, E. Conducción de agua a Santander. *Rev. Obras Públicas* **1878**, *28*, 2–5.
5. Borregón López-Peñalver, A. Abastecimiento provisional de aguas en Valladolid. *Rev. Obras Públicas* **1880**, *28*, 2–5.
6. Plasencia-Lozano, P. El Proyecto de Abastecimiento de Cáceres realizado por García Faria en 1895. *Rev. Obras Públicas* **2011**, *158*, 55–62.
7. Merdinger, C.J. Water supply through the Ages: Part I. Early History. *Mil. Eng.* **1955**, *318*, 280–285.
8. Houghton, J. Abastecimiento de aguas en varias poblaciones de los Estados Unidos. *Rev. Obras Públicas* **1855**, *3*, 225–226.
9. Editorial Office. Consumo de agua de Nueva York. *Rev. Obras Públicas* **1876**, *24*, 74.
10. Editorial Office. Abastecimiento de aguas para grandes poblaciones. *Rev. Obras Públicas* **1899**, *46*, 459–460.
11. Del Valle Arana, L. Abastecimiento de agua en Lisboa. *Rev. Obras Públicas* **1856**, *4*, 133–136.
12. Editorial Office. Nota sobre la distribución de las aguas de Berlín. *Rev. Obras Públicas* **1857**, *5*, 124–127.
13. Churruca Brunet, E. Abastecimiento de aguas de Chicago. *Rev. Obras Públicas* **1874**, *22*, 109–116.
14. Monterde, A. Consumo de agua en las grandes poblaciones. *Rev. Obras Públicas* **1863**, *11*, 87–88.
15. Blasco Esquivias, B. Toledo y Madrid: Sistemas de captación y uso del agua para servicio doméstico en la Edad Moderna. In *Patrimonio Cultural Vinculado con el Agua. Paisaje, Urbanismo, Arte, Ingeniería y Turismo*; Editora Regional de Extremadura: Mérida, Mexico, 2014; pp. 267–279
16. Melosi, M.V. *Precious Commodity: Providing Water for America's Cities*; University of Pittsburgh Press: Pittsburgh, PA, USA, 2011.
17. Hassan, F. *Water, History for Our Times. IHP Essays on Water History*; UNESCO: Paris, France, 2011; Volume 2.
18. Blake, N.M. *Water for the Cities: A History of the Urban Water Supply Problem in the United States*; University Microfilm International: Ann Arbor, MI, USA, 1995.
19. Angelakis, A.; Zheng, X.Y. Evolution of water supply, sanitation, wastewater, and stormwater technologies globally. *Water* **2015**, *7*, 455–463. [CrossRef]
20. Vilar, D.D.; Fonseca, F.P. The waters of Cantareira: The hydraulic heritage of the city of Sao Paolo. In *Hydraulic Heritage in Ibero-America*; Francisco da Silva, C., Vieira, A., José Manuel Lopes, C., Jesus, N.-G., Eds.; Nova Science Publisher: Hauppauge, NY, USA, 2020; pp. 127–158.
21. Billington, D.P.; Jackson, D.C.; Melosi, M.V. *The History of Large Federal Dams: Planning, Design and Construction in the Era of Big Dams*; US Department of the Interior, Bureau of Reclamations: Washington, DC, USA, 2005.

22. Jackson, D.C. *Dams*; Routledge: Oxfordshire, UK; Taylor & Francis Group: Abingdon, UK, 1998.
23. Astwood, K.; Fell, G. *Karoki Water Supply Dams and Reservoirs*; Engineering Heritage New Zealand: Wellington, New Zealand, 2012.
24. Douet, J. *Temples of Steam. Waterworks Architecture in the Steam Age*; Bristol Polytechnic: Bristol, UK, 1992.
25. Vonka, M.; Kořinek, R. Chimney Reservoirs: Unique Technical Structures from the first half of the 20th century in the Czech Republic. *Acta Polytech.* **2018**, *58*, 155–164. [CrossRef]
26. Bandrés Mariscal, C.; Robador González, M.D.; Albardonedo Freire, A. Analysis and structure of water distribution system in the distribution depot of the Carmona Gate as a way of supplying water to the Royal Alcazar of Seville. In *Conserving Cultural Heritage, Proceedings of the 3rd International Congress on Science and Technology for the Conservation of Cultural Heritage, Cádiz, Spain, 21–24 May 2017*; Mosquera, M., Almoraima, M.L., Eds.; Taylor & Francis Group: Oxfordshire, UK, 2018; pp. 209–212.
27. Koeppel, G. *Water for Gotham: A History*; Princeton University Press: Princeton, NJ, USA, 2001.
28. Baker, M.N.; Taras, M. *The Quest for Pure Water: The History of Water Purification from the Earliest Records to the Twentieth Century*; American Water Works Association: Denver, CO, USA, 1981.
29. Şener, İ. Assessment of Izmir Halkapinar Water Pump Station for Its Conservation as Industrial Heritage. Ph.D. Thesis, School of Engineering and Sciences of Izmir Institute of Technology, Izmir, Turkey, 2019.
30. Voudouris, K.; Valipour, M.; Kaiafa, A.; Zheng, X.Y.; Kumar, R.; Zanier, K.; Kolokytha, E.; Angelakis, A. Evolution of water wells focusing on Balkan and Asian civilizations. *Water Supply* **2019**, *19*, 347–364. [CrossRef]
31. Ursino, N.; Pozzato, L. Heritage-Based Water Harvesting Solutions. *Water* **2019**, *11*, 924. [CrossRef]
32. Douet, J. *The Water Industry as World Heritage*; TICCIH Thematic Paper; 2018; Available online: https://ticcih.org/wp-content/uploads/2018/05/TICCIH-Water-Report.pdf (accessed on 16 May 2022).
33. Hunter Burkett, M. Silent and Unseen: Stewardship of Water Infraestructural Heritage. In *Adaptative Strategies for Water Heritage. Past, Present and Future*; Springer International Publishing: Berlin/Heidelberg, Germany, 2020; pp. 21–39.
34. Blazquez Herrero, C. *El Acueducto de Aguas Livres en Lisboa*; Agua Ibérica: Aragon, Spain, 2019; Available online: https://www.aguaiberica.com/2019/06/18/el-acueducto-de-aguas-livres-en-lisboa/ (accessed on 15 June 2022).
35. Ruiz, R. Coordinator. Análisis y definición de estrategias para la caracterización, recuperación y puesta en valor del patrimonio de las obras públicas. Una aproximación desde la escala territorial (PID2019-105877RA-I00). In *I+D+I Project*; Ministerio de Ciencia; Innovación y Universidades: Madrid, Spain, 2020–2022.
36. Rueda Laffond, J.C. *El Agua en Madrid. Datos Para Historia del Canal de Isabel II (1851–1930)*; Fundación Empresa Pública: Madrid, Spain, 1994.
37. Martínez Vázquez de Parga, R. *Historia del Canal de Isabel II*; Fundación Canal de Isabel II: Madrid, Spain, 2001.
38. Fernández Casado, C. *Acueductos Romanos en España*; Instituto Eduardo Torroja: Madrid, Spain, 1972.
39. García Grinda, J.L. Arquitectura e ingenios hidráulicos: Orígenes y presencia medieval. *Bibl. Estud. E Investig.* **2009**, *24*, 143–168.
40. Plasencia Lozano, P. L´acquedotto "Arcos de San Antón" di Plasencia. In *Il Valore Dell'acqua nel Patrimonio dei beni Culturali Attraverso la Lettura di Alcuni Episodi Architettonici, Urbani e Territoriai. Gli Acquedotti e le Fontane a Roma dal XVI al XIX Secolo*; Aracne: Roma, Italy, 2014; pp. 203–218.
41. Pelegrín Garrido, M.C. Obras hidráulicas históricas de Lorca. *Alberca* **2006**, *4*, 165–171.
42. Camacho Martínez, R. Los problemas del agua en Málaga en el siglo XVIII: El Acueducto de San Telmo y su valoración actual. In *Paisajes Modelados por el Agua: Entre el Arte y la Ingeniería*; Editora Regional de Extremadura: Mérida, Spain, 2012; pp. 41–61.
43. Torroja, E. Acueducto de Alloz. *Inf. Construcción* **1962**, *14*, 134–148. [CrossRef]
44. Nobile, M.R. Fontane e acquedotti nella Sicilia tra XV e XVII secolo. In *Patrimonio Cultural Vinculado con el Agua. Paisaje, Urbanismo, Arte, Ingeniería y Turismo*; Editora Regional de Extremadura: Mérida, Spain, 2014.
45. Trindale, L. A água nas cidades portugueseas entre os séculos XIV e XVI. In *Patrimonio Cultural Vinculado con el Agua. Paisaje, Urbanismo, Arte, Ingeniería y Turismo*; Editora Regional de Extremadura: Mérida, Spain, 2014; pp. 367–380.
46. de Jesús Martínez Pelcastre, E. Herencia constructiva en la fábrica de acueductos novohispanos en el siglo XVI en el estado de Hidalgo. In *Tercer Congreso Internacional Hispanoamericano de Historia de la Construcción*; Instituto Juan de Herrera: Madrid, Spain, 2019; pp. 319–330.
47. Cabrera Aceves, J. El acueducto histórico de Valladolid, hoy Morelia, México. Nuevos acercamientos a su función hidráulica y estereotomía. In Proceedings of the Noveno Congreso Nacional y Primer Congreso Internacional Hispanoamericano de Historia de la Construcción, Segovia, Spain, 13–17 October 2015; pp. 289–299.
48. Sánchez de Toca, J. Juicio del Sr. Sánchez de Toca. *Rev. Obras Públicas* **1925**, *Tomo I*, 20–23.
49. Gavira, C. De la cultura del agua a la técnica hidráulica: El canal de Isabel II. In *Ciudad, Territorio y Patrimonio. Materiales de Investigación*; Castrillo Romón, M., Ed.; Instituto Universitario de Urbanística: Valladolid, Spain, 2000.
50. de Isabel, C., II. Plan de obras para mejorar y completar el abastecimiento de aguas de Madrid. *Rev. Obras Públicas* **1907**, *Tomo I*, 483–487.
51. Ortega de Miguel, E.; Sanz Mulas, A. A public sector multinational company: The case of Canal de Isabel II. *Util. Policy* **2007**, *15*, 143–150. [CrossRef]
52. Rubio Gavilán, A.; Hernández Lamas, P.; Bernabéu Larena, J. Water Supplies build the cities: The Canal de Isabel II as origin of the metropolis if Madrid. In *Back to the Sense of the City, Proceedings of the 11th International Congress Virtual Cities and Territories, 6–8 July 2016*; 11 VCT.; Institute of Urban Design, Faculty of Architecture, Cracow University of Technology: Kraków, Poland, 2016.

53. Marcos, J. El Canal, Que Aguirre Quería Privatizar, Ganó 225 Millones en 2014 El Pais (Madrid). Available online: http://ccaa.elpais.com/ccaa/2015/06/20/madrid/1434822230_248284.html (accessed on 21 June 2021).
54. Harvey, D.C. Heritage Past and Heritage Presents: Temporality, meaning and the scope of heritage studies. *Int. J. Herit. Stud.* **2001**, *7*, 319–338. [CrossRef]
55. Aguiló Alonso, M. *El Agua en Madrid*; Diputación de Madrid. Area de Urbanismo y Ordenación Territorial: Madrid, Spain, 1983.
56. Fundación Canal. About Us. Available online: https://www.fundacioncanal.com/en/about-us/ (accessed on 24 April 2022).
57. Sáenz Ruidrejo, F. Presentación. In *Descripciones de los Canales Imperial de Aragón y Real de Tauste (1976)*; Ministerio de Fomento: Madrid, Spain, 1984.
58. Plasencia Lozano, P. Apuntes sobre la relación entre la obra pública y el paisaje. *Carreteras. Rev. Técnica Asoc. Española Carret.* **2017**, *212*, 20–27.
59. Lynch, K. *The Image of the City*; The MIT Press: Cambridge, UK, 1960.
60. Waterman, T. *Principios Básicos de la Arquitectura del Paisaje*; Nerea: San Sebastián, Spain, 2009.
61. Nárdiz Ortiz, C. *El Paisaje en la Ingeniería*; Ministerio de Fomento: Madrid, Spain, 2019.
62. Español Echániz, I. *La Carretera en el Paisaje. Criterios Para su Planificación, Trazado y Proyecto*; Junta de Andalucía: Seville, Spain, 2008.
63. Plasencia Lozano, P. Alconétar, paisaje cultural del al ingeniería. Una propuesta de ordenación territorial. In *Paisajes Modelados por el Agua: Entre el Arte y la Ingeniería*; Editora Regional de Extremadura: Badajoz, Spain, 2012; pp. 187–205.
64. Plasencia Lozano, P. Miradores a la obra pública y la obra pública como mirador. In *El Patrimonio de las Obras Públicas. Del Puente Romano de Alcántara al Diálogo con la Actualidad*; Sial Pigmalión: Madrid, Spain, 2022; pp. 429–462.
65. Machuca Santa Cruz, L.; Machuca Casares, B. Caminito del Rey: "el proyecto ha sido un desafio". *PH Boletín Inst. Andal. Patrim. Histórico* **2016**, *24*, 19–21. [CrossRef]
66. Leno Cerro, F. La evaluación de los recursos turísticos: El caso del Canal de Castilla. *Treb. Geogr.* **1990**, *43*, 135.
67. AevvBoss. Declaración para una "Red Verde Europea". Asociación Europea de Vías Verdes, 15-03-2016. Available online: https://www.aevv-egwa.org/es/declaracion-para-una-red-verde-europea/ (accessed on 15 May 2022).
68. Bargón García, M.; Plasencia Lozano, P. Las vías verdes en Asturias. La reutilización de una infraestructura ferroviaria obsoleta como parques lineales urbanos y regionales. In *Restauro: Temi Contemporanei per un Confronto Dialettico*; Università degli Studi di Firenze: Firenze, Italy, 2020; pp. 428–437.
69. Medina Lasansky, D. Introducción. In *Arquitectura y Turismo. Percepción, Representación y Lugar*; Gustavo Gili: Barcelona, Spain, 2006; pp. 15–27.
70. Rossi, A. *L ´Architettura Della Città*; Marsilio Editore: Padua, Italy, 1966.
71. Plasencia Lozano, P. El puente sobre el Tajo de Eiffel en España. *Inf. Construcción* **2018**, *70*, e268. [CrossRef]
72. Sainz-Jimenez, C.; González, J.M. Aerobiology and cultural heritage: Some reflections and future challenges. *Aerobiologia* **2007**, *23*, 89–90. [CrossRef]
73. Sainz-Jimenez, C. Biodeterioration vs biodegradation: The role of microorganisms in the removal of pollutants deposited on historic buidlings. *Int. Biodeterior. Biodegrad.* **1997**, *40*, 225–232. [CrossRef]

Article

Ontology-Driven Cultural Heritage Conservation: A Case of *The Analects of Confucius*

Fengxiang Wang [1,2], Tong Wei [1,2,*] and Jun Wang [1,2]

1 The Department of Information Management, Peking University, Beijing 100871, China; wfximer@pku.edu.cn (F.W.); junwang@pku.edu.cn (J.W.)
2 The Center of Digital Humanities, Peking University, Beijing 100871, China
* Correspondence: weitong@pku.edu.cn

Abstract: Confucianism, recognized as the belief system of Chinese, is one of the most important intangible cultural heritages of China. The main ideas of its founder, Confucius, are written in *The Analects of Confucius*. However, its scattered chapters and the obscurity of ancient Chinese have prevented many people from understanding it. In order to overcome this difficulty, it needs some modern ways to reveal the vague connotation of Confucianism. This paper aims to describe how to construct the Lunyu ontology in which all concepts are abstract within the core scope, i.e., morality of Confucianism. The key task of this project lies in identifying essential characteristics, a notion that is compliant with the ISO principles on Terminology (ISO 1087 and 704), according to which a concept is defined as a combination of essential characteristics. This paper proposed an approach in the practice of identifying essential characteristics of abstract concepts from different meanings of its Chinese terms in *The Analects of Confucius*. With this work, Lunyu ontology established a semantic, formal, and explicit representation system for concepts of Confucianism, and the new proposed approach provides a useful reference for other researchers.

Keywords: cultural heritage; ontology; Confucianism; *The Analects of Confucius*; essential characteristics; protégé

Citation: Wang, F.; Wei, T.; Wang, J. Ontology-Driven Cultural Heritage Conservation: A Case of *The Analects of Confucius*. *Appl. Sci.* **2022**, *12*, 287. https://doi.org/10.3390/app12010287

Academic Editor: Cesareo Saiz-Jimenez

Received: 13 November 2021
Accepted: 24 December 2021
Published: 28 December 2021

1. Introduction

Cultural Heritage is a civilization's memory of a country or a society, as well as the wealth shared by people all over the world. It records the thoughts and activities of the ancients who lived in a certain area and at a certain time. Cultural heritage is divided into the tangible cultural heritage and intangible cultural heritage (ICH). According to *Convention for the Safeguarding of the Intangible Cultural Heritage* [1] reported by UNSECO, "Intangible Cultural Heritage means the practices, representations, expressions, knowledge, skills—as well as the instruments, objects, artifacts and cultural spaces associated therewith—that communities, groups and, in some cases, individuals recognize as part of their cultural heritage." As the most important one among the rich ICH of China, Confucianism has been transmitted from generation to generation for over 2200 years and has been interpreted by thinkers in different eras, including thinkers in the present. Literally, the reason to recognize it as the most essential ICH of China is that Confucianism is the cultural foundation of Chinese society since the Qin dynasty (221 BCE–207 BCE) until the beginning of the 20th century. The main content of Confucian culture is condensed in the book named *The Analects of Confucius* (Chinese: Lunyu), which was written during the Spring and Autumn Period and the Warring States Period (770 BCE–221 BCE). The difficulty of understanding *The Analects of Confucius* lies in two aspects. One is that this book is a collection of dialogues; thus, it is difficult to understand its meaning without context. Second, there are great differences between ancient Chinese and modern Chinese. Therefore, compared with the requirement of other cultural heritages, which is to be preserved, Confucianism needs to be represented in more modern ways such that people with different cultural backgrounds can understand it better.

People in different dynasties of China all faced the problems mentioned above more or less and attempted to solve them. In Chinese history, many scholars annotated *The Analects of Confucius*. One of the most famous scholars is Zhu Xi, who lived in the Southern Song Dynasty, and his contributions will be referenced in this study in later chapters. In the modern era, the form of research on *The Analects of Confucius* varies from annotation to illustration, translation, etc. However, although much effort has been paid, it is still hard to clearly define Confucianism. Since the challenge to comprehend *The Analects of Confucius* derives from the illogicality of the text and the barrier of the ancient Chinese, this study attempts to reshape its core content—the morality system of Confucianism, based on ontology to propose an approach to solve this problem. Ontology is a modeling tool that could be used to describe concepts of information systems on semantic and knowledge levels [2]. Conversely, the Confucian morality system is embodied in the concept of some single characters that are independent in form but related in connotation. From the perspective of requirement and function, the formalization and clarity of ontology are exactly in line with the purpose of reshaping ideas in *The Analects of Confucius*. The outcome of this study, Lunyu ontology, aims to devote contributions to both providing a new way to learn about Confucianism and expanding the application of this methodology.

The rest of this paper is organized as follows: Section 2 briefly introduces the history of Confucianism and describes the domain of research, Section 3 puts forwards the objectives, Section 4 reviews the state of the art, Section 5 is dedicated to our contribution to ontology building methodology relying on a meaning analysis of Chinese terms and the ISO principles on Terminology, Section 6 presents the Lunyu ontology in Protégé, and the final section presents the evaluation of the Lunyu ontology.

2. Foundation of Research

2.1. Brief History of Confucianism

Confucianism is one of the most influential philosophies in the history of China, and it has existed for over 2500 years. It is a belief system rather than a religion, which focuses on the importance of personal ethics and morality. Confucius, the creator of Confucianism, is the most famous philosopher and teacher in China history and his thoughts on ethics, good behavior, and moral character were written down by his disciples in several books, the most important being *The Analects of Confucius* [3,4].

After Confucius, Mencius (Chinese: Mengzi), a fourth-century BCE Chinese thinker, enriched Confucianism by his theory that all human begins share innate goodness that either can be cultivated through education and self-discipline or squandered through neglect and negative influences, but never lost altogether [5].

When time came to the Southern Song Dynasty, another famous Confucian Zhu Xi created the supreme synthesis of Son-Ming Dynasty (960–1628 CE) Neo-Confucianism, which is known as *lixue* or *daoxue*. He redefined the Confucian tradition and brought the focus of Confucianism back to the moral cultivation of people which is the original focus of Confucianism by selecting the essential classical Confucian texts including *The Analects of Confucius* then compiling them with annotations [6]. This is why we referenced his contribution.

About 400 years later, Wang Yangming, who studied Zhu Zi seriously in his teenage years, developed his contribution to Neo-Confucianism. Wang Yangming's thought broke through the limitations of *lixue* by arguing for a different philosophical interpretation and cultivation of the *xin* or mind-heart, such that his distinctive philosophy is called *xinxue* or the teaching of the mind-heart. *Xinxue* and *lixue* are the two most significant movements of Neo-Confucianism. Until the end of China's feudal dynasty, Neo-Confucianism dominated the political field of China [7,8].

2.2. Description about Data Resource

Obviously, all content of Confucianism is too complex and profound; thus, this study selected the most core part of it, the morality system, to promote our work. The domain of this research can be outlined as follow four constraints:

1. All concepts, therefore, all terms, come from *The Analects of Confucius*.
2. Every concept is related to Confucian morality.
3. Meanings of terms which points to the concepts are referenced from *Shisanjing Dictionary* [9] and *The Analects of Confucius Variorum* [10].
4. Meanings of terms irrelevant to Confucian morality are not considered.

The first constraint is based on the fact that Confucian culture experienced a process of continuous derivation and scholars in different times made different interpretations of Confucianism. *The Analects of Confucius* mainly records the dialogues between Confucius and his disciples, some of the events of his life, and disciple's comments on Confucius; thus, it is not a philosophical book that logically introduces the thought system of Confucius. If any book other than *the Analects of Confucian* was involved, there may be ambiguity in the connotation of a specific concept. Even if the scope was limited to such a book, there is no worry if some important concepts have been missed, because *The Analects of Confucius* comparably concentrated embodies the political views, ethical thoughts, moral concepts, and educational principles of Confucius and Confucian school. *The Analects of Confucius* seems to be short in length, but actually, as mentioned above, it covers many aspects of Confucian Culture.

Confucius advocated telling rather than writing; thus, *The Analects of Confucius* did not describe or explain Confucius's thoughts, and even if they recorded what Confucius said, they did not record the context and background of the dialogues. Therefore, each chapter of *The Analects of Confucius* is only a few sentences long and almost independent from each other. This is why the second limitation is that concepts should be related to morality.

In particular, ancient Chinese is dominated by monosyllabic words, that is, a word represents an independent and complete meaning; thus, all concepts in our ontology appear in the form of a single word in *The Analects of Confucius*. These particularities of this book caused another problem for this study, that is, it is impossible to find the characteristics of these concepts in the original text. It led to the third condition that *Shisanjing Dictionary* and *The Analects of Confucius Variorum* provide meanings of terms. Actually, the entries in the *Shisanjing Dictionary* already contain almost all the meaningful words (i.e., characters) in *The Analects of Confucius* and list all their explanations that occurred in this book. While some of them are still not detailed or comprehensive enough and were replenished by Zhu Xi's annotation in *The Analects of Confucius Variorum*. Of course, meaning items of terms referring to moral concepts have been eliminated if they are not within the scope of Confucian morality, as the fourth constraint says.

To sum up, we obtained all terms from *The Analects of Confucius* with the guide of experts. In order to ensure the accuracy of the comprehension of Confucianism, we acquired the meanings of each term from the *Shisanjing Dictionary*, a dictionary specially compiled for 13 Confucian classics, and Zhu Xi's *The Analects of Confucius Variorum* of The Southern Song Dynasty (960–1279) as a complement.

3. Objective

The goal of this study includes (1) to build a knowledge representation of Confucianism at the aspect of morality and ethic in the form of an open ontology in a W3C standard and (2) to build a bilingual (Chinese-English) terminological e-dictionary for users who want to know the thoughts of Confucianism. This article aims to present the first goal that aims to build a knowledge base based on ontology for the second goal. It is an attempt to express Chinese philosophical ideas in a new way from traditional ones where Chinese philosophy tends to use vague and suggestive language when expressing ideas. With the building of Lunyu ontology, thoughts of Confucianism will be represented in a formal and explicit form. Moreover, we hope to accumulate some experience in the process of constructing the ontology to contribute to the methodology.

Competency questions are used to specify the requirements of the ontology [11]. A question such as "What are the Chinese terms of concept and their equivalents in English?" raises the problem of modeling the linguistic dimension of the terminology. Questions such as "What are the concepts that express character?" stress the hierarchy of classes. While "What are the definitions of the "filiality" in English and Chinese?" particularly affects ontology modeling because it directly relates to the focus of the ontology. Answering these questions requires an explicit representation of essential characteristics.

Table 1 presents some of the competency questions taken into account for the implementation of the Lunyu ontology. The table specifies the classes and relations required to answer these queries.

Table 1. The three competency questions.

CQ	Competency Questions	Class	Relation
1	What are the Chinese terms of concept and their equivalents in English?	Concept	Rdfs:subClassOf, skos:prefLabel
2	What are the concepts that express character?	Concept, Character	Rdfs:subClassOf
3	What are the definitions of the "filiality" in English and Chinese?	Concept, Filiality	Rdfs:subClassOf, skos:definition

4. State of the Art

This chapter will present some current work about ontology and its application in cultural heritage. A general introduction of ontology and ontology building methodologies is to be made initially. Then we are going to take some examples to show how ontology is applied in cultural heritage conservation. In the last part, standards and models available will be introduced.

Ontology originated in philosophy and extended to other fields with the rapid development of computer science. After Nech et al. proposed the definition of ontology in artificial intelligence [12], its definition was expanded and refined to the point where it is now the most common definition which is "an ontology is a formal, explicit specification of a shared conceptualization" including four layers of meaning: conceptual model, explicit specification, formal, and shared [13,14]. Ontology construction is a systematic engineering, but there is no absolutely correct methodology for developing ontologies. The characteristic of domain knowledge is also a key factor to be considered during ontology design. Nonetheless, ontology construction still follows certain basic steps. There are various methods to develop a domain ontology, including Skeletal Methodology [15], IDEF5 [16], TOVE [17], SENSUS [18], and so on. The Ontology Development 101 [19] method developed by Stanford University School of Medicine is the most mature and widely used ontology development method at present.

Ontology has a wide range of application scenarios in the domain of cultural heritage and could be used in different stages in cultural heritage protection. Ontology for capturing and processing 3D cultural heritage objects could enhance the operability of 3D scanning and the control of data quality [20]. Mohammed Maree et al. proposed Holy-Land ontology, which attempts to encode knowledge about Palestine's cultural heritage, aiming to provide multi-language semantic retrieval support for their cultural heritage system [21]. Compared to tangible cultural heritage, ICH is abstract and requires more detailed consideration in ontology design. Correspondingly, an appropriate ontology model can be of great benefit in representing abstract transactions. Ontology of Terengganu Brassware Craft completely records the knowledge about this craft and the process of constructing it presenting the experience in developing a domain ontology for ICH [22]. Similarly, ontology-based knowledge and understanding obtained from experts and knowledgeable members improved the inheritance of culture [23]. By building ontologies beforehand, entities and relationships could be extracted from the structured or unstructured text about ICH [24,25]. In some cases, existing ontologies can be reused to organize and classify new cultural heritage data collection. The degree of reuse depends on what class and properties those ontologies have. As for an example of representing an idea system with ontology, we have not yet found one, although there are not many studies that introduce ontology into ICH.

The digitization of the ICH is much more complicated than the general cultural heritage; thus, it requires the extension of the standard ontology for digitization of cultural heritage [26]. For this reason, it would be helpful to reference some public resources when developing a specific domain ontology of ICH. The W3C standard OWL (Web Ontology Language) [27] can be used to explicitly represent the meaning of terms in vocabularies and the relationships between those terms with good machine interpretability. SKOS (Simple Knowledge Organization System) [28] is a data model for sharing and linking knowledge organization systems on the Web. It captures the similarity of many knowledge organization systems and makes it explicit; thus, it is widely used in knowledge organization. DC (Dublin Core) [29] is a metadata schema (Schemas are machine-processable specifications that define the structure and syntax of metadata specifications in a formal schema language.) based on 15 essential properties to describe online and physical resources and it is also cited in this study. OTContainer proposed a term-guided method for ontology construction, which is applied in TAO CI ontology [30]. We will adopt this method in our research. There are also specific standards for cultural heritage. The CIDOC CRM [31] ontology is an official standard whose ISO standard can be found at ISO 21127: 2014. It provides the basic classes and relations devised for the cultural heritage. EDM [32] defined formal specification of the classes and properties that could be used in Europeana which develops expertise, tools, and policies to empower the cultural heritage sector in its digital transformation, whereas both are not suitable for our ontology modeling because we need to build a "granular" ontology to represent concepts of *The Analects of Confucius*, which means that we extract knowledge at a much deeper and more detailed level and build the ontology from it. The knowledge is specific and has the particularity of the field. Nevertheless, classes CIDOC and EDM defined are too general and common to be transplanted to our ontology.

5. The Methodology of Lunyu Ontology

Ontology building follows a lifecycle of several stages [33]. Some of them have to be specialized, others have to be newly introduced to take into account the specificities of the domain. The theory of concept underpinning the ontology can also strongly impact the methodology of engineering it. Following the ISO principles on Terminology where "a term is a verbal designation of a concept" and "a concept is a unique combination of (essential) characteristics" (An "essential characteristic" is a characteristic (abstraction of a property) of a concept and is indispensable to understanding that concept (ISO 1087). Essential characteristics correspond to rigid predicates in DL [34] and to rigid properties in the OntoClean method [35]. Unlike essential characteristics, which define the concept, "descriptive characteristics" own values which describe the current state of an object, e.g., weight, color, etc.) (ISO 1087, ISO 704), we adopt a "term-and-characteristic" guided method derived from previous work [36–38]. Identifying essential characteristics becomes the first goal to achieve.

The problem of identifying essential characteristics is a new and central phase of our methodology. For concrete concepts, this phase is aimed at identifying differences between objects (chair with leg versus chair without a leg). For abstract concepts, this phase is aimed at identifying differences between meanings of concepts because the concepts do not have a physical entity corresponding to it, e.g., the meaning of the term "礼" (ritual) is a hierarchical system that regulates the external behavior. Thus, the essential characteristics were {/system/, /hierarchical/, /external/, /behavior/}.

The term-and-characteristic guided methodology includes seven steps, each aiming at different tasks. Step 1: identify the scope of the domain and the objectives; Step 2: identify terms and objects; Step 3: identify essential characteristics; Step 4: define concepts;

Step 5: build ontology using one of the available tools; Step 6: integration of other resources; Step 7: evaluation.

This chapter is dedicated to step 3 (identifying essential characteristics) and step 4 (combining essential characteristics into concepts). The first step has been presented in the previous chapter 3. There is no phase of identification of terms and objects, since the experts know the terms to be defined. The Lunyu ontology engineering in the OWL phase as well as the evaluation phase will be presented in dedicated chapters.

5.1. Identifying Essential Characteristics of Abstract Concepts

Identifying differences between meanings is a useful means toward identifying essential characteristics. Because this paper focuses on the concepts related to personal morality in *The Analects of Confucius*, the differences can be an object that is described by a concept (e.g., for oneself, for others), object's identity (parents, brother, friends, monarch), concept's type (character, ability, need, attitude, system, and morality). Thus, the presence or the absence of differences can be interpreted as essential characteristics. For example, the meaning of concepts could be divided into analysis axis: person, resource, identity, binding, range, action, hierarchy, performance and type (Figure 1).

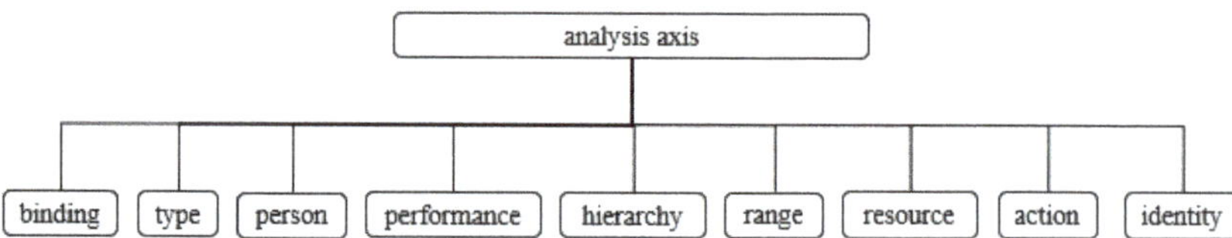

Figure 1. The analysis axis.

From the concept type point of view, the concepts could be divided into concept for attitude, concept for character, concept for ability, concept for need, concept for system, and concept for morality corresponding to the essential characteristic/character/, /ability/, /need/, /attitude/, /system/, and /morality/ (Figure 2). These essential characteristics are exclusive to each other.

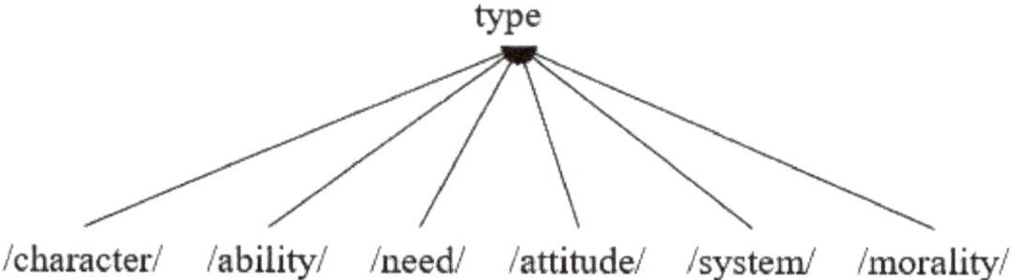

Figure 2. The essential characteristics of type analysis axis.

5.2. Defining Concepts Based on Essential Characteristics' Combination

From the terminology point of view, a concept is defined as a unique combination of essential characteristics (ISO 1087). However, not any combination of essential characteristics defines a meaningful concept from the expert's point of view. For the expert, concepts of interest are those that are named in a natural language. Hence, a concept is a set of essential characteristics stable enough to be named in a given language (even if some concepts, without any designation in natural language, can be introduced for organizational purposes of the conceptual system). Terms can be then considered as guidelines for identifying domain concepts to be defined from the expert point of view. For example, the term "信", "trust" in English, denotes the set of essential characteristics {/morality/, /realize/, /promise/, /others/, /honesty/}. Based on this formal definition, the definition in natural language is then: "Morality that realizes promise to others and expressed as honest".

6. The Lunyu Ontology

The specific implementation method of the "term-and-characteristics" guided method in Protégé was presented in the literary [37,38]. For example, concepts are defined as named classes in Protégé. Terms are represented as labels (skos:prefLabel, skos:altLabel, skos:definition). Relations, such as "hasRange". "hasObject" are represented as object properties. Implementing essential characteristics is a slightly more complex process. Since essential characteristics correspond to rigid predicates, they cannot be directly expressed into description logic. Reference [39] introduced different ways to implement essential characteristics in Protégé. The first way is to implement essential characteristics as classes. For example, the essential characteristics corresponding to the action of concepts are subclasses of the Action class. Owning an essential characteristic for a concept (class) is represented as a restriction of an object property whose range is the class associated with the essential characteristic. For example, owing the essential characteristic/control/will be translated into the restriction of the "has action" object property: "has action" some Control.

However, the essential characteristics of type are special in our ontology. They were defined as the subclasses of the Concept class. In other words, the essential characteristics of types were translated as classes, which is an "is-a" relationship between owning essential characteristics for a concept (class) and essential characteristics of types (classes).

6.1. Class

The Lunyu ontology includes 100 classes, 8 object properties, 204 logical axioms. Although *The Analects of Confucius* includes different concepts, this paper focused on the concepts related to morality.

Let us consider the concept denoted by the term "礼" (ritual), denotes the set of essential characteristics: {/external/,/hierarchy/,/normalize/,/behavior system/}, which was defined in OWL as following (Figure 3):

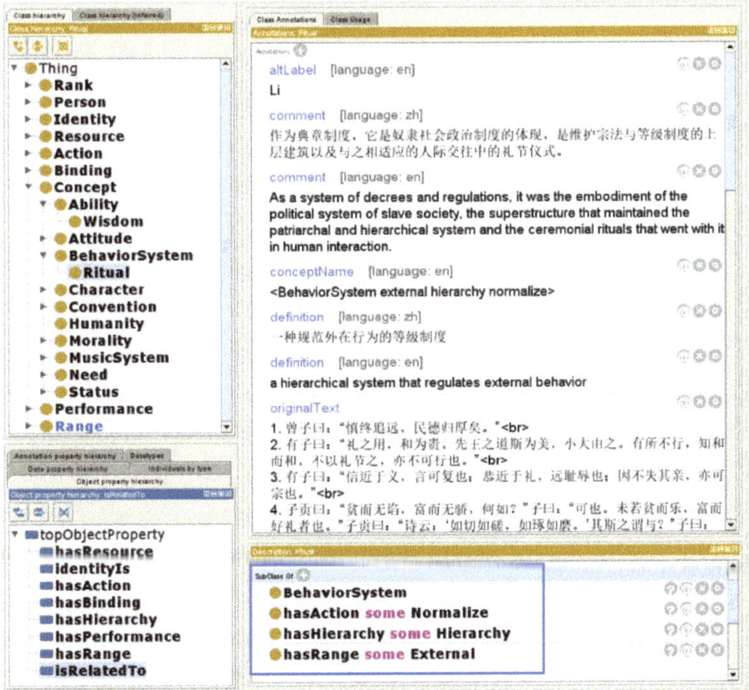

Figure 3. The "Ritual" as a class.

Li ≡ BehaviorSystem ⊓ ∃hasAction. Normalize ⊓ ∃hasRange. External ⊓ ∃hasHierarchy. Hierarchy.

6.2. Properties

There were eight object properties, which included hasResource, identityIs, has Action, hasBinding, hasHierarchy, hasPerformance, hasRange, isRelatedTo (Table 2). There are no data properties because the concept is an abstract concept, whose attributes are defined as the essential characteristics. Thus, we did not need data properties to describe the concept in Lunyu ontology.

Table 2. The object properties.

Object Properties	Domain	Range
hasResource	Concept	Resource
identityIs	Concept	Identity
hasAction	Concept	Action
hasBinding	Concept	Binding
hasHierarchy	Concept	Rank
hasPerformance	Concept	Performance
hasRange	Concept	Range
isRelatedTo	Concept	Person

6.3. Annotation

In Lunyu ontology, we reuse the RDFS, DC, OTContainer, and SKOS vocabularies to express metadata information. For example, dc:publisher, dc:license, dc:creator, skos:prefLabel, skos:altLabel, skos:definition, rdfs:comment. OTContainer:conceptName.

7. Evaluation

Evaluation is the last stage of building ontology, whose goal is "to assess the quality and correctness of the obtained ontology" [40]. In this paper, we used the OOPS! [41], OntoMetrics [42] and queried the ontology against the Competency Questions defined in chapter 3.

When we submitted Lunyu ontology to OOPS!, it only detected minor pitfalls (e.g., P08 "Missing annotations", P13 "Inverse relationship not explicitly declared", P36 "URI contains file extension"). The minor pitfalls are not a problem.

When we submitted Lunyu ontology to OntoMetrics, Table 3 shows the result of schema metrics and knowledge base metrics about ontology clarity and conciseness.

Table 3. The Lunyu ontology advanced metrics.

Metric	Value
Attribute richness	0.0
Inheritance richness	1.84
Relationship richness	0.05641
Class/relation ratio	0.512821
Average population	0
Class richness	0

Most of the scores are low, that is due to:

1. The implementation of essential characteristics in Description Logics. In our work, the essential characteristics were translated as the classes without any attributes (attribute richness). Thus, the value of the attribute richness is a "0".
2. The goal of Lunyu ontology is the classification and definition of concepts; they did not represent the relationships between concepts in *The Analects of Confucius* and concepts defined in other literature (relationship richness, class/relation ration).

3. The Lunyu ontology is about the abstraction concept, which does not correspond to the physical entity in the real world. Thus, we could not define any individuals in Lunyu ontology (average population, class richness). Depending on the calculation formula of the average population and class richness in OntoMetrics, the values of the average population and class richness are "0".

Evaluation of criteria strongly depends on the aims of the ontology and the choices made to regard to its implementation: "a good ontology does not perform equally well with regards to all criteria" [43].

We should note that, about our objectives of classification and terminology, the Lunyu ontology well covers the domain concept (classification), and each concept is clearly defined as a unique combination of essential characteristics (terminology). It should provide the domain knowledge to a bilingual terminology e-dictionary (http://dh.ketrc.com/lunyu/index.html, accessed on 25 November 2021). Thus, the low values of OntoMetrics do not affect the construction of a bilingual terminology e-dictionary.

The last validation concerns the answers to the Competency Questions. All of them are satisfied. The three competency questions were translated into SPARQL as following. Figure 4 is the returned results.

	name_en	name_zh
21	"cautious"@en	"慎"@zh
22	"Ti"@en	"悌"@zh
23	"reverent"@en	"恭"@zh
24	"reciprocity"@en	"恕"@zh
25	"character"@en	"性格"@zh
26	"attitude"@en	"态度"@zh
27	"loyalty"@en	"忠"@zh
28	"tolerate"@en	"忍"@zh
29	"morals"@en	"德"@zh
30	"Kuan"@en	"宽"@zh

a). The result of Q1

	Name
1	"Kuan"@en
2	"Liang"@en
3	"agile"@en
4	"brave"@en
5	"cautious"@en
6	"faithful"@en
7	"gentleness"@en
8	"reticence"@en
9	"simplicity"@en
10	"steadfastness"@en

b). The result of Q2

	definition_en	definition_zh
1	"a morality that realizes promises to others and expressed as honest"@en	"一种实现对他人承诺的品德，表现为诚实的。"@zh

c). The result of Q3

Figure 4. The result of Sparql.

1. What are the Chinese terms of concept and their equivalents in English?

PREFIX xsd: <http://www.w3.org/2001/XMLSchema#>
PREFIX skos: <http://www.w3.org/2004/02/skos/core#>
PREFIX owl: <http://www.w3.org/2002/07/owl#>
PREFIX rdf: <http://www.w3.org/1999/02/22-rdf-syntaxns#>
PREFIX rdfs: <http://www.w3.org/2000/01/rdf-schema#>
PREFIX otc:<http://www.dh.ketrc.com/otcontainer/data/otc.owl#>

```
PREFIX lunyu:<http://www.dh.ketrc.com/ChineseCulture/data/lunyu.owl#>
PREFIX skos:<http://www.w3.org/2004/02/skos/skos#>
SELECT ?name_en ?name_zh
WHERE {
?x rdfs:subClassOf* skos:Concept.
?x skos:prefLabel ?name_en.
?x skos:prefLabel ?name_zh.
FILTER(Lang(?name_en) = 'en')
FILTER(Lang(?name_zh) = 'zh')
```

2. What are the concepts that express character?

```
PREFIX rdfs: <http://www.w3.org/2000/01/rdf-schema#>
PREFIX otc:<http://www.dh.ketrc.com/otcontainer/data/otc.owl#>
PREFIX lunyu:<http://www.dh.ketrc.com/ChineseCulture/data/lunyu.owl#>
PREFIX skos:<http://www.w3.org/2004/02/skos/skos#>
SELECT DISTINCT ?Name
WHERE {
?x rdfs:subClassOf lunyu:Character.
?x skos:prefLabel ?Name
}
ORDER BY ?Name
```

3. What are the definitions of the "filiality" in English and Chinese?

```
PREFIX xsd: <http://www.w3.org/2001/XMLSchema#>
PREFIX skos: <http://www.w3.org/2004/02/skos/core#>
PREFIX owl: <http://www.w3.org/2002/07/owl#>
PREFIX rdf: <http://www.w3.org/1999/02/22-rdf-syntaxns#>
PREFIX rdfs: <http://www.w3.org/2000/01/rdf-schema#>
PREFIX otc:<http://www.dh.ketrc.com/otcontainer/data/otc.owl#>
PREFIX lunyu:<http://www.dh.ketrc.com/ChineseCulture/data/lunyu.owl#>
PREFIX skos:<http://www.w3.org/2004/02/skos/skos#>
SELECT DISTINCT ?definition_en ?definition_zh
WHERE {
?x skos:prefLabel "trust"@en.
?x skos:definition ?definition_zh.
?x skos:definition ?definition_en.
FILTER(Lang(?definition_zh)='zh')
FILTER(Lang(?definition_en)='en')
```

In our research, concepts referring to morality in *The Analects of Confucius* are translated into a clearly-defined format, and then a bilingual terminology e-dictionary is developed. This work will provide practical value on at least two levels. For anyone who wants to learn the thought in *The Analects of Confucius*, they can avoid the difficulties caused by the chaotic organization of the text and easily gain access to the connotation of each concept and the potential relationship between them. Especially for those who are not familiar with ancient and Confucianism, a clear definition with modern expression can help them quickly understand the concept. Conversely, the bilingual terminology e-dictionary could support cross-language retrieval such that even people who do not speak Chinese can accurately acquire knowledge about Confucianism. With all the work we have performed, we hope the Lunyu ontology, with the e-dictionary based on it, will be helpful to anyone interested in Confucianism.

8. Conclusions

Our publication proposed Lunyu ontology to formally represent concepts concerning morality in *The Analects of Confucius*. In this research, we explored how to develop a class hierarchy from discrete and abstract ideas. We also adopted a term-and-characteristic guided methodology derived from taking into account the ISO principles on Terminology, whereby "a term is a verbal designation of a concept" and "a concept is a unique combination of (essential) characteristics".

The problem of identifying essential characteristics is a new and central phase of our methodology. Usually, the essential characteristics of the concrete concept are extracted from differences between objects, but for abstract concept, they come from the differences between meanings in this study. This method may provide reference for developing ontology based on abstract concepts.

Lunyu ontology was published at: http://www.dh.ketrc.com/ChineseCulture/data/lunyu.owl, accessed on 25 November 2021.

In our future work, we would like to enrich our work in two aspects: (i) expand the scope of concepts to other topics, such as politics, learning and life; (ii) develop a multilingual e-dictionary for our ontology. At present, we limited the range of concepts to the ethics topic, which reflects only a single slice of Confucianism. It will make our ontology more valuable to enrich the content of it. The multilingual e-dictionary will provide a user-friendly entry point to access our outcome. Another issue that deserves to be researched is that is there a common method or process or workflow to identify essential characteristics of abstract concepts. It will be a methodological discussion.

Author Contributions: Conceptualization, T.W. and F.W.; methodology, T.W.; software, T.W.; validation, T.W.; formal analysis, T.W.; resources, F.W.; data curation, F.W.; writing—original draft preparation, T.W. and F.W.; writing—review and editing, T.W. and F.W.; supervision, J.W.; project administration, F.W.; funding acquisition, J.W. All authors have read and agreed to the published version of the manuscript.

Funding: This research was funded by the NSFC project "the Construction of the Knowledge Graph for the History of Chinese Confucianism" grant number 72010107003.

Conflicts of Interest: The authors declare no conflict of interest.

References

1. United Nations Educational, Scientific and Cultural Organization. Available online: https://ich.unesco.org/en/convention (accessed on 23 October 2021).
2. Hu, J.; Lv, Y.; Zhang, M. The ontology design of intangible cultural heritage based on CIDOC CRM. *Int. J. u- e-Serv. Sci. Technol.* **2014**, *7*, 261–274. [CrossRef]
3. National Geographic Society. Available online: https://www.nationalgeographic.org/encyclopedia/confucianism/ (accessed on 25 October 2021).
4. Asia Society. Available online: https://asiasociety.org/education/confucianism (accessed on 25 October 2021).
5. Richey, J. Mencius (c. 372—289 B.C.E.). Available online: https://iep.utm.edu/mencius/ (accessed on 25 October 2021).
6. Thompson, K.O. Zhu Xi (Chu Hsi, 1130—1200). Available online: https://iep.utm.edu/zhu-xi/ (accessed on 25 October 2021).
7. Kim, Y. Wang Yangming (1472—1529). Available online: https://iep.utm.edu/wangyang/ (accessed on 25 October 2021).
8. Berthrong, J.H. Neo-Confucian Philosophy. Available online: https://iep.utm.edu/neo-conf/ (accessed on 25 October 2021).
9. Xuelin, L. *Shisanjing Dictionary*; Shanxi People's Press: Shanxi, China, 2002.
10. Xi, Z. *The Analects of Confucius Variorums*; The Commercial Press: Beijing, China, 2015.
11. Ren, Y.; Parvizi, A.; Mellish, C.; Pan, J.Z.; Deemter, K.V.; Stevens, R. Towards Competency Question-Driven Ontology Authoring. In Proceedings of 11th International Conference, ESWC 2014, Anissaras, Crete, Greece, 25–29 May 2014; pp. 752–767.
12. Neches, R.; Fikes, R.E.; Finin, T.; Gruber, T.; Patil, R.; Senator, T.; Swartout, W.R. Enabling technology for knowledge sharing. *AI Mag.* **1990**, *18*, 36–56.
13. Gruber, T.R. A translation approach to portable ontology specifications. *Knowl. Acquis.* **1993**, *5*, 199–220. [CrossRef]
14. Studer, R.; Benjamins, V.R.; Fensel, D. Knowledge engineering: Principles and methods. *Data Knowl. Eng.* **1998**, *25*, 161–197. [CrossRef]
15. Uschold, M.; Gruninger, M. Ontologies: Principles, methods and applications. *Knowl. Eng. Rev.* **1996**, *11*, 93–136. [CrossRef]
16. Menzel, C.P.; Mayer, R.J. *IDEF5 Ontology Description Capture Method: Concept Paper*; Technical Report; Knowledge Based Systems Laboratory Texas A&M University: College Station, TX, USA, 1990.
17. Tham, K.D.; Fox, M.S.; Gruninger, M. A Cost Ontology for Enterprise Modelling. In Proceedings of 3rd IEEE Workshop on Enabling Technologies: Infrastructure for Collaborative Enterprises, Morgantown, WV, USA, 17–19 April 1994; pp. 197–210.

18. Swartout, B.; Patil, R.; Knight, K.; Russ, T. Toward Distributed Use of Large-Scale Ontologies. In *AAAI Spring Symposium on Ontological Engineering*; Stanford University: Stanford, CA, USA, 1997.
19. Noy, N.F.; McGuinness, D.L. Ontology Development 101: A Guide to Creating Your First Ontology. *Knowl. Syst. Lab.* **2001**, *32*, 1–25.
20. Homburg, T.; Cramer, A.; Raddatz, L.; Mara, H. Metadata schema and ontology for capturing and processing of 3D cultural heritage objects. *Herit. Sci.* **2021**, *9*, 1–19. [CrossRef]
21. Maree, M.; Rattrout, A.; Altawil, M.; Belkhatir, M. Multi-modality search and recommendation on palestinian cultural heritage Based on the holy-land ontology and extrinsic semantic resources. *J. Comput. Cult. Heritage.* **2021**, *14*, 1–23. [CrossRef]
22. Isa, W.M.W.; Zin, N.A.M.; Rosdi, F.; Sarim, H.M.; Saad, S. Ontology Modeling of Intangible Cultural Heritage Terengganu Brassware Craft. In Proceedings of the 2019 International Conference on Electrical Engineering and Informatics, Bandung, Indonesia, 9–10 July 2019; pp. 1–6.
23. Pramartha, C.; Davis, J.G. Digital Preservation of Cultural Heritage: Balinese Kulkul Artefact and Practices. In Proceedings of the 6th International Conference, EuroMed 2016, Nicosia, Cyprus, 31 October–5 November 2016; pp. 491–500.
24. Araújo, C.; Martini, R.G.; Henriques, P.R.; Almeida, J.J. Annotated Documents and Expanded CIDOC-CRM Ontology in the Automatic Construction of a Virtual Museum. In *Developments and Advances in Intelligent Systems and Applications*; Rocha, Á., Reis, L.P., Eds.; Springer: Cham, Switzerland, 2018; Volume 718, pp. 91–110.
25. Dou, J.; Qin, J.; Jin, Z.; Li, Z. Knowledge graph based on domain ontology and natural language processing technology for Chinese intangible cultural heritage. *J. Vis. Lang. Comp.* **2018**, *48*, 19–28. [CrossRef]
26. Stewart, R.; Zhelev, Y.; Monova-Zheleva, M. Development of Digital Collections of Intangible Cultural Heritage Objects—Base Ontology. In Proceedings of the 2021 International Conference on Digital Presentation and Preservation of Cultural and Scientific Heritage, Burgas, Bulgaria, 23–25 September 2021; pp. 51–56.
27. W3C-OWL Web Ontology Language Overview. Available online: https://www.w3.org/TR/2004/REC-owl-features-20040210/ (accessed on 29 October 2021).
28. W3C-SKOS Simple Knowledge Organization System Reference. Available online: https://www.w3.org/TR/skos-reference/ (accessed on 29 October 2021).
29. Dublin Core Metadata Initiative. Available online: https://dublincore.org/schemas/ (accessed on 29 October 2021).
30. OTContainer. Available online: http://www.dh.ketrc.com/otcontainer/data/OTContainer.owl# (accessed on 29 October 2021).
31. CIDOC CRM. Available online: http://www.cidoc-crm.org/ (accessed on 29 October 2021).
32. Europeana Pro. Available online: https://pro.europeana.eu/page/edm-documentation (accessed on 29 October 2021).
33. Fernández-López, M.; Gómez-Pérez, A.; Juristo, N. Methontology: From ontological art towards ontological engineering. *Eng. Workshop Ontol. Eng.* **1997**, 33–40. Available online: https://www.researchgate.net/publication/50236211_METHONTOLOGY_from_ontological_art_towards_ontological_engineering (accessed on 29 October 2021).
34. Guarino, N.; Guizzardi, G. In the defense of ontological foundations for conceptual modeling. *Scand. J. Inf. Systems.* **2006**, *18*, 1.
35. Guarino, N.; Welty, C.A. An Overview of OntoClean. In *Handbook on Ontologies*; Staab, S., Studer, R., Eds.; Springer: Berlin/Heidelberg, Germany, 2009; pp. 201–220.
36. Roche, C.; Papadopoulou, M. Mind the Gap: Ontology Authoring for Humanists. In Proceedings of the 1st International Workshop for Digital Humanities and their Social Analysis (WODHSA)-Episode V: The Styrian Autumn of Ontology, Medical University of Graz, Graz, Austria, 23–25 September 2019.
37. Wei, T.; Roche, C.; Papadopoulou, M.; Jia, Y. Using ISO and semantic web standard for building a multilingual terminology e-dictionary: A use case of Chinese ceramic vases. *J. Inf. Sci.* **2021**, *1*, 1–16. [CrossRef]
38. Wei, T.; Roche, C.; Papadopoulou, M.; Jia, Y. An Ontology of Chinese Ceramic Vases. In Proceedings of the 12th International Conference on Knowledge Engineering and Ontology Development, Budapest, Hungary, 2–4 November 2020; pp. 53–63.
39. Spies, M.; Roche, C. Aristotelian Ontologies and OWL Modeling. In *Handbook of Ontologies for Business Interaction*; Rittgen, P., Ed.; IGI Global: Hershey, PA, USA, 2008; pp. 21–33.
40. Sabou, M.; Fernandez, M. Ontology (network) Evaluation. In *Ontology Engineering in a Networked World*; Springer: Berlin/Heidelberg, Germany, 2012; pp. 193–212.
41. Poveda-Villalón, M.; Gómez-Pérez, A.; Suárez-Figueroa, M.C. Oops! (ontology pitfall scanner!): An on-line tool for ontology evaluation. *Int. J. Semant. Web Inf. Syst.* **2014**, *10*, 7–34. [CrossRef]
42. Lantow, B. OntoMetrics: Putting Metrics into Use for Ontology Evaluation. In Proceedings of the 8th International Conference on Knowledge Engineering and Ontology Development, Porto, Portugal, 9–11 November 2016; pp. 186–191.
43. Vrandečić, D. Ontology Evaluation. In *Handbook on Ontologies*; Staab, S., Studer, R., Eds.; Springer: Berlin/Heidelberg, Germany, 2009; pp. 293–313.

MDPI
St. Alban-Anlage 66
4052 Basel
Switzerland
Tel. +41 61 683 77 34
Fax +41 61 302 89 18
www.mdpi.com

Applied Sciences Editorial Office
E-mail: applsci@mdpi.com
www.mdpi.com/journal/applsci

www.ingramcontent.com/pod-product-compliance
Lightning Source LLC
LaVergne TN
LVHW070930170726
843515LV00005B/907

* 9 7 8 3 0 3 6 5 6 8 1 8 8 *